Intelligent Construction Monitoring Systems:

Real-Time Safety, Environmental Prediction, and Risk Management

Aquil Mirza Mohammed
https://orcid.org/0000-0001-7756-4363
The Hong Kong Polytechnic University, Hong Kong

Hewa Majeed Zangana
https://orcid.org/0000-0001-7909-254X
Duhok Polytechnic University, Iraq

IGI Global
Scientific Publishing
Publishing Tomorrow's Research Today

Vice President of Editorial	Melissa Wagner
Director of Acquisitions	Mikaela Felty
Director of Book Development	Jocelynn Hessler
Production Manager	Mike Brehm
Cover Design	Jose Rosado

Published in the United States of America by
IGI Global Scientific Publishing
701 East Chocolate Avenue
Hershey, PA, 17033, USA
Tel: 717-533-8845 | Fax: 717-533-7115
Website: https://www.igi-global.com E-mail: cust@igi-global.com

Library of Congress Cataloging-in-Publication Data

LCCN: (CIP Data Pending)
ISBN13: 9798337392455
Isbn13Softcover: 9798337392462
EISBN13: 9798337392479
British Cataloguing in Publication Data
A Cataloguing in Publication record for this book is available from the British Library.

Table of Contents

Detailed Table of Contents

mainly relies on manual checks and basic methods, which have problems like limited coverage, slow response, and biased judgments, lacking real-time protection. This paper introduces an integrated construction safety and environmental monitoring system. It combines computer vision, IoT sensors, and data analytics within a PyQt5 framework. The system uses image processing to detect safety helmets in real time and sensors to monitor environmental parameters like temperature and humidity. Data is transmitted via MQTT protocol, with multi-layer visualization for simultaneous monitoring of helmet compliance, temperature, and humidity. It also conducts intelligent analysis based on predefined thresholds, including automatic report generation, temporal pattern analysis, and predictive risk assessment. Tests show the system runs well, providing a foundation for future expansion.

Qizheng Zhu, The Hong Kong Polytechnic University, Hong Kong
Changyu Zhang, The Hong Kong Polytechnic University, Hong Kong
Zhenming Xu, The Hong Kong Polytechnic University, Hong Kong
Qinyan Wang, The Hong Kong Polytechnic University, Hong Kong
Haoting Wan, The Hong Kong Polytechnic University, Hong Kong
Aquil Mirza Mohammed, The Hong Kong Polytechnic University, Hong Kong & School of Low-Altitude Digital Intelligence, Liaoning Technical University, Huludao, China
Cong Wu, Liaoning Technical University, Huludao, China

The current automatic safety helmet detection system does not include real-time environmental data such as temperature, humidity and wind speed. This information is very important for a comprehensive assessment of the risks at the work site. This study aims to combine real-time helmet compliance monitoring with an advanced environmental forecasting framework. It solves the limitations of traditional methods, which rely on manual supervision and do not include environmental data. We have created a complete cycle from monitoring to warning, using a module design connected through Internet of Things communication. It includes reliable data transmission, dynamic visualization, multi-level alerts, environmental assessment and security storage. The environmental prediction part uses wavelet decomposition to clean up noise sensor data, uses long- and short-term memory (LSTM) networks for accurate time series prediction, and Gaussian process regression (GPR) to provide probability prediction and measurement uncertainty.

Modular integrated construction (MiC) offers faster, safer, and more sustainable building, but on-site environmental factors can reduce connection quality and productivity. This study presents a real-time risk management system using IoT sensors; thermometers, hygrometers, and anemometers; networked via MQTT to a central dashboard. The system continuously monitors temperature, humidity, and wind speed, comparing data to activity-specific thresholds for key MiC tasks. Risks are classified into three levels, triggering instant alerts and recommendations from “proceed with caution” to “stop work.” Simulations confirm reliable detection of hazards like high humidity, extreme temperatures, and unsafe winds. With AI helmet detection, the system reached 80% safety compliance, reducing misalignments and rework. This framework effectively mitigates environmental impacts, and future predictive analytics could further enhance proactive management.

In today's fast-paced work environments, fatigue among workers can decrease productivity and health. To address this, we developed a real-time fatigue detection system using Artificial Intelligence of Things (AIoT). It integrates sensors—including cameras, simulated physiological sensors, environmental monitors, and logs—to assess workers' cognitive and physical states. Key indicators such as eye aspect ratio (EAR), head pose, and behaviors are extracted. A weighted fusion algorithm combines these to generate a fatigue score, triggering alerts when thresholds are exceeded. This non-intrusive system enables timely interventions, promoting well-being and productivity. Achieving 86% accuracy, it effectively monitors fatigue via vision, physiological, and behavioral data, offering personalized feedback. Ultimately, this AIoT solution shows great promise for improving worker health and efficiency.

concept viability, with dataset scale identified as the primary constraint. This work contributes to proactive construction safety through continuous, automated PPE compliance verification.

Head injuries at industrial sites often occur because workers don't wear safety helmets. Manual supervision is error-prone and lacks continuous feedback. To address this, a Smart Safety Helmet Compliance Monitoring System uses computer vision, IoT, and real-time visualization. The system employs an MQTT client that requests helmet detection from a remote AI service, which returns results with labeled images. A Python application processes these responses, tracking helmet-wearing behavior over time and visualizing compliance trends via dynamic graphs and live feeds. Results show the system accurately measures helmet compliance, distinguishes compliant from non-compliant events, and enhances situational awareness. This modular framework demonstrates the effective integration of AI and IoT for safety training and industrial safety improvements.

Construction sites face critical safety challenges from environmental hazards, structural risks, and PPE non-compliance, yet traditional manual monitoring remains inefficient and error-prone. This paper presents a Python-based Smart Construction Monitoring Simulator demonstrating AIoT principles for comprehensive safety management without physical infrastructure. The simulator integrates six modules: sensor data simulation (temperature, humidity, structural strain, PPE compliance), MQTT communication, threshold analytics, real-time visualization, intelligent alerting, and logging. Systematic validation demonstrates 94.12% detection accuracy, 96.3% environmental alert precision, and stable 72-hour operation processing

Preface

INTRODUCTION

The construction industry is undergoing a profound digital transformation driven by rapid advances in intelligent monitoring technologies, artificial intelligence (AI), the Internet of Things (IoT), digital twins, and data-driven decision-making systems. As Editors of this volume, we observe that traditional construction monitoring approaches—largely dependent on manual inspections and reactive safety measures—are increasingly inadequate for addressing the growing complexity, scale, and risk of modern construction projects. In response to these challenges, *Intelligent Construction Monitoring Systems (ICMS)* have emerged as a critical paradigm for enabling real-time safety assurance, environmental prediction, and proactive risk management across the entire construction lifecycle.

Intelligent Construction Monitoring Systems integrate heterogeneous data sources, including smart sensors, drones, wearable devices, Building Information Modeling (BIM), and cyber-physical systems, to continuously monitor site conditions, worker behavior, equipment performance, and environmental factors. When coupled with advanced analytics, machine learning, and predictive modeling, these systems can identify hazards before accidents occur, forecast environmental impacts, optimize resource utilization, and support evidence-based decision-making. Such capabilities are essential for improving occupational safety, minimizing project delays, reducing cost overruns, and enhancing sustainability in construction projects.

Real-time safety monitoring has become a major research and industrial priority due to the persistently high rate of construction-related accidents and fatalities worldwide. Intelligent systems capable of detecting unsafe behaviors, equipment malfunctions, and structural anomalies represent a paradigm shift from reactive incident response toward proactive prevention. In parallel, environmental prediction—covering air quality, noise, vibration, weather impacts, energy use, and carbon emissions—plays a crucial role in regulatory compliance and sustainable construction practices. Risk management, strengthened by predictive analytics, simulation, and optimization, enables stakeholders to anticipate uncertainties and mitigate operational, financial, and safety-related risks more effectively.

This edited reference volume aims to provide a comprehensive and interdisciplinary platform for researchers, practitioners, and policymakers to explore the latest theories, methodologies, technologies, and real-world applications of Intelligent Construction Monitoring Systems. The chapters collectively highlight cutting-edge research, practical case studies, and emerging trends that demonstrate how intelligent monitoring can transform construction safety, environmental management, and risk governance. By bringing together contributions from diverse disciplinary perspectives, the book seeks to bridge the gap between academic research and industry practice while offering actionable insights for the future of smart, safe, and sustainable construction.

The primary objective of this volume is to advance the state of knowledge and practice in ICMS by consolidating emerging research, innovative technologies, and practical implementations that address the limitations of traditional, reactive monitoring approaches. Specifically, the book presents state-of-the-art methodologies involving AI, machine learning, IoT, digital twins, computer vision, and cyber-physical systems; advances real-time safety management through proactive, data-driven frameworks; enhances environmental prediction and sustainability through intelligent monitoring and forecasting models; strengthens risk management and decision-making using predictive and simulation-based approaches; and bridges theory and practice through real-world case studies and industrial applications. Equally important, the volume fosters interdisciplinary collaboration among construction engineering, computer science, data science, project management, and policy studies.

This book is intended for a broad and interdisciplinary audience, including academic researchers and scholars, graduate and postgraduate students, industry professionals, technology developers, policymakers, consultants, and sustainability experts. By addressing both theoretical advancements and practical applications, the volume serves as a comprehensive reference for readers seeking to understand, design, evaluate, and implement next-generation construction monitoring systems.

We hope that this book will not only serve as a valuable scholarly reference but also inspire innovation, collaboration, and practical adoption of intelligent monitoring solutions. Ultimately, the contributions in this volume aim to shape the future of construction by promoting safer worksites, reduced environmental impact, and more effective risk governance through intelligent, real-time monitoring systems.

ORGANIZATION OF THE BOOK

Chapter 1: Intelligent Construction Monitoring Systems: Foundations, Architectures, and Future Directions

In this opening chapter, the Editors provide a conceptual and technological foundation for Intelligent Construction Monitoring Systems (ICMS). The chapter situates ICMS within the broader digital transformation of the construction industry and systematically examines the core architectures that enable real-time monitoring, prediction, and decision-making. Key enabling technologies—such as computer vision, machine learning, IoT platforms, digital twins, and cloud–edge computing—are discussed in terms of their roles and integration strategies. By outlining current capabilities alongside emerging research directions, this chapter establishes a unifying framework that guides the subsequent contributions and highlights future opportunities for intelligent, proactive, and scalable construction monitoring solutions.

Chapter 2: Dual-Dimension Monitoring and Data Management via MQTT—Temperature-Humidity and Helmet Safety

This chapter presents an integrated safety and environmental monitoring system that addresses the limitations of traditional, manual construction supervision. The Editors highlight how the authors combine computer vision–based helmet detection with IoT-enabled temperature and humidity sensing within a unified data management framework supported by the MQTT protocol. Through real-time visualization, intelligent threshold-based analysis, and automated reporting, the chapter demonstrates how dual-dimension monitoring can improve responsiveness, reduce bias, and enhance on-site safety awareness. The work provides a practical and extensible foundation for intelligent monitoring systems that simultaneously address worker protection and environmental conditions.

Chapter 3: An Integrated System for Real-Time Safety Helmet Monitoring and Advanced Environmental Prediction

Building upon helmet compliance monitoring, this chapter advances the discussion by integrating real-time environmental prediction into construction safety systems. The Editors emphasize the chapter's end-to-end design philosophy, which links monitoring, data transmission, visualization, and warning mechanisms through an IoT-enabled modular architecture. Notably, advanced predictive techniques—including wavelet decomposition, long short-term memory (LSTM) networks, and

Gaussian process regression—are employed to deliver both accurate forecasts and uncertainty-aware risk assessments. This contribution illustrates how combining compliance detection with predictive environmental intelligence can significantly enhance proactive risk management on construction sites.

Chapter 4: Real-Time Environmental Risk Management and Safety Monitoring System for Modular Construction

Focusing on modular integrated construction (MiC), this chapter addresses the unique environmental risks that affect quality, productivity, and safety in off-site and on-site assembly processes. The Editors note how the proposed IoT-based framework continuously monitors temperature, humidity, and wind speed, classifying risks into actionable levels tied to specific construction activities. By integrating real-time alerts, decision recommendations, and AI-based helmet detection, the system demonstrates tangible improvements in safety compliance and construction quality. This chapter underscores the importance of context-aware environmental thresholds and real-time decision support in modern construction methods.

Chapter 5: Real-Time AIoT-Based Fatigue Detection System for Enhancing Workers' Well-Being and Learning Efficiency

This chapter broadens the scope of intelligent monitoring beyond physical hazards to include human factors, particularly worker fatigue. The Editors highlight the innovative use of AIoT to fuse visual, physiological, behavioral, and environmental data into a unified fatigue assessment framework. Through non-intrusive sensing and weighted data fusion, the system delivers real-time fatigue scores and alerts, enabling timely interventions that promote worker well-being and productivity. The chapter demonstrates how intelligent monitoring systems can support not only safety compliance but also long-term health, learning efficiency, and workforce sustainability.

Chapter 6: Leveraging OWL-ViT and MQTT for Real-Time Construction Safety—An Integrated AIoT Monitoring Framework

In this chapter, the Editors draw attention to the application of advanced vision-language models and real-time communication protocols to enhance construction safety oversight. The proposed AIoT framework leverages OWL-ViT for automated helmet detection and MQTT for efficient client–server data exchange, enabling continuous monitoring of PPE compliance alongside environmental factors such as temperature, humidity, and noise. Experimental results demonstrate the system's operational viability and potential to reduce accident risks. This chapter exemplifies how cutting-edge deep learning models can be effectively embedded within practical, real-time construction monitoring infrastructures.

Chapter 7: Multi-Class Personal Protective Equipment Detection for Construction Safety Using YOLOv11

Addressing the broader challenge of PPE compliance, this chapter introduces a YOLOv11-based system capable of detecting multiple classes of protective equipment in real time. The Editors emphasize the chapter's contribution to proactive safety management through automated, high-frame-rate detection and alerting at construction site entry points. While performance varies across PPE categories, the results validate the feasibility of multi-class detection and identify data scalability as a key research challenge. This chapter contributes valuable insights into the design, evaluation, and deployment of deep learning–based PPE monitoring systems in real-world construction environments.

Chapter 8: An Integrated IoT Framework for Real-Time PPE Compliance Monitoring and Visual Analytics in Construction Sites

This chapter focuses on the integration of computer vision, IoT communication, and visual analytics to improve helmet compliance monitoring. The Editors note the system's modular design, in which an MQTT-based client interacts with a remote AI service to obtain detection results that are then analyzed and visualized over time. By transforming raw detection outputs into actionable compliance trends and situational awareness dashboards, the chapter demonstrates how intelligent visualization enhances both safety supervision and training. The work highlights the value of combining real-time analytics with intuitive visual feedback in construction safety management.

Chapter 9: A Python-Based AIoT Simulation Platform for Multi-Parameter Construction Safety Monitoring

Recognizing the challenges of deploying physical monitoring infrastructure, this chapter introduces a simulation-based AIoT platform for construction safety research and education. The Editors emphasize the platform's modular architecture, which replicates real-world workflows involving sensor data generation, MQTT communication, analytics, visualization, alerting, and logging. With strong validation results and long-duration stability, the simulator serves as both a research testbed and an educational tool. This chapter demonstrates how simulation can accelerate innovation, experimentation, and learning in intelligent construction monitoring without the constraints of physical deployment.

Chapter 10: Future Trends in Intelligent Construction Monitoring Technologies

Concluding the volume, this chapter looks ahead to the evolving landscape of intelligent construction monitoring. The Editors highlight emerging trends in advanced sensing, AI-driven analytics, automated reporting, and adaptive risk management, while also addressing critical challenges such as data interoperability, cybersecurity, and ethical considerations. By synthesizing technological advances with policy and governance perspectives, this chapter provides a forward-looking roadmap for researchers, practitioners, and decision-makers. It reinforces the book's central message: that intelligent, integrated, and responsible monitoring systems are essential to achieving safer, more efficient, and more sustainable construction practices in the future.

CONCLUSION

As Editors of this volume, we conclude by reaffirming the central premise that Intelligent Construction Monitoring Systems represent a transformative shift in how safety, environmental performance, and risk are managed across the construction lifecycle. The chapters in this edited reference book collectively demonstrate that construction monitoring is no longer limited to reactive inspections or isolated technological solutions. Instead, it is evolving into an integrated, intelligent, and adaptive ecosystem driven by real-time data, advanced analytics, and interdisciplinary innovation.

The contributions in this volume illustrate how emerging technologies—such as artificial intelligence, Internet of Things platforms, computer vision, digital twins, predictive modeling, and AIoT frameworks—can be orchestrated to deliver actionable insights and proactive interventions. From foundational system architectures to real-time PPE compliance, environmental prediction, fatigue detection, modular construction risk management, simulation-based platforms, and future-oriented perspectives, the chapters highlight both the maturity and the rapid evolution of intelligent monitoring solutions. Together, they underscore the importance of integrating human, environmental, and operational dimensions to achieve meaningful improvements in construction safety, quality, and sustainability.

Equally important, this book bridges theory and practice by presenting real-world case studies, applied frameworks, and validated systems alongside methodological and conceptual advances. By doing so, it addresses persistent challenges in construction, including high accident rates, environmental uncertainty, fragmented data flows, and limited decision support. The interdisciplinary nature of the contributions reflects the reality that effective construction monitoring requires collaboration among engineers, computer scientists, data analysts, project managers, technology developers, and policymakers.

Looking forward, the insights presented in this volume point toward a future in which construction sites function as intelligent, interconnected environments capable of self-monitoring, learning, and adaptation. While challenges related to scalability, interoperability, cybersecurity, ethics, and workforce adoption remain, the research and applications showcased in this book provide a strong foundation for continued progress. We hope that this volume will inspire further research, foster cross-sector collaboration, and support the responsible deployment of intelligent monitoring systems worldwide.

Ultimately, this edited reference book is intended to serve as both a scholarly benchmark and a practical guide for advancing Intelligent Construction Monitoring Systems. By promoting safer worksites, enhanced environmental stewardship, and more effective risk governance, the contributions in this volume collectively contribute to shaping a smarter, safer, and more sustainable future for the global construction industry.

Chapter 1
Intelligent Construction Monitoring Systems:
Foundations, Architectures, and Future Directions

Hewa Majeed Zangana
https://orcid.org/0000-0001-7909-254X
Duhok Polytechnic University, Iraq

Anuradha Rangarajan
https://orcid.org/0009-0006-7179-5404
Illinois Institute of Technology, USA

Aquil Mirza Mohammed
Hong Kong Polytechnic University, Hong Kong

Faisal Kaleem
Metro State Applied Innovation Institute, USA

Marwan Omar
https://orcid.org/0009-0000-7534-5490
Illinois Institute of Technology, USA

Jamal N. Al-Karaki
https://orcid.org/0009-0000-7833-3970
Zayed University, UAE

Xinwei Cao
Jiangnan University, China

ABSTRACT

The rapid digital transformation of the construction industry has driven the adoption of intelligent construction monitoring systems to enhance safety, productivity, and risk management in complex and dynamic project environments. These systems integrate advanced sensing technologies, artificial intelligence, Internet of Things (IoT) platforms, and data analytics to enable real-time observation, prediction, and decision-making throughout the construction lifecycle. This chapter presents a comprehensive overview of the foundations, system architectures, and emerging trends in intelligent construction monitoring systems. It examines key enabling technologies, including computer vision, machine learning, digital twins, and

DOI: 10.4018/979-8-3373-9245-5.ch001

cloud–edge computing frameworks, and discusses how these components are orchestrated to support real-time safety monitoring, environmental impact prediction, and proactive risk mitigation.

1. INTRODUCTION

The construction industry is undergoing a profound transformation driven by rapid advances in digital technologies, artificial intelligence (AI), and cyber-physical integration. Traditionally characterized by fragmented workflows, safety risks, cost overruns, and environmental uncertainty, construction environments are increasingly adopting intelligent monitoring systems to enable real-time visibility, predictive decision-making, and proactive risk mitigation. Intelligent Construction Monitoring Systems (ICMS) represent a convergence of sensing, communication, data analytics, and intelligent reasoning technologies that collectively support safer, more efficient, and more resilient construction processes (Irani & Kamal, 2014; Liu et al., 2022).

At the foundation of ICMS lies the evolution of intelligent systems research within the construction domain. Early efforts focused on rule-based expert systems and basic automation, but recent developments have shifted toward data-driven intelligence enabled by machine learning, deep learning, and distributed computing architectures (Lu, 2019; Sarker, 2021; Taye, 2023). These advances have been accelerated by the proliferation of wireless sensor networks (WSNs), smart sensors, and embedded systems capable of continuously capturing structural, environmental, and operational data from construction sites (Abdelhafidh et al., 2019; Preethichandra et al., 2023; Sonko et al., 2024).

The integration of Internet of Things (IoT) technologies has further expanded the scope and capabilities of construction monitoring. IoT-enabled architectures allow heterogeneous devices—such as wearable sensors, unmanned aerial vehicles (UAVs), ground vehicles, and smart equipment—to communicate seamlessly and operate within unified monitoring frameworks (Shafique et al., 2020; Saleem et al., 2019; Rane et al., 2023). These systems increasingly align with broader cyber-physical system (CPS) paradigms and smart city infrastructures, where construction sites function as intelligent nodes within urban ecosystems (Juma & Shaalan, 2020; Vinueza-Martinez et al., 2024; Mazzetto, 2024).

Artificial intelligence plays a central role in transforming raw construction data into actionable insights. Machine learning and deep learning models enable automated damage detection, structural health monitoring, safety risk prediction, and progress tracking with unprecedented accuracy (Mondal & Chen, 2022; Barbhuiya & Das, 2025; Wojciechowska et al., 2024). Computer vision techniques, often deployed through UAVs and fixed cameras, support real-time hazard identification, worker behavior analysis, and asset condition assessment (Telli et al., 2023; Wang et al., 2025). Similar AI-driven paradigms have proven effective in domains such as healthcare, manufacturing, agriculture, and transportation, offering transferable insights for construction monitoring applications (Li et al., 2022; Wu et al., 2025; Mahmud et al., 2025).

Recent advances in Building Information Modeling (BIM) and digital twin technologies further strengthen intelligent monitoring capabilities. The integration of BIM with AI enables dynamic synchronization between physical construction activities and virtual models, facilitating predictive scheduling, quality assurance, and lifecycle risk management (Pan & Zhang, 2023; Liu et al., 2022). Digital twins extend this concept by enabling continuous simulation, scenario analysis, and environmental impact forecasting, aligning construction monitoring with sustainability and resilience objectives (Mazzetto, 2024; Rane, 2023).

Emerging communication technologies such as 5G and future 6G networks are also reshaping ICMS architectures. Ultra-low latency, high bandwidth, and massive device connectivity enable real-time monitoring at scale, supporting applications such as autonomous equipment coordination, edge intelligence, and immersive visualization (Chowdhury et al., 2020; Shafique et al., 2020). These developments are closely aligned with Industry 4.0 and Industry 5.0 paradigms, which emphasize human-centric automation, interoperability, and intelligent decision support across industrial systems (Xu et al., 2018; Xiang et al., 2023; Verma et al., 2022).

Beyond technical innovation, intelligent construction monitoring systems must address critical challenges related to security, privacy, trust, and governance. As construction sites become data-intensive and interconnected, vulnerabilities related to cyberattacks, authentication, and data misuse become increasingly significant (Aouedi et al., 2024; Shende et al., 2024; Zangana, 2025). Emerging frameworks such as Trust, Risk, and Security Management (TRiSM) and model context protocols highlight the importance of explainability, accountability, and ethical AI deployment in intelligent systems (Ray, 2025a; Ray, 2025b). Blockchain technologies are also being explored to enhance data integrity, transparency, and decentralized trust in construction monitoring ecosystems (Verma et al., 2022).

Furthermore, interdisciplinary innovations—including robotics, agentic AI, quantum computing, and large language models—are expected to influence the next generation of intelligent monitoring systems. Autonomous robots and UAVs enhance site inspection and hazardous task execution, while agent-based AI systems support adaptive coordination and decision-making (Ryalat et al., 2025; Kostopoulos et al., 2025). Although quantum computing applications in construction remain exploratory, insights from other sectors suggest long-term potential for optimization and large-scale simulation (Bukkarayasamudram et al., 2025).

In this context, this chapter aims to provide a structured and comprehensive introduction to intelligent construction monitoring systems by synthesizing foundational concepts, architectural paradigms, and emerging research directions. Drawing on advancements across AI, IoT, CPS, smart cities, and Industry 5.0, the chapter establishes a conceptual bridge between theory and practice. By doing so, it supports researchers, practitioners, and policymakers in understanding how intelligent monitoring systems can enable real-time safety assurance, environmental prediction, and risk-aware decision-making in modern construction projects (Adebayo et al., 2025; Ali et al., 2024).

2. TRADITIONAL CONSTRUCTION MONITORING: LIMITATIONS AND CHALLENGES

Traditional construction monitoring has historically relied on manual inspections, periodic reporting, and fragmented technological tools to oversee project progress, safety compliance, and structural integrity. While these approaches laid the groundwork for modern construction management practices, they are increasingly inadequate in addressing the complexity, scale, and dynamism of contemporary construction projects. The growing demand for real-time safety assurance, environmental sustainability, and proactive risk management has exposed fundamental limitations in conventional monitoring paradigms (Irani & Kamal, 2014; Liu et al., 2022).

2.1 Dependence on Manual and Periodic Inspections

One of the most significant limitations of traditional construction monitoring is its heavy dependence on manual inspections and human judgment. Site supervisors and engineers typically conduct visual inspections at scheduled intervals, which limits the ability to detect rapid changes, emerging hazards, or early-stage structural damage. Such practices are time-consuming, labor-intensive, and prone to subjectivity and human error (Mondal & Chen, 2022; Barbhuiya & Das, 2025). In contrast to continuous data-driven monitoring, periodic assessments fail to capture transient events such as sudden material degradation, unsafe worker behavior, or environmental fluctuations.

Moreover, manual monitoring methods struggle to scale across large or geographically distributed construction sites, leading to inconsistent data quality and delayed decision-making. These challenges are exacerbated in complex projects such as high-rise buildings, infrastructure corridors, and heritage structures, where comprehensive real-time visibility is essential (Wojciechowska et al., 2024).

2.2 Limited Sensing and Data Collection Capabilities

Traditional monitoring systems often employ isolated sensors or basic instrumentation without integrated networking or intelligence. Early wireless sensor network (WSN) deployments, while innovative for their time, were constrained by limited energy efficiency, data transmission reliability, and interoperability (Abdelhafidh et al., 2019). These systems typically operated in silos, providing localized measurements without holistic situational awareness.

The absence of smart sensors capable of adaptive sensing, self-calibration, and intelligent data fusion further limits traditional monitoring effectiveness. As highlighted by Preethichandra et al. (2023), legacy sensor-based systems are often unable to support predictive maintenance or lifecycle asset management due to insufficient data granularity and contextual awareness. Similar limitations have been observed across other cyber-physical domains, including smart grids and autonomous systems, underscoring the need for more integrated and intelligent sensing infrastructures (Saleem et al., 2019; Sonko et al., 2024).

2.3 Lack of Real-Time Data Processing and Predictive Intelligence

Traditional construction monitoring primarily focuses on descriptive and retrospective analysis rather than predictive or prescriptive intelligence. Data, when collected, is often processed offline using static analytical methods, preventing timely responses to emerging risks. This limitation contrasts sharply with modern AI-driven approaches that leverage machine learning and deep learning to identify patterns, anomalies, and future risk trajectories (Lu, 2019; Sarker, 2021; Taye, 2023).

Without intelligent analytics, traditional systems are unable to perform automated damage detection, safety risk forecasting, or environmental impact prediction. Studies in civil infrastructure monitoring consistently demonstrate that conventional methods fail to detect early-stage failures that could otherwise be identified using AI-based techniques (Mondal & Chen, 2022; Barbhuiya & Das, 2025). Similar gaps have been documented in other data-intensive fields such as healthcare, agriculture, and transportation, where legacy systems lack adaptive learning capabilities (Li et al., 2022; Wu et al., 2025; Mahmud et al., 2025).

2.4 Fragmentation and Poor System Integration

Another critical challenge of traditional construction monitoring lies in fragmented system architectures. Monitoring tools for safety, structural health, scheduling, and environmental assessment are often deployed independently, resulting in data silos and limited cross-functional coordination. This fragmentation undermines holistic decision-making and prevents seamless integration with broader project management platforms such as Building Information Modeling (BIM) systems (Pan & Zhang, 2023).

The absence of integrated cyber-physical system (CPS) frameworks further constrains interoperability between physical construction assets and digital representations. As construction sites increasingly function as components of smart cities and urban digital ecosystems, traditional monitoring approaches struggle to align with CPS and smart city architectures (Juma & Shaalan, 2020; Vinueza-Martinez et al., 2024; Mazzetto, 2024).

2.5 Communication and Connectivity Constraints

Conventional monitoring systems are also limited by communication infrastructure constraints. Many traditional setups rely on wired networks or low-bandwidth wireless technologies that cannot support high-frequency data transmission, real-time video analytics, or large-scale device connectivity. These limitations hinder the deployment of advanced monitoring tools such as UAV-based inspection, autonomous equipment coordination, and immersive visualization (Telli et al., 2023; Wang et al., 2025).

With the emergence of 5G and anticipated 6G communication systems, the gap between traditional and modern monitoring approaches has become increasingly pronounced. Legacy systems are ill-equipped to leverage ultra-low latency, massive machine-type communication, and edge intelligence capabilities envisioned for next-generation construction environments (Chowdhury et al., 2020; Shafique et al., 2020).

2.6 Security, Privacy, and Trust Deficiencies

Security and trust represent major shortcomings of traditional construction monitoring systems. Early monitoring frameworks were not designed with cybersecurity, authentication, or privacy preservation as core requirements. As a result, data integrity, access control, and system resilience remain vulnerable, particularly when systems are retrofitted with network connectivity (Aouedi et al., 2024; Shende et al., 2024).

The lack of structured trust, risk, and security management frameworks further limits the reliability of traditional systems. Emerging concepts such as TRiSM and model context protocols emphasize explainability, accountability, and governance—dimensions largely absent in legacy monitoring architectures (Ray, 2025a; Ray, 2025b). These issues are especially critical in safety-critical construction environments, where compromised data can lead to severe human and financial consequences (Zangana, 2025).

2.7 Inability to Support Emerging Construction Paradigms

Traditional monitoring approaches are fundamentally misaligned with emerging paradigms such as Industry 4.0, Industry 5.0, and intelligent automation. Modern construction increasingly requires human-centric, resilient, and collaborative systems that integrate AI, robotics, IoT, and blockchain technologies (Xu et al., 2018; Xiang et al., 2023; Verma et al., 2022). However, legacy monitoring frameworks lack the flexibility and intelligence required to support these transformations.

Advanced applications such as agentic AI, autonomous robotics, and collaborative UAV–ground systems remain largely incompatible with traditional monitoring infrastructures (Ryalat et al., 2025; Kostopoulos et al., 2025; Wang et al., 2025). Even forward-looking technologies such as quantum computing, while still exploratory in construction, highlight the widening gap between conventional monitoring methods and future computational paradigms (Bukkarayasamudram et al., 2025).

2.8 Summary of Key Challenges

In summary, traditional construction monitoring systems face critical challenges related to manual dependency, limited sensing and intelligence, fragmented architectures, communication constraints, security vulnerabilities, and misalignment with emerging digital paradigms. These limitations hinder the ability of construction projects to achieve real-time safety assurance, predictive risk management, and sustainable performance. The shortcomings identified in this section provide a strong justification for the transition toward intelligent construction monitoring systems, which are explored in subsequent sections of this chapter (Adebayo et al., 2025; Ali et al., 2024; Rane, 2023).

To clearly illustrate the transition from conventional construction monitoring approaches to intelligent, data-driven systems, Figure 1 presents a conceptual comparison between traditional and intelligent construction monitoring paradigms. The diagram highlights the key technological, architectural, and functional differences that motivate the shift toward intelligent construction monitoring systems (ICMS).

Figure 1. Evolution from traditional to intelligent construction monitoring systems

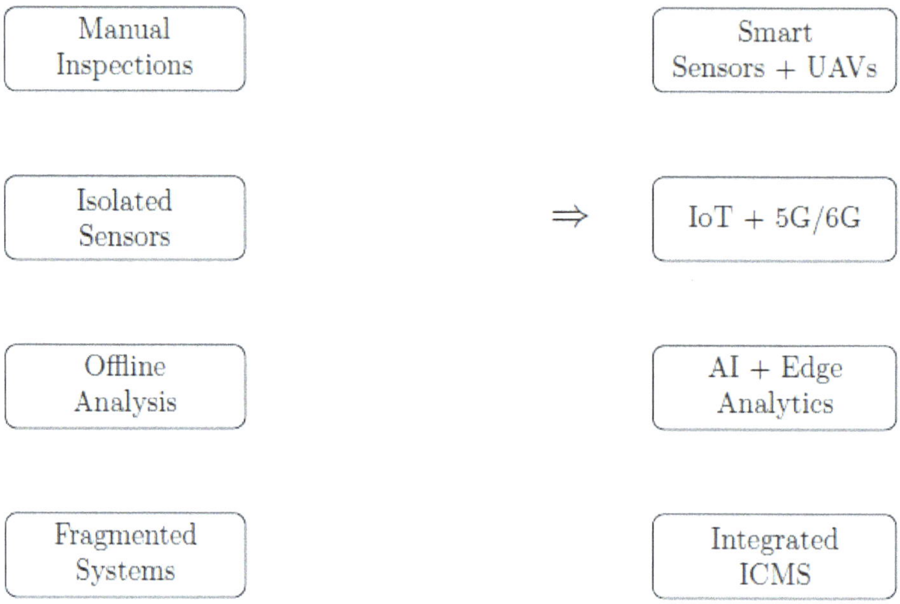

3. FOUNDATIONS OF INTELLIGENT CONSTRUCTION MONITORING SYSTEMS (ICMS)

Intelligent Construction Monitoring Systems (ICMS) represent a paradigm shift from reactive and fragmented oversight mechanisms toward integrated, data-driven, and autonomous construction management frameworks. The foundations of ICMS are rooted in advances across artificial intelligence, Internet of Things (IoT), cyber-physical systems (CPS), sensing technologies, and digital infrastructure. Together, these foundations enable continuous situational awareness, predictive analytics, and adaptive decision-making across the construction lifecycle (Irani & Kamal, 2014; Liu et al., 2022).

3.1 Evolution from Conventional Monitoring to Intelligent Systems

The conceptual origins of ICMS can be traced to early intelligent systems research in the construction industry, which initially focused on expert systems and decision-support tools for scheduling, cost estimation, and quality control (Irani &

Kamal, 2014). While these systems provided structured reasoning capabilities, they lacked real-time data integration and adaptive learning.

The emergence of machine learning and deep learning significantly expanded the intelligence layer of construction monitoring. Modern ICMS leverage supervised, unsupervised, and reinforcement learning models to interpret complex, high-dimensional data streams generated by sensors, images, and operational logs (Lu, 2019; Sarker, 2021; Taye, 2023). These capabilities enable automated damage detection, anomaly identification, and predictive maintenance, which are foundational to intelligent monitoring in both building and infrastructure projects (Mondal & Chen, 2022; Barbhuiya & Das, 2025).

3.2 Sensing Infrastructure and Data Acquisition Foundations

Robust sensing infrastructure forms the physical backbone of ICMS. Wireless sensor networks (WSNs) enable distributed data acquisition across structural, environmental, and operational dimensions, including strain, vibration, temperature, humidity, and equipment performance (Abdelhafidh et al., 2019). Unlike traditional sensors, smart sensors in ICMS support adaptive sampling, local preprocessing, and event-driven data transmission, significantly improving efficiency and reliability (Preethichandra et al., 2023).

Advances in embedded systems and autonomous platforms further enhance data acquisition capabilities. UAVs, ground robots, and mobile sensor units provide flexible, non-intrusive monitoring of large or hazardous construction environments, enabling visual inspection, progress tracking, and safety assessment from multiple perspectives (Telli et al., 2023; Wang et al., 2025; Sonko et al., 2024). These technologies mirror developments in autonomous transportation and robotics research, reinforcing their relevance to construction monitoring (Ryalat et al., 2025).

3.3 Internet of Things and Cyber-Physical Integration

The Internet of Things serves as a core enabler of ICMS by facilitating seamless connectivity among heterogeneous sensing devices, equipment, and digital platforms. IoT-based architectures support real-time data exchange, system scalability, and interoperability, enabling construction sites to function as dynamic cyber-physical environments (Shafique et al., 2020; Saleem et al., 2019).

Intelligent IoT extends these capabilities by embedding AI-driven analytics, security mechanisms, and context-awareness directly into networked devices (Aouedi et al., 2024). Through CPS integration, ICMS bridge the physical construction environment with computational intelligence, enabling feedback loops between sensing, analysis, and control actions (Juma & Shaalan, 2020). Such integration aligns construction monitoring with broader smart city ecosystems and urban digital infrastructures (Vinueza-Martinez et al., 2024; Mazzetto, 2024).

3.4 Communication Technologies and Edge–Cloud Computing

Advanced communication technologies are foundational to the real-time operation of ICMS. Emerging 5G and future 6G networks provide ultra-low latency, high reliability, and massive device connectivity, enabling continuous monitoring, high-resolution video analytics, and autonomous system coordination (Chowdhury et al., 2020; Shafique et al., 2020).

Edge and cloud computing architectures complement these communication capabilities by distributing intelligence across multiple computational layers. Edge computing enables low-latency processing near data sources, supporting time-critical safety applications, while cloud platforms facilitate large-scale analytics, historical data storage, and cross-project learning (Deka et al., 2024). This layered architecture enhances system resilience and scalability, mirroring best practices observed in healthcare, transportation, and industrial automation systems (Mahmud et al., 2025).

3.5 Artificial Intelligence and Data Intelligence Foundations

AI constitutes the analytical core of ICMS, transforming raw sensor data into actionable insights. Machine learning and deep learning models support image-based inspection, structural health assessment, safety risk prediction, and progress monitoring (Pan & Zhang, 2023; Adebayo et al., 2025). Computer vision techniques, widely applied in domains such as medical imaging and manufacturing, are increasingly adapted for construction monitoring to detect defects, unsafe behaviors, and environmental hazards (Li et al., 2022; Wu et al., 2025).

Recent advancements in agentic AI and large language models further extend the intelligence layer of ICMS. Agent-based systems enable autonomous coordination among monitoring components, while language-based models facilitate human–machine interaction, knowledge extraction, and decision support (Kostopoulos et al., 2025; Mahmud et al., 2025). These developments support a transition from passive monitoring toward proactive and self-adaptive construction environments.

3.6 Digital Twins and Information Modeling Foundations

Digital twin technology provides a critical foundation for integrating real-time monitoring data with virtual representations of construction assets. By synchronizing sensor inputs with BIM models, ICMS enable continuous comparison between planned and actual construction states, supporting predictive scheduling, quality control, and risk assessment (Pan & Zhang, 2023; Liu et al., 2022).

Digital twins also support scenario simulation and environmental impact prediction, aligning construction monitoring with sustainability and resilience objectives. Their growing adoption in smart cities and infrastructure management highlights their importance as a foundational component of next-generation ICMS (Mazzetto, 2024; Wojciechowska et al., 2024).

3.7 Security, Trust, and Governance Foundations

As ICMS become increasingly interconnected and data-intensive, security and trust emerge as foundational requirements rather than optional features. Intelligent monitoring systems must protect sensitive data, ensure system integrity, and support trustworthy decision-making. IoT security, authentication mechanisms, and privacy-preserving analytics are therefore integral to ICMS design (Aouedi et al., 2024; Shende et al., 2024).

Emerging governance frameworks such as Trust, Risk, and Security Management (TRiSM) and model context protocols emphasize explainability, accountability, and risk-aware AI deployment (Ray, 2025a; Ray, 2025b). Blockchain technologies further contribute to trust by enabling tamper-resistant data sharing and decentralized coordination, particularly in multi-stakeholder construction environments (Verma et al., 2022). These approaches align with defense-in-depth cybersecurity principles, which are increasingly relevant to safety-critical construction systems (Zangana, 2025).

3.8 Alignment with Industry 4.0 and Industry 5.0 Paradigms

ICMS are fundamentally aligned with Industry 4.0 principles, including automation, interoperability, and data-driven optimization (Xu et al., 2018). However, emerging Industry 5.0 concepts place greater emphasis on human-centric design, resilience, and ethical intelligence, further shaping ICMS foundations (Xiang et al., 2023).

Technologies such as robotics, autonomous systems, and blockchain enable collaborative and transparent construction ecosystems, while exploratory paradigms such as quantum computing highlight future opportunities for large-scale optimization and simulation (Ryalat et al., 2025; Verma et al., 2022; Bukkarayasamudram et al., 2025). These foundations collectively position ICMS as a cornerstone of intelligent, sustainable, and risk-aware construction practices (Ali et al., 2024; Rane, 2023; Rane et al., 2023).

4. SYSTEM ARCHITECTURES FOR INTELLIGENT CONSTRUCTION MONITORING SYSTEMS (ICMS)

Intelligent Construction Monitoring Systems (ICMS) rely on multi-layered, cyber–physical architectures that integrate sensing, communication, computation, intelligence, and decision-support components to enable real-time safety assurance, environmental prediction, and risk management. Modern ICMS architectures are influenced by advances in intelligent systems research within construction, Industry 4.0 and 5.0 paradigms, cyber-physical systems, and intelligent Internet of Things (IoT) frameworks (Irani & Kamal, 2014; Xu et al., 2018; Xiang et al., 2023). This section presents a detailed architectural analysis of ICMS, covering layered designs, data acquisition mechanisms, communication infrastructures, intelligence layers, integration platforms, and security-aware architectural considerations.

4.1 Layered Architecture of ICMS

Most ICMS implementations adopt a layered architecture to ensure modularity, scalability, and interoperability. A typical ICMS consists of perception, communication, middleware, intelligence, and application layers, similar to architectures proposed for wireless sensor networks and intelligent IoT systems (Abdelhafidh et al., 2019; Aouedi et al., 2024).

Figure 2 illustrates the layered architectural structure of Intelligent Construction Monitoring Systems (ICMS). The model highlights how sensing, communication, intelligence, and application layers interact to enable real-time safety monitoring, predictive analytics, and integrated decision support.

Figure 2. Layered architecture of intelligent construction monitoring systems

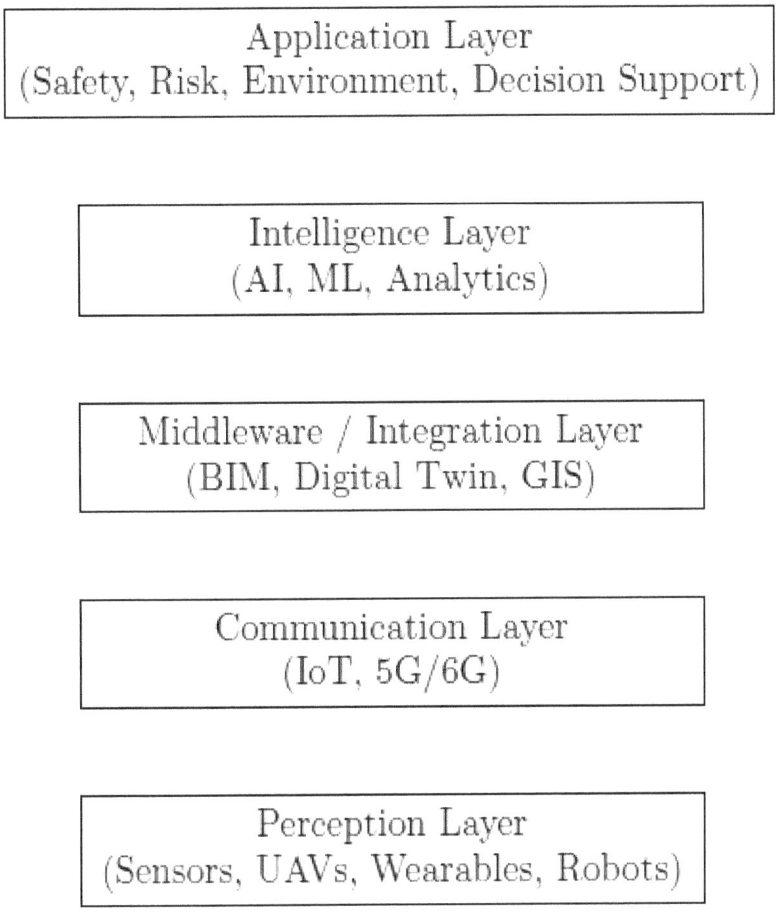

The perception layer includes heterogeneous sensing devices such as smart sensors, embedded systems, unmanned aerial vehicles (UAVs), and robotic platforms that capture real-time construction data related to structural integrity, worker behavior, equipment usage, and environmental conditions (Preethichandra et al., 2023; Telli et al., 2023; Ryalat et al., 2025). These sensors form the foundation for structural health monitoring and damage detection in concrete and civil infrastructure systems (Barbhuiya & Das, 2025; Mondal & Chen, 2022).

The communication layer ensures reliable data transmission between field devices and computational platforms. IoT-based networking architectures, including 5G- and emerging 6G-enabled systems, are increasingly adopted to support low-latency, high-bandwidth, and massive device connectivity requirements in construction environments (Shafique et al., 2020; Chowdhury et al., 2020). These communication models align with architectures used in smart grids and smart cities, reinforcing cross-domain convergence (Saleem et al., 2019; Juma & Shaalan, 2020).

4.2 Edge, Cloud, and Serverless Computing Architectures

To handle the volume, velocity, and variety of construction data, ICMS architectures increasingly integrate edge–cloud–serverless computing paradigms. Edge computing enables on-site processing for time-critical safety applications, such as fall detection or equipment collision avoidance, reducing dependency on centralized resources (Rane et al., 2023). Cloud platforms, in contrast, provide scalable storage, long-term analytics, and model training capabilities for predictive risk assessment and environmental forecasting (Rane, 2023).

Recent architectural trends borrow concepts from serverless computing models used in smart healthcare systems, enabling event-driven data processing and elastic resource allocation (Deka et al., 2024). These architectures improve cost efficiency and adaptability, particularly for large-scale or geographically distributed construction projects.

4.3 Intelligence and Analytics Layer

At the core of ICMS lies the intelligence layer, where artificial intelligence (AI), machine learning (ML), and deep learning models transform raw data into actionable insights. Deep learning architectures are widely applied for vision-based safety monitoring, damage detection, and anomaly recognition, leveraging advances in convolutional and hybrid neural networks (Sarker, 2021; Taye, 2023).

The evolution of AI models and workflows has significantly influenced ICMS architecture design, enabling predictive and prescriptive analytics rather than reactive monitoring (Lu, 2019; Adebayo et al., 2025). AI-driven damage detection systems and infrastructure health monitoring frameworks illustrate how intelligence can be embedded into system architectures to enable continuous assessment and early warning mechanisms (Barbhuiya & Das, 2025; Liu et al., 2022).

Emerging architectural enhancements include the integration of agentic AI and large language models (LLMs) to support adaptive decision-making, automated reporting, and human–machine collaboration within construction monitoring platforms (Kostopoulos et al., 2025; Mahmud et al., 2025). These intelligent agents improve system autonomy and contextual understanding across complex construction scenarios.

4.4 Integration with BIM, Digital Twins, and GIS

Modern ICMS architectures increasingly incorporate Building Information Modeling (BIM), digital twins, and Geographic Information Systems (GIS) as integrative platforms. BIM-AI integration enables real-time synchronization between physical construction activities and digital representations, supporting progress tracking, risk visualization, and resource optimization (Pan & Zhang, 2023).

Digital twin architectures extend BIM by enabling continuous bi-directional data exchange between physical assets and virtual models, allowing predictive simulation and scenario analysis (Mazzetto, 2024). GIS-based WebGIS architectures further enhance spatial awareness by contextualizing construction data within geographic and environmental dimensions, supporting large-scale monitoring and urban infrastructure management (Vinueza-Martinez et al., 2024).

4.5 Autonomous Systems and Robotic Integration

ICMS architectures increasingly support autonomous and semi-autonomous systems, including UAVs, ground vehicles, and robotic inspection platforms. UAV-based sensing architectures provide flexible and high-resolution data acquisition for site inspection, progress monitoring, and hazard detection (Telli et al., 2023; Wang et al., 2025). These systems share architectural principles with embedded systems in autonomous vehicles, emphasizing real-time processing, sensor fusion, and fault tolerance (Sonko et al., 2024).

Robotic and cyber-physical system integration further enhances ICMS capabilities by enabling automated inspection and intervention, aligning with broader trends in robotics research and smart city development (Juma & Shaalan, 2020; Ryalat et al., 2025).

4.6 Security, Trust, and Governance in ICMS Architectures

As ICMS architectures become more interconnected and intelligent, security, trust, and governance emerge as critical architectural requirements. Blockchain-enabled architectures are increasingly proposed to ensure data integrity, traceability, and decentralized trust across construction stakeholders (Verma et al., 2022). These approaches align with Industry 5.0 visions emphasizing human-centric, resilient, and trustworthy systems (Xiang et al., 2023).

Advanced authentication mechanisms based on deep learning further enhance device-level security in heterogeneous ICMS environments (Shende et al., 2024). Additionally, Trust, Risk, and Security Management (TRiSM) frameworks and model context protocols provide architectural guidance for managing AI risks, transparency, and lifecycle governance in intelligent monitoring systems (Ray, 2025a; Ray, 2025b).

From a strategic perspective, defense-in-depth architectural principles are increasingly adopted to protect ICMS against cyber threats, ensuring layered security across sensing, communication, intelligence, and application components (Zangana, 2025).

4.7 Future-Ready Architectural Trends

Looking forward, ICMS architectures are expected to evolve toward hyper-connected, adaptive, and human-centric systems. Emerging technologies such as quantum computing, although currently explored in domains like healthcare, offer long-term potential for optimizing large-scale construction analytics and complex decision problems (Bukkarayasamudram et al., 2025). Cross-domain learning from agriculture, manufacturing, and smart infrastructure further informs scalable and resilient architectural designs (Wu et al., 2025; Xiang et al., 2023).

Overall, system architectures for ICMS represent a convergence of intelligent sensing, advanced communication, AI-driven analytics, and secure integration frameworks. These architectures provide the structural backbone required to realize real-time safety monitoring, environmental prediction, and proactive risk management in next-generation construction environments (Ali et al., 2024; Wojciechowska et al., 2024).

5. DATA ACQUISITION AND SENSOR TECHNOLOGIES

Data acquisition and sensor technologies constitute the foundational layer of Intelligent Construction Monitoring Systems (ICMS), enabling continuous, real-time observation of physical, environmental, and operational conditions across construction sites. Advances in intelligent sensing, embedded systems, wireless communication, and cyber–physical integration have significantly expanded the scope, accuracy, and reliability of construction data collection (Irani & Kamal, 2014; Liu et al., 2022). This section provides an in-depth examination of sensor typologies, data acquisition mechanisms, intelligent sensing frameworks, and emerging trends shaping next-generation ICMS.

To illustrate the diversity of sensing technologies used in intelligent construction monitoring, Figure 3 presents the sensor ecosystem responsible for data acquisition in ICMS. The diagram shows how environmental, structural, visual, and operational data are collected using heterogeneous sensing platforms.

Figure 3. Data acquisition and sensor ecosystem in ICMS

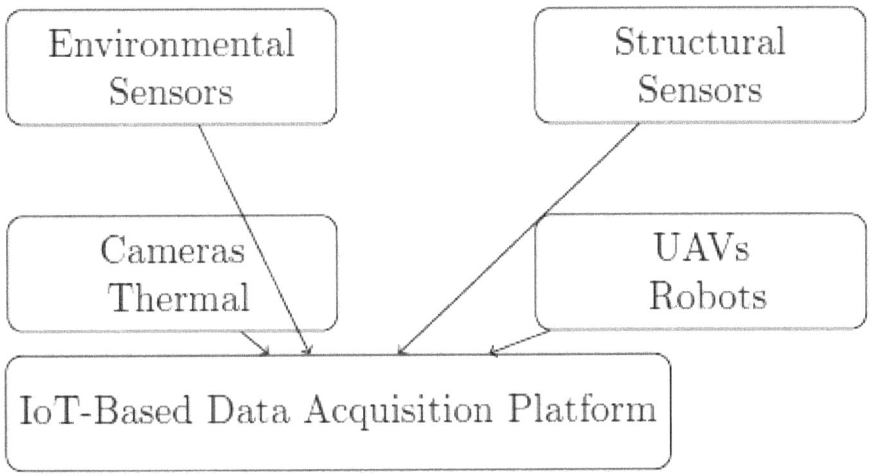

5.1 Wireless Sensor Networks and Smart Sensing Infrastructure

Wireless Sensor Networks (WSNs) represent a core component of ICMS data acquisition architectures. These networks consist of distributed sensor nodes capable of sensing, processing, and wirelessly transmitting data related to structural behavior, environmental parameters, and site operations (Abdelhafidh et al., 2019). In construction contexts, WSNs support scalable monitoring of temperature, humidity, vibration, strain, noise, and air quality, enabling proactive safety and environmental management.

The integration of **smart sensors**—equipped with embedded processing and communication capabilities—has further enhanced data fidelity and autonomy in monitoring systems (Preethichandra et al., 2023). These sensors support self-calibration, adaptive sampling, and localized decision-making, reducing data redundancy and communication overhead. Such capabilities are critical for long-term structural health monitoring and asset management in intelligent construction environments (Mondal & Chen, 2022).

5.2 IoT-Enabled Data Acquisition Frameworks

The convergence of WSNs with the Internet of Things (IoT) has enabled interoperable, intelligent data acquisition frameworks that connect heterogeneous devices across construction ecosystems. Intelligent IoT architectures facilitate seamless integration of sensors, actuators, and analytics platforms, enabling real-time data exchange and system-wide situational awareness (Aouedi et al., 2024).

IoT-enabled data acquisition supports the monitoring of construction machinery, worker safety equipment, and environmental conditions, aligning with broader trends in smart and sustainable architecture, engineering, and construction (AEC) systems (Rane, 2023; Rane et al., 2023). These frameworks leverage advances in 5G and emerging 6G communication technologies to achieve ultra-low latency, high reliability, and massive device connectivity in dense construction environments (Shafique et al., 2020; Chowdhury et al., 2020).

5.3 Vision-Based and Imaging Sensors

Vision-based sensors, including RGB cameras, thermal imagers, and multispectral imaging devices, play a critical role in capturing rich visual data for safety monitoring, progress tracking, and defect detection. Advances in computer vision and deep learning have significantly enhanced the interpretability and utility of visual data in construction monitoring systems (Sarker, 2021; Taye, 2023).

Vision-based sensing approaches in ICMS draw methodological inspiration from medical and industrial imaging domains, where AI-driven image analysis has demonstrated high accuracy in pattern recognition and anomaly detection (Li et al., 2022). In construction, these techniques are applied to detect cracks, surface degradation, unsafe worker behaviors, and equipment misuse, supporting automated and scalable inspection processes (Barbhuiya & Das, 2025).

5.4 Structural Health Monitoring Sensors

Structural Health Monitoring (SHM) systems rely on specialized sensors such as strain gauges, accelerometers, fiber optic sensors, and acoustic emission devices to assess the integrity and performance of construction assets. These sensors enable continuous monitoring of load responses, vibration characteristics, and damage progression in concrete and civil infrastructure systems (Preethichandra et al., 2023; Mondal & Chen, 2022).

AI-enhanced SHM architectures integrate sensor data with machine learning models to enable early damage detection and predictive maintenance (Barbhuiya & Das, 2025). Such systems are particularly valuable for large-scale and complex construction projects, where manual inspection is costly, time-consuming, and prone to human error (Liu et al., 2022).

5.5 Embedded Systems, Robotics, and Autonomous Platforms

Embedded sensing systems form the backbone of autonomous monitoring platforms in ICMS, including robotic inspection units, unmanned aerial vehicles (UAVs), and ground-based autonomous systems. Embedded processors enable real-time sensor fusion, local analytics, and adaptive control, supporting autonomous data acquisition in hazardous or inaccessible construction zones (Sonko et al., 2024).

UAV-based sensing systems provide flexible and high-resolution data acquisition for site surveying, progress monitoring, and environmental assessment (Telli et al., 2023). These platforms are increasingly integrated with ground vehicles and robotic systems to support coordinated monitoring and emergency response operations (Wang et al., 2025). The adoption of robotics in construction monitoring aligns with broader research trends in autonomous systems and robotics education (Ryalat et al., 2025).

5.6 BIM, Digital Twins, and GIS-Based Data Integration

Sensor-generated data gains significant value when integrated with digital representations such as Building Information Modeling (BIM), digital twins, and Geographic Information Systems (GIS). BIM-integrated sensing enables real-time synchronization between physical construction activities and digital project models, enhancing accuracy in progress tracking and risk assessment (Pan & Zhang, 2023).

Digital twins extend sensor integration by enabling continuous bi-directional data flows between physical assets and virtual models, supporting simulation-based decision-making and predictive analytics (Mazzetto, 2024). GIS-based WebGIS architectures further contextualize sensor data spatially, enabling large-scale environmental monitoring and infrastructure planning (Vinueza-Martinez et al., 2024). These integrated data acquisition ecosystems are particularly valuable in urban and heritage construction contexts (Wojciechowska et al., 2024).

5.7 Intelligent Analytics and Context-Aware Data Acquisition

Modern ICMS increasingly incorporate **context-aware sensing** and intelligent analytics to optimize data acquisition strategies. Machine learning and deep learning models dynamically adjust sampling rates, sensor activation, and data transmission based on contextual conditions and risk levels (Lu, 2019; Sarker, 2021). This adaptive sensing paradigm improves system efficiency while maintaining high monitoring fidelity.

The emergence of agentic AI and large language models further enhances contextual interpretation of sensor data, enabling intelligent summarization, anomaly explanation, and decision support (Kostopoulos et al., 2025; Mahmud et al., 2025).

These approaches align with AI evolution trends across construction project management and intelligent systems research (Adebayo et al., 2025; Ali et al., 2024).

5.8 Security, Trust, and Governance in Sensor Data Acquisition

As sensor networks expand, ensuring the security and trustworthiness of acquired data becomes a critical challenge. Blockchain-enabled data acquisition architectures provide immutable logging, decentralized trust, and secure data sharing among construction stakeholders (Verma et al., 2022). These approaches are aligned with Industry 5.0 principles emphasizing resilience, transparency, and human-centric design (Xiang et al., 2023).

Advanced authentication mechanisms based on deep learning further protect sensor nodes and smart devices from unauthorized access and cyber threats (Shende et al., 2024). Trust, Risk, and Security Management (TRiSM) frameworks and model context protocols provide governance mechanisms for managing AI-driven data pipelines and ensuring compliance with ethical and operational standards (Ray, 2025a; Ray, 2025b). Defense-in-depth strategies further reinforce multi-layered protection across sensing, communication, and analytics layers (Zangana, 2025).

5.9 Cross-Domain Influences and Emerging Directions

Data acquisition technologies in ICMS increasingly benefit from cross-domain innovation. Lessons from smart grids, healthcare systems, agriculture, and advanced manufacturing inform scalable, resilient sensing architectures (Saleem et al., 2019; Deka et al., 2024; Wu et al., 2025). Industry 4.0 and 5.0 paradigms further influence sensor integration strategies, emphasizing interoperability, sustainability, and human–machine collaboration (Xu et al., 2018; Xiang et al., 2023).

Looking forward, emerging computational paradigms such as quantum computing, although currently explored primarily in healthcare, may offer future opportunities for optimizing large-scale sensor data processing and complex decision-making in construction monitoring systems (Bukkarayasamudram et al., 2025).

6. INTELLIGENT DATA PROCESSING AND ANALYTICS

Intelligent data processing and analytics form the cognitive core of Intelligent Construction Monitoring Systems (ICMS), transforming raw, heterogeneous sensor data into actionable knowledge that supports real-time safety assurance, environmental prediction, and proactive risk management. The evolution of intelligent analytics in construction has been shaped by advances in artificial intelligence (AI), machine learning (ML), deep learning (DL), big data technologies, and cyber–physical systems, enabling a shift from descriptive monitoring toward predictive and prescriptive intelligence (Irani & Kamal, 2014; Lu, 2019; Liu et al., 2022). This section examines data processing pipelines, AI-driven analytics, context-aware intelligence, integration frameworks, and future-ready computational paradigms within ICMS.

Figure 4 illustrates the intelligent data processing pipeline used in ICMS, showing how raw sensor data is transformed into actionable knowledge through preprocessing, AI analytics, and decision support mechanisms.

Figure 4. Intelligent data processing pipeline in ICMS

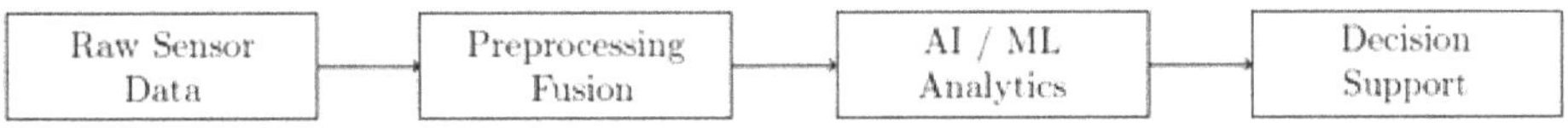

6.1 Data Preprocessing and Fusion in ICMS

Construction monitoring environments generate large volumes of noisy, incomplete, and heterogeneous data from distributed sensors, vision systems, and autonomous platforms. Effective data preprocessing—including filtering, normalization, synchronization, and anomaly removal—is therefore essential for reliable analytics (Abdelhafidh et al., 2019). Wireless sensor network data often require temporal alignment and redundancy reduction to address communication constraints and sensor failures.

Data fusion techniques integrate multi-source data streams, combining physical sensor readings, visual data, and contextual information to enhance situational awareness and analytical robustness (Preethichandra et al., 2023). Such fusion is particularly important in cyber–physical construction systems, where environmental, structural, and human-related data must be jointly analyzed to assess safety and operational risks (Juma & Shaalan, 2020; Liu et al., 2022).

6.2 Machine Learning and Deep Learning for Construction Analytics

Machine learning and deep learning techniques underpin most intelligent analytics capabilities in ICMS. Supervised, unsupervised, and reinforcement learning models are applied to detect patterns, classify events, and predict future states of construction processes (Sarker, 2021; Taye, 2023). These methods support applications such as hazard detection, productivity analysis, equipment monitoring, and schedule deviation forecasting.

Deep learning architectures, including convolutional and recurrent neural networks, are particularly effective for analyzing high-dimensional data such as images, videos, and time-series sensor streams (Li et al., 2022). In construction monitoring, DL-based models enable automated detection of structural defects, unsafe worker behaviors, and equipment anomalies, significantly reducing reliance on manual inspection (Barbhuiya & Das, 2025; Mondal & Chen, 2022).

6.3 AI-Driven Structural Health and Damage Analytics

Intelligent data analytics play a central role in structural health monitoring (SHM) and damage detection within ICMS. AI-enhanced SHM systems leverage sensor data to identify early signs of deterioration, fatigue, and failure in concrete and civil infrastructure (Mondal & Chen, 2022). These analytics integrate vibration analysis, strain monitoring, and acoustic emission data with ML models to achieve high detection accuracy and robustness (Preethichandra et al., 2023).

Recent advances in AI-driven damage detection emphasize hybrid analytical pipelines that combine physics-based models with data-driven learning, improving interpretability and reliability in safety-critical construction environments (Barbhuiya & Das, 2025). Such approaches align with broader trends in intelligent construction automation and monitoring (Liu et al., 2022).

6.4 Big Data Analytics and Cloud–Edge Intelligence

The scale and complexity of construction monitoring data necessitate big data analytics frameworks capable of handling high volume, velocity, and variety. Cloud-based analytics platforms enable large-scale data storage, model training, and long-term trend analysis, supporting strategic decision-making in construction project management (Rane, 2023; Adebayo et al., 2025).

Edge intelligence complements cloud analytics by enabling real-time data processing near data sources, reducing latency and bandwidth requirements for time-critical safety applications (Rane et al., 2023). These architectures are influenced by serverless computing paradigms used in smart healthcare systems, enabling event-driven analytics and elastic resource allocation (Deka et al., 2024). The combination of edge, cloud, and serverless computing enhances scalability and responsiveness in ICMS.

6.5 Vision Analytics and Autonomous Monitoring Systems

Vision-based analytics represent a major component of intelligent data processing in ICMS. AI-powered computer vision models analyze images and videos captured by fixed cameras, UAVs, and robotic platforms to monitor construction progress, detect hazards, and assess compliance with safety regulations (Telli et al., 2023; Sonko et al., 2024).

UAV-based analytics enable rapid and flexible data acquisition and processing, supporting applications such as site mapping, environmental assessment, and post-incident analysis (Wang et al., 2025). These autonomous monitoring systems integrate embedded intelligence for real-time inference, aligning with broader research trends in robotics and autonomous systems (Ryalat et al., 2025).

6.6 BIM, Digital Twins, and GIS-Driven Analytics

Intelligent analytics in ICMS increasingly rely on integration with BIM, digital twins, and GIS platforms to contextualize data and enhance decision support. BIM-integrated analytics enable real-time comparison between planned and actual construction states, supporting progress tracking, quality control, and risk assessment (Pan & Zhang, 2023).

Digital twin analytics extend these capabilities by enabling continuous simulation and predictive modeling based on live sensor data, supporting scenario analysis and proactive risk mitigation (Mazzetto, 2024). GIS-driven analytics further enrich ICMS by incorporating spatial and environmental context, enabling large-scale infrastructure monitoring and urban construction management (Vinueza-Martinez et al., 2024). These integrated analytics frameworks are particularly valuable for heritage and complex building monitoring applications (Wojciechowska et al., 2024).

6.7 Context-Aware and Agentic Intelligence

Recent advances in AI emphasize the development of context-aware and agentic intelligence, enabling ICMS to adapt dynamically to changing construction conditions. Context-aware analytics leverage environmental, temporal, and operational cues to adjust model behavior, data prioritization, and alert thresholds (Lu, 2019; Aouedi et al., 2024).

Agentic AI systems further enhance autonomy by enabling intelligent agents to reason, plan, and collaborate with human stakeholders (Kostopoulos et al., 2025). The integration of large language models (LLMs) into intelligent transportation and monitoring systems demonstrates the potential of natural language reasoning and explainable analytics for construction decision support (Mahmud et al., 2025).

These developments align with evolving AI trends in construction and industrial applications (Ali et al., 2024).

6.8 Security-Aware and Trustworthy Analytics

As ICMS analytics become increasingly autonomous and data-driven, ensuring security, trust, and governance is essential. Blockchain-enabled analytics frameworks support secure data sharing, traceability, and decentralized trust among construction stakeholders (Verma et al., 2022). These approaches align with Industry 5.0 principles emphasizing human-centric, resilient, and trustworthy intelligent systems (Xiang et al., 2023).

Advanced authentication mechanisms based on deep learning protect data pipelines and analytics services from unauthorized access and cyber threats (Shende et al., 2024). Trust, Risk, and Security Management (TRiSM) frameworks provide structured governance for AI analytics, addressing transparency, accountability, and lifecycle risks (Ray, 2025a). Model context protocols further enhance analytics reliability by ensuring consistent contextual interpretation across AI components (Ray, 2025b). Defense-in-depth strategies reinforce multi-layered protection across sensing, processing, and analytics layers (Zangana, 2025).

6.9 Cross-Domain Influences and Emerging Paradigms

Intelligent data processing in ICMS increasingly benefits from cross-domain innovation. Techniques developed for smart grids, agriculture, healthcare, and advanced manufacturing inform scalable analytics architectures and optimization strategies (Saleem et al., 2019; Wu et al., 2025). Industry 4.0 and 5.0 paradigms further influence analytics design, emphasizing interoperability, sustainability, and human–AI collaboration (Xu et al., 2018; Xiang et al., 2023).

Looking forward, emerging computational paradigms such as quantum computing—currently explored primarily in healthcare analytics—may offer future opportunities for solving complex optimization and prediction problems in large-scale construction monitoring systems (Bukkarayasamudram et al., 2025). Together, these developments point toward increasingly autonomous, adaptive, and trustworthy intelligent analytics ecosystems for next-generation ICMS.

7. REAL-TIME SAFETY AND RISK MONITORING FRAMEWORKS

Real-time safety and risk monitoring frameworks form the operational core of intelligent construction monitoring systems, enabling proactive hazard identification, continuous situational awareness, and predictive decision-making in complex construction environments. These frameworks integrate sensing technologies, artificial intelligence (AI), cyber–physical systems, and data-driven analytics to detect unsafe conditions, assess risk levels, and support timely interventions. As construction sites evolve into highly digitized and interconnected ecosystems, the importance of robust, scalable, and intelligent monitoring frameworks has become increasingly pronounced (Irani & Kamal, 2014; Liu et al., 2022).

To demonstrate how real-time safety monitoring operates in intelligent construction environments, Figure 5 presents a flowchart of the safety and risk monitoring process, from sensing and data analysis to alert generation and response actions.

Figure 5. Real-time safety and risk monitoring flow in ICMS

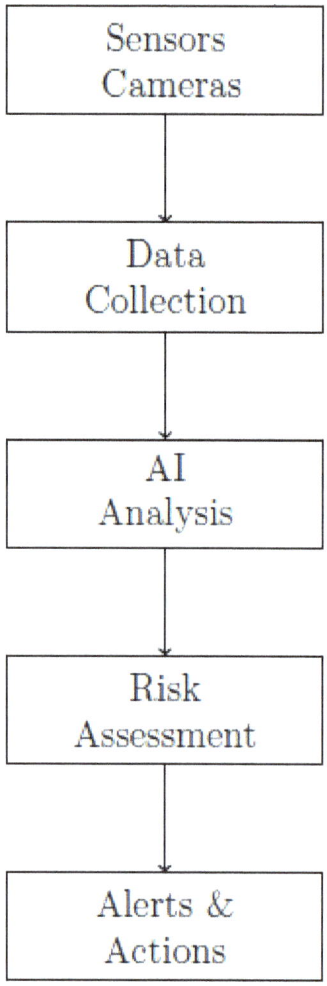

7.1 Conceptual Foundations of Real-Time Monitoring Frameworks

At their foundation, real-time safety and risk monitoring frameworks are designed to continuously collect, process, and interpret multi-source data from construction sites. Wireless sensor networks (WSNs), embedded devices, and smart sensors serve as primary data acquisition layers, capturing environmental, structural, and operational parameters such as vibration, temperature, air quality, equipment status, and worker location (Abdelhafidh et al., 2019; Preethichandra et al., 2023). These sensing infrastructures are often embedded within cyber–physical systems that tightly couple physical processes with computational intelligence, enabling real-time feedback loops between monitoring and control mechanisms (Juma & Shaalan, 2020).

The evolution of intelligent monitoring frameworks has been strongly influenced by advances in AI and machine learning, which allow systems to move beyond reactive alerting toward predictive and prescriptive safety management. Deep learning architectures, in particular, have demonstrated strong capabilities in recognizing complex patterns, anomalies, and safety-related events from high-dimensional data streams such as video feeds and sensor time series (Sarker, 2021; Taye, 2023). These capabilities are critical for addressing the dynamic and uncertain nature of construction sites, where risks emerge from constantly changing interactions among workers, machinery, and the environment.

7.2 Architectural Layers of Real-Time Safety Monitoring Systems

Modern real-time safety and risk monitoring frameworks typically adopt multi-layered architectures to ensure scalability, reliability, and interoperability. The perception layer consists of IoT-enabled sensors, wearable devices, unmanned aerial vehicles (UAVs), and computer vision systems that capture real-time data from the construction environment (Shafique et al., 2020; Telli et al., 2023). UAVs, in particular, have gained prominence for monitoring hazardous or inaccessible areas, enabling rapid risk assessment during emergencies and complex operations (Wang et al., 2025).

The communication layer ensures low-latency and reliable data transmission, increasingly leveraging next-generation wireless technologies such as 5G and emerging 6G networks to support ultra-reliable and low-latency communications (Chowdhury et al., 2020). These capabilities are essential for real-time safety applications, where delays in hazard detection or response can lead to severe consequences.

At the intelligence layer, AI-driven analytics engines process incoming data streams using machine learning, deep learning, and agent-based models to identify unsafe conditions, predict potential incidents, and generate actionable insights. Recent research highlights the growing role of agentic AI systems capable of autonomous reasoning, adaptation, and coordination across multiple monitoring tasks (Kostopoulos et al., 2025). Integration with Building Information Modeling (BIM) and digital twins further enhances contextual awareness by aligning real-time data with spatial, temporal, and semantic representations of construction assets (Pan & Zhang, 2023; Mazzetto, 2024).

7.3 AI-Driven Risk Detection and Predictive Safety Analytics

AI-powered risk detection mechanisms play a central role in transforming raw construction data into meaningful safety intelligence. Computer vision techniques enable automated detection of unsafe worker behaviors, missing personal protective equipment, and proximity risks involving heavy machinery (Mondal & Chen, 2022). Similarly, AI-based structural health monitoring systems leverage sensor fusion and deep learning models to detect early signs of material degradation, damage, or instability in concrete and steel structures (Barbhuiya & Das, 2025).

Predictive analytics further extend the capabilities of real-time frameworks by forecasting accident likelihood and risk escalation based on historical patterns and real-time indicators. These predictive functions draw on advancements in machine learning workflows and model architectures that have been successfully applied across domains such as healthcare, manufacturing, and smart infrastructure (Lu, 2019; Li et al., 2022; Xiang et al., 2023). Cross-domain insights from intelligent healthcare systems and autonomous vehicles highlight the importance of explainability, robustness, and real-time inference in safety-critical applications (Deka et al., 2024; Sonko et al., 2024).

7.4 Integrated Risk Management and Trustworthy AI Considerations

As safety monitoring frameworks become increasingly autonomous, ensuring trust, transparency, and resilience has emerged as a critical concern. Trustworthy AI principles, including model interpretability, data governance, and ethical decision-making, are essential for gaining stakeholder confidence and ensuring responsible deployment in construction environments (Ray, 2025a). Frameworks such as Trust, Risk, and Security Management (TRiSM) provide structured approaches for managing AI-related risks, including model bias, data leakage, and adversarial attacks (Ray, 2025b).

Cybersecurity considerations are particularly critical given the interconnected nature of construction monitoring systems. Blockchain-enabled architectures have been proposed to enhance data integrity, access control, and auditability across distributed monitoring components (Verma et al., 2022). Defense-in-depth strategies further strengthen system resilience by integrating multi-layered security controls across sensing, communication, and analytics layers (Zangana, 2025).

7.5 Emerging Technologies and Future-Oriented Frameworks

Emerging technologies are expected to significantly reshape real-time safety and risk monitoring frameworks in construction. The convergence of IoT, big data, and AI continues to drive innovation in smart construction ecosystems, enabling more holistic and adaptive risk management strategies (Rane, 2023; Rane et al., 2023). Advances in robotics and autonomous systems further support automated inspection, hazard mitigation, and emergency response operations (Ryalat et al., 2025).

Additionally, quantum computing has been identified as a long-term enabler for solving complex optimization and risk assessment problems that are currently computationally infeasible, with early insights drawn from applications in healthcare and intelligent systems research (Bukkarayasamudram et al., 2025). The integration of large language models (LLMs) into intelligent transportation and monitoring systems also opens new possibilities for natural-language interaction, decision support, and cross-system coordination in safety-critical contexts (Mahmud et al., 2025).

7.6 Toward Holistic and Sustainable Safety Monitoring Ecosystems

The future of real-time safety and risk monitoring frameworks lies in their ability to function as holistic ecosystems rather than isolated solutions. Integration with geographic information systems (GIS), smart grids, and urban digital twins enables construction safety monitoring to align with broader smart city and sustainability objectives (Saleem et al., 2019; Vinueza-Martinez et al., 2024; Wojciechowska et al., 2024). Such convergence supports not only immediate risk mitigation but also long-term resilience, environmental protection, and asset lifecycle management.

In summary, real-time safety and risk monitoring frameworks represent a critical pillar of intelligent construction monitoring systems. By integrating advanced sensing, AI-driven analytics, trustworthy architectures, and emerging technologies, these frameworks enable proactive, data-driven, and resilient safety management practices that are essential for the future of smart and sustainable construction.

8. ENVIRONMENTAL MONITORING AND PREDICTION SYSTEMS

Environmental monitoring and prediction systems constitute a critical pillar of intelligent construction monitoring systems, particularly in the context of sustainable development, regulatory compliance, and risk mitigation. Construction activities significantly influence air quality, noise levels, water resources, soil stability, and surrounding ecosystems. Consequently, the integration of intelligent environmental monitoring frameworks has become essential for minimizing adverse impacts while enabling real-time awareness and predictive decision-making (Irani & Kamal, 2014; Liu et al., 2022).

Recent advancements in artificial intelligence (AI), Internet of Things (IoT), cyber–physical systems, and data analytics have transformed traditional environmental monitoring approaches into proactive, adaptive, and predictive systems. These intelligent frameworks not only collect environmental data but also analyze trends, forecast future conditions, and support informed interventions across the construction lifecycle (Lu, 2019; Ali et al., 2024).

8.1 Foundations of Environmental Monitoring in Intelligent Construction

Environmental monitoring in construction environments relies on distributed sensing infrastructures capable of capturing real-time data related to air pollutants, noise emissions, vibration, temperature, humidity, and water quality. Wireless sensor networks (WSNs) serve as foundational components, offering scalable, energy-efficient, and flexible architectures for continuous environmental observation (Abdelhafidh et al., 2019). The evolution of smart sensors has further enhanced monitoring precision, reliability, and integration with asset management and decision-support systems (Preethichandra et al., 2023).

These sensing systems are embedded within cyber–physical environments that tightly couple physical construction processes with digital intelligence, enabling real-time feedback and adaptive control (Juma & Shaalan, 2020). From an Industry 4.0 perspective, environmental monitoring systems align with the broader digitalization of construction workflows, facilitating data-driven sustainability and compliance strategies (Xu et al., 2018; Rane, 2023).

8.2 IoT-Enabled Environmental Data Acquisition and Communication

IoT technologies play a central role in enabling real-time environmental monitoring across complex and geographically dispersed construction sites. Intelligent IoT architectures integrate heterogeneous sensors, edge devices, and cloud platforms to support continuous data acquisition and analysis (Aouedi et al., 2024). These architectures must address challenges related to scalability, energy efficiency, data security, and interoperability, particularly in dynamic construction environments (Shafique et al., 2020).

Advanced wireless communication technologies, including 5G and emerging 6G networks, provide the ultra-low latency and high reliability required for real-time environmental monitoring and prediction (Chowdhury et al., 2020). Similar communication paradigms have been successfully applied in smart grids and urban infrastructure monitoring, offering valuable insights for construction-related environmental systems (Saleem et al., 2019).

8.3 AI-Driven Environmental Prediction and Analytics

AI-driven analytics significantly enhance the capabilities of environmental monitoring systems by enabling predictive modeling, anomaly detection, and scenario analysis. Machine learning and deep learning techniques are widely employed to model complex environmental dynamics, identify pollution patterns, and forecast future conditions based on historical and real-time data (Sarker, 2021; Taye, 2023). These approaches allow construction managers to anticipate environmental risks rather than merely responding to threshold-based alerts.

Deep learning architectures, originally developed for domains such as medical image analysis and civil infrastructure monitoring, have demonstrated strong potential for extracting meaningful insights from high-dimensional environmental datasets (Li et al., 2022; Mondal & Chen, 2022). In construction contexts, these techniques support predictive assessments of dust dispersion, noise propagation, and vibration-induced structural impacts, thereby improving environmental risk management (Barbhuiya & Das, 2025).

8.4 Integration with BIM, Digital Twins, and GIS Platforms

The integration of environmental monitoring systems with Building Information Modeling (BIM), digital twins, and geographic information systems (GIS) represents a major advancement in intelligent construction management. BIM-based platforms provide spatial and semantic context for environmental data, enabling more accurate impact assessments and visualization-driven decision-making (Pan & Zhang, 2023). Digital twins further extend these capabilities by synchronizing real-time environmental data with virtual replicas of construction sites and surrounding environments (Mazzetto, 2024).

GIS and WebGIS architectures enable large-scale environmental analysis by incorporating spatial data, regulatory boundaries, and urban context into monitoring frameworks (Vinueza-Martinez et al., 2024). Such integration is particularly valuable for construction projects located in dense urban areas or environmentally sensitive zones, where predictive environmental modeling supports compliance with sustainability and planning regulations (Wojciechowska et al., 2024).

8.5 Autonomous Platforms and Aerial Environmental Monitoring

Unmanned aerial vehicles (UAVs) and autonomous robotic platforms have emerged as powerful tools for environmental monitoring in construction. UAVs enable rapid assessment of air quality, thermal conditions, and site-wide environmental impacts, particularly in hazardous or hard-to-reach areas (Telli et al., 2023). Collaborative UAV–ground vehicle systems further enhance monitoring coverage and responsiveness during emergencies and extreme environmental events (Wang et al., 2025).

The integration of embedded systems, robotics, and AI enables autonomous environmental sensing and data interpretation, reducing reliance on manual inspections and improving monitoring frequency (Sonko et al., 2024; Ryalat et al., 2025). Similar approaches in agri-food engineering and advanced manufacturing provide transferable insights into scalable and adaptive environmental monitoring solutions (Wu et al., 2025; Xiang et al., 2023).

8.6 Trustworthy, Secure, and Sustainable Environmental Monitoring Frameworks

As environmental monitoring systems increasingly rely on autonomous intelligence and interconnected infrastructures, trust, security, and resilience become paramount. Trustworthy AI principles, including transparency, explainability, and robustness, are essential for ensuring confidence in environmental predictions and decision-support outputs (Ray, 2025a). Model context protocols further support consistent and reliable AI behavior across distributed monitoring components (Ray, 2025b).

Cybersecurity considerations are particularly critical given the sensitivity of environmental data and the potential consequences of system compromise. Blockchain-based architectures and Industry 5.0 paradigms offer promising solutions for enhancing data integrity, traceability, and access control in environmental monitoring systems (Verma et al., 2022). Defense-in-depth strategies provide a comprehensive security framework by integrating multiple layers of protection across sensing, communication, and analytics components (Zangana, 2025).

8.7 Emerging Trends and Future Directions in Environmental Prediction Systems

Emerging technologies continue to reshape the future of environmental monitoring and prediction in construction. The convergence of AI, IoT, and big data analytics supports more adaptive and context-aware environmental intelligence, enabling real-time optimization of construction processes to minimize ecological impact (Adebayo et al., 2025; Rane et al., 2023). Agentic AI systems further enhance autonomy by enabling monitoring platforms to reason, learn, and coordinate actions across complex environments (Kostopoulos et al., 2025).

Long-term research directions also point toward the potential role of quantum computing in solving large-scale environmental optimization and prediction problems, drawing inspiration from early applications in healthcare and intelligent systems research (Bukkarayasamudram et al., 2025). Additionally, the integration of large language models into intelligent monitoring platforms opens new avenues for natural-language querying, reporting, and cross-domain knowledge integration (Mahmud et al., 2025).

9. INTEGRATION WITH BIM AND DIGITAL TWINS

The integration of Building Information Modeling (BIM) and digital twin technologies represents a transformative advancement in intelligent construction monitoring systems. These technologies enable a unified digital representation of physical construction assets, processes, and environments, facilitating real-time monitoring, predictive analytics, and informed decision-making. By combining BIM's structured data models with the dynamic, data-driven capabilities of digital twins, construction projects can achieve enhanced situational awareness, operational efficiency, safety assurance, and environmental sustainability (Irani & Kamal, 2014; Liu et al., 2022).

Recent progress in artificial intelligence (AI), IoT-enabled sensing, cyber–physical systems, and advanced analytics has accelerated the adoption of BIM–digital twin frameworks across the construction lifecycle, from planning and execution to operation and asset management (Lu, 2019; Ali et al., 2024). This section examines architectural foundations, enabling technologies, data integration mechanisms, and emerging research directions associated with BIM and digital twin integration in intelligent construction monitoring systems.

9.1 Conceptual Foundations of BIM and Digital Twins in Construction

BIM provides a centralized, object-oriented digital representation of construction assets, incorporating geometric, semantic, and relational information across project phases. Traditionally used for design coordination and clash detection, BIM has evolved into a data-rich platform capable of supporting intelligent monitoring and analytics when integrated with real-time data streams (Pan & Zhang, 2023). Digital twins extend BIM by enabling continuous synchronization between physical assets and their virtual counterparts, thereby supporting dynamic state estimation, predictive simulation, and adaptive control (Mazzetto, 2024).

Within intelligent construction contexts, BIM and digital twins operate as cyber–physical systems that link physical construction environments with digital intelligence layers. This integration supports closed-loop feedback mechanisms, enabling proactive risk mitigation and performance optimization (Juma & Shaalan, 2020). Similar paradigms have been successfully adopted in Industry 4.0 and smart city ecosystems, offering transferable insights for construction monitoring applications (Xu et al., 2018).

9.2 IoT and Sensor Integration for BIM-Driven Digital Twins

The effectiveness of BIM–digital twin integration depends heavily on reliable, real-time data acquisition from construction sites. Wireless sensor networks and smart IoT devices form the backbone of this data layer, capturing information related to structural behavior, environmental conditions, equipment status, and worker safety (Abdelhafidh et al., 2019; Preethichandra et al., 2023). These sensor networks enable continuous updating of digital twins, ensuring alignment between virtual models and physical reality.

Advanced IoT architectures emphasize scalability, interoperability, and secure data transmission, which are essential for integrating heterogeneous sensor data into BIM environments (Aouedi et al., 2024). Emerging wireless communication technologies, including 5G and future 6G networks, further enhance the responsiveness and reliability of BIM–digital twin systems by supporting ultra-low-latency data exchange (Chowdhury et al., 2020). Comparable IoT-enabled architectures in smart grids and autonomous systems highlight best practices for resilient data integration (Saleem et al., 2019; Sonko et al., 2024).

9.3 AI-Enhanced Analytics within BIM and Digital Twin Platforms

Artificial intelligence plays a pivotal role in extracting actionable insights from the vast volumes of data generated by BIM-integrated digital twins. Machine learning and deep learning algorithms enable pattern recognition, anomaly detection, and predictive modeling across structural, environmental, and operational dimensions (Sarker, 2021; Taye, 2023). These capabilities allow digital twins to move beyond passive visualization tools toward autonomous decision-support systems.

AI-driven damage detection and structural health monitoring techniques have demonstrated strong potential for integration with BIM-based digital twins, enabling early identification of defects and degradation in construction assets (Barbhuiya & Das, 2025; Mondal & Chen, 2022). Similar advances in image-based analytics and pattern recognition from healthcare and manufacturing domains further validate the robustness of these AI methodologies for construction monitoring (Li et al., 2022; Xiang et al., 2023).

9.4 Spatial Intelligence through GIS and Urban Digital Twins

The integration of BIM and digital twins with geographic information systems (GIS) enhances spatial intelligence and contextual awareness in construction monitoring. GIS platforms provide geospatial context, regulatory overlays, and environmental constraints that complement BIM's asset-level detail (Vinueza-Martinez et al., 2024). This integration is particularly valuable for large-scale infrastructure projects and urban developments, where construction activities interact closely with surrounding environments.

Urban digital twins extend this concept by linking individual construction digital twins to broader city-scale models, enabling holistic analysis of environmental impacts, mobility patterns, and infrastructure resilience (Mazzetto, 2024). Applications in historical building management further demonstrate the value of integrated digital twins for preserving cultural assets while supporting intelligent monitoring and maintenance strategies (Wojciechowska et al., 2024).

9.5 Autonomous Systems and Aerial Data Integration

Unmanned aerial vehicles (UAVs) and autonomous robotic systems contribute significantly to BIM–digital twin ecosystems by providing high-resolution spatial and environmental data. UAV-based sensing supports rapid site inspections, progress tracking, and environmental assessments, which can be directly integrated into BIM and digital twin platforms for near-real-time updates (Telli et al., 2023). Collaborative UAV–ground vehicle systems further enhance monitoring coverage and responsiveness, particularly in emergency and disaster scenarios (Wang et al., 2025).

The integration of robotics, embedded systems, and AI enables autonomous data collection and interpretation, reducing manual intervention and improving monitoring accuracy (Ryalat et al., 2025). Insights from agri-food engineering and autonomous mobility research highlight scalable approaches for integrating autonomous platforms into digital twin ecosystems (Wu et al., 2025).

9.6 Trust, Security, and Governance in BIM–Digital Twin Integration

As BIM and digital twin systems become increasingly interconnected and autonomous, ensuring trust, security, and governance becomes critical. Trustworthy AI frameworks emphasize transparency, explainability, and robustness, which are essential for gaining stakeholder confidence in AI-driven BIM analytics (Ray, 2025a). Model context protocols further support consistent AI behavior across distributed BIM–digital twin components, enabling reliable multi-agent coordination (Ray, 2025b).

Cybersecurity risks associated with data integrity, unauthorized access, and system manipulation necessitate comprehensive protection strategies. Blockchain-based architectures and Industry 5.0 paradigms offer promising solutions for enhancing data traceability, access control, and collaborative governance in BIM–digital twin ecosystems (Verma et al., 2022). Defense-in-depth approaches provide layered security mechanisms across sensing, communication, and analytics layers, ensuring resilience against evolving cyber threats (Zangana, 2025).

9.7 Emerging Trends and Future Research Directions

Future research on BIM and digital twin integration in intelligent construction monitoring is increasingly oriented toward autonomy, intelligence, and sustainability. The convergence of AI, IoT, and big data analytics supports more adaptive and predictive construction management practices, enabling proactive decision-making and lifecycle optimization (Adebayo et al., 2025; Rane, 2023; Rane et al., 2023). Agentic AI systems introduce new possibilities for self-coordinating digital twins capable of reasoning, learning, and negotiating actions across complex construction environments (Kostopoulos et al., 2025).

Longer-term perspectives suggest potential applications of quantum computing for large-scale optimization and simulation within digital twin environments, inspired by early research in other intelligent systems domains (Bukkarayasamudram et al., 2025). Additionally, the integration of large language models into BIM platforms offers novel interfaces for querying, reporting, and collaborative decision-making, further enhancing the accessibility and usability of intelligent construction monitoring systems (Mahmud et al., 2025).

10. CHALLENGES AND IMPLEMENTATION ISSUES

The deployment of intelligent construction monitoring systems, integrating AI, IoT, BIM, and digital twin technologies, offers unprecedented opportunities for enhanced project management, structural health monitoring, and operational efficiency. However, these systems face a range of challenges and implementation issues spanning technical, organizational, and socio-ethical domains. Understanding these challenges is essential for both researchers and practitioners to design robust, scalable, and sustainable intelligent monitoring solutions (Irani & Kamal, 2014; Liu et al., 2022).

10.1 Technical Challenges

10.1.1 Data Acquisition and Sensor Reliability

The reliability and accuracy of data collected through IoT devices and wireless sensor networks remain critical concerns. Inconsistent sensor performance, environmental interference, and network failures can compromise the fidelity of digital twin models and AI-driven analytics (Abdelhafidh et al., 2019; Preethichandra et al., 2023). Ensuring the continuous operation of heterogeneous sensor networks requires advanced fault-tolerant architectures, robust calibration methods, and redundancy mechanisms (Aouedi et al., 2024; Shafique et al., 2020).

10.1.2 Integration and Interoperability

Integrating multiple data streams from BIM, IoT, UAVs, and autonomous devices presents significant interoperability challenges. The diverse data formats, protocols, and update frequencies require sophisticated middleware and standardization frameworks to maintain coherence across platforms (Pan & Zhang, 2023; Rane et al., 2023). Additionally, synchronizing real-time operational data with static BIM models demands high-performance computing capabilities and efficient data management pipelines (Mazzetto, 2024; Saleem et al., 2019).

10.1.3 AI Model Limitations

While AI algorithms offer predictive and prescriptive capabilities, they are susceptible to data bias, overfitting, and generalization errors. Large-scale construction environments often introduce noisy, incomplete, or non-stationary datasets that challenge conventional machine learning and deep learning models (Sarker, 2021; Taye, 2023; Shende et al., 2024). Moreover, AI-driven damage detection and monitoring systems must account for uncertainties in sensor measurements and structural behavior, requiring probabilistic or hybrid modeling approaches (Barbhuiya & Das, 2025; Mondal & Chen, 2022).

10.1.4 Cybersecurity and Privacy

The increasing interconnectivity of AI-enabled BIM and digital twin systems heightens vulnerability to cyberattacks. Unauthorized access, data tampering, and ransomware threats can disrupt monitoring systems and compromise safety-critical decisions (Zangana, 2025; Verma et al., 2022). Ensuring data confidentiality and integrity across distributed IoT networks and cloud-based platforms necessitates robust encryption, blockchain integration, and layered security strategies (Aouedi et al., 2024; Chowdhury et al., 2020).

10.2 Operational Challenges

10.2.1 Scalability and Computational Demand

Scaling intelligent monitoring solutions to large infrastructure projects introduces high computational and storage demands. Real-time analytics, 3D digital twin updates, and continuous AI model training require advanced computing infrastructures, including edge computing and cloud–edge hybrid models (Wu et al., 2025; Bukkarayasamudram et al., 2025; Mahmud et al., 2025). Resource-constrained environments may experience latency issues that reduce the efficacy of real-time decision-making.

10.2.2 Maintenance and Lifecycle Management

Long-term operation of digital twins and sensor networks involves significant maintenance efforts. Sensor calibration, firmware updates, and hardware replacement are ongoing tasks that impact operational continuity and cost-efficiency (Abdelhafidh et al., 2019; Preethichandra et al., 2023). Furthermore, ensuring consistency between the physical asset and its digital twin throughout the lifecycle is critical for maintaining the validity of predictive and prescriptive analytics (Pan & Zhang, 2023; Rane, 2023).

10.3 Organizational and Implementation Challenges

10.3.1 Knowledge and Skill Gaps

Effective implementation of AI-integrated BIM and digital twin systems requires multidisciplinary expertise spanning civil engineering, computer science, and data analytics (Adebayo et al., 2025; Ali et al., 2024). Many construction organizations lack sufficient in-house technical capacity, necessitating investments in training, knowledge transfer, and collaboration with technology providers (Irani & Kamal, 2014; Liu et al., 2022).

10.3.2 Change Management and Adoption

Integrating intelligent monitoring systems into existing construction workflows often encounters resistance due to organizational inertia, perceived complexity, and fear of job displacement (Rane et al., 2023; Lu, 2019). Effective change management strategies, including stakeholder engagement, incremental implementation, and demonstrable value delivery, are essential for successful adoption (Pan & Zhang, 2023; Mondal & Chen, 2022).

10.4 Ethical, Legal, and Regulatory Challenges

10.4.1 Data Governance and Privacy Compliance

AI-driven monitoring systems rely heavily on data collection, raising concerns regarding privacy, consent, and compliance with local and international regulations (Aouedi et al., 2024; Vinueza-Martinez et al., 2024). Implementing GDPR-compliant practices, anonymization techniques, and ethical guidelines is critical to mitigate legal risks and maintain stakeholder trust.

10.4.2 Accountability and Liability

The increasing autonomy of AI algorithms and digital twin systems complicates accountability in construction projects. Determining liability in case of system failure, structural damage, or misprediction requires clear frameworks for human oversight, auditability, and explainability (Ray, 2025a; Ray, 2025b; Kostopoulos et al., 2025). Ethical considerations around AI decision-making, transparency, and fairness are integral to responsible deployment (Lu, 2019; Mahmud et al., 2025).

10.5 Emerging Implementation Solutions

Several strategies are being explored to address these challenges. Cyber–physical co-design, edge–cloud hybrid architectures, blockchain-enabled trust mechanisms, and AI-enhanced predictive maintenance frameworks have shown promise in increasing resilience, scalability, and reliability (Shafique et al., 2020; Verma et al., 2022; Wu et al., 2025). Furthermore, standards for interoperability, data modeling, and communication protocols are gradually maturing, facilitating smoother integration of heterogeneous systems (Abdelhafidh et al., 2019; Rane, 2023; Mazzetto, 2024). Advanced robotics, UAVs, and autonomous data collection mechanisms offer additional pathways to reduce human dependency and operational bottlenecks (Telli et al., 2023; Wang et al., 2025; Ryalat et al., 2025).

10.6 Summary

The implementation of intelligent monitoring systems in construction, while transformative, is fraught with multifaceted challenges spanning technical, operational, organizational, and ethical dimensions. Addressing these challenges requires a combination of advanced technological solutions, robust governance mechanisms, interdisciplinary expertise, and forward-looking research. Future efforts should focus on scalable AI models, resilient IoT infrastructures, explainable decision-making, and secure data management, ultimately enabling the realization of fully autonomous, efficient, and trustworthy intelligent construction monitoring ecosystems (Lu, 2019; Mondal & Chen, 2022; Zangana, 2025).

11. FUTURE DIRECTIONS AND EMERGING TRENDS

The integration of artificial intelligence (AI), Internet of Things (IoT), Building Information Modeling (BIM), digital twins, and other advanced technologies is reshaping the architecture, engineering, and construction (AEC) industry. As research and practice evolve, several emerging trends and future directions are anticipated, addressing technological, operational, and strategic dimensions of intelligent systems (Lu, 2019; Irani & Kamal, 2014).

11.1 AI and Machine Learning Evolution

The future of AI in construction and civil infrastructure is expected to emphasize enhanced predictive capabilities, hybrid modeling approaches, and real-time analytics. The integration of deep learning and large language models (LLMs) into intelligent monitoring and decision-making systems promises improved accuracy in structural health monitoring, construction project planning, and autonomous site management (Mahmud et al., 2025; Taye, 2023; Sarker, 2021). Research is moving towards explainable AI (XAI), ensuring that models provide transparent reasoning to support decision-making in safety-critical environments (Barbhuiya & Das, 2025; Li et al., 2022).

Emerging approaches are also exploring the combination of AI with reinforcement learning and agentic systems for autonomous task execution in construction and education-related applications (Kostopoulos et al., 2025; Mondal & Chen, 2022).

11.2 Advanced IoT and Smart Sensor Networks

Next-generation IoT-enabled systems are expected to facilitate ubiquitous sensing, real-time data aggregation, and adaptive decision-making (Aouedi et al., 2024; Shafique et al., 2020; Saleem et al., 2019). Wireless sensor networks (WSNs) and embedded systems are evolving to support low-latency communication, fault tolerance, and predictive maintenance of structures (Abdelhafidh et al., 2019; Sonko et al., 2024). Smart sensors will play a critical role in bridging the physical and digital domains, enabling more accurate and dynamic digital twin representations (Preethichandra et al., 2023; Rane et al., 2023).

The integration of IoT with 5G and emerging 6G networks will enhance connectivity, allowing high-bandwidth data streams to support real-time monitoring, autonomous construction robotics, and UAV operations (Chowdhury et al., 2020; Telli et al., 2023).

11.3 BIM and Digital Twin Integration

BIM and digital twin technologies are poised to become the backbone of future smart construction systems. Their integration with AI, IoT, and cloud-based platforms will enable predictive asset management, automated construction planning, and lifecycle optimization (Pan & Zhang, 2023; Mazzetto, 2024). Digital twins will evolve from static 3D models to real-time dynamic systems, capable of simulating complex structural behaviors, environmental interactions, and operational scenarios (Wojciechowska et al., 2024; Liu et al., 2022).

Future research will focus on achieving seamless interoperability among heterogeneous data sources, improving data fidelity, and developing standardized protocols for digital twin deployment (Rane, 2023; Rane, Choudhary, & Rane, 2023).

11.4 Cybersecurity, Privacy, and Ethical Considerations

As intelligent systems become increasingly autonomous, ensuring cybersecurity and data privacy is a critical research priority. Blockchain, defense-in-depth strategies, and secure multi-agent architectures are emerging as key enablers for safe and resilient digital infrastructures (Zangana, 2025; Verma et al., 2022). Ethical considerations, including fairness, accountability, and transparency of AI decisions, are gaining attention, especially in applications with safety-critical outcomes, such as autonomous construction robotics and UAV operations (Ray, 2025a; Ray, 2025b).

Future directions emphasize the development of explainable, auditable AI frameworks and privacy-preserving data analytics to support regulatory compliance and public trust (Aouedi et al., 2024; Vinueza-Martinez et al., 2024).

11.5 Emerging Technologies in Industry 5.0 and Smart Systems

The convergence of AI, quantum computing, robotics, and advanced manufacturing under Industry 5.0 principles is expected to drive intelligent automation, collaborative robotics, and human–AI synergy (Bukkarayasamudram et al., 2025; Xiang et al., 2023). Quantum computing could revolutionize optimization and predictive analytics in construction, healthcare, and logistics, enabling faster and more complex simulations (Bukkarayasamudram et al., 2025).

Additionally, UAV–ground vehicle collaborative systems and autonomous delivery platforms will play an increasing role in emergency response, precision agriculture, and large-scale infrastructure management (Wang et al., 2025; Telli et al., 2023). Embedded intelligence in autonomous systems will allow adaptive control, anomaly detection, and self-optimization across smart infrastructures (Sonko et al., 2024; Wu et al., 2025).

11.6 Smart Cities and Urban Digital Twins

Urban digital twins are emerging as strategic tools for city planning, infrastructure optimization, and sustainability monitoring. The fusion of GIS, BIM, and IoT data enables predictive modeling for traffic management, energy consumption, and disaster response (Mazzetto, 2024; Vinueza-Martinez et al., 2024). Cyber–physical integration will facilitate real-time monitoring of city assets, with AI predicting maintenance needs and resource allocation (Juma & Shaalan, 2020; Rane, 2023).

Future research is likely to focus on enhancing scalability, multi-city integration, and cross-domain interoperability to create resilient and intelligent urban ecosystems (Pan & Zhang, 2023; Xu, Xu, & Li, 2018).

11.7 Smart Healthcare and AI-Enabled Services

Intelligent AI-driven systems in healthcare are evolving towards server-less architectures, distributed analytics, and predictive diagnostics (Deka et al., 2024; Bukkarayasamudram et al., 2025). Wearable devices, IoT-enabled patient monitoring, and AI-based image analysis will enhance personalized healthcare and early intervention strategies (Li et al., 2022; Shende et al., 2024). Integration with digital twins for patient or hospital infrastructure models will enable scenario simulations for capacity planning and emergency management (Mahmud et al., 2025; Ryalat et al., 2025).

11.8 Future Research Priorities

Several research avenues are emerging as priorities:

1. Scalable AI models capable of handling multimodal, heterogeneous datasets in real-time applications (Sarker, 2021; Lu, 2019).
2. Robust IoT infrastructures with low-latency, fault-tolerant, and energy-efficient operations (Abdelhafidh et al., 2019; Aouedi et al., 2024).
3. Explainable and auditable AI to support transparency, accountability, and ethical compliance (Ray, 2025a; Kostopoulos et al., 2025).
4. Interoperable BIM and digital twin frameworks for multi-disciplinary collaboration and lifecycle management (Pan & Zhang, 2023; Mazzetto, 2024).
5. Integration of emerging technologies such as quantum computing, Industry 5.0 systems, autonomous vehicles, and UAVs to enhance decision-making and operational resilience (Xiang et al., 2023; Wang et al., 2025; Bukkarayasamudram et al., 2025).
6. Smart city and healthcare applications, focusing on predictive modeling, resource optimization, and real-time monitoring for sustainable and resilient infrastructure (Juma & Shaalan, 2020; Deka et al., 2024; Vinueza-Martinez et al., 2024).

11.9 Summary

The future of intelligent systems in construction, infrastructure, and smart cities lies in the convergence of AI, IoT, BIM, digital twins, and emerging computational technologies. Emphasis on explainable AI, secure and interoperable architectures, and autonomous systems is expected to redefine operational efficiency, sustainability, and resilience. Researchers and practitioners must address the technological, ethical, and regulatory challenges while embracing opportunities in Industry 5.0, urban digital twins, and AI-enabled healthcare to realize the full potential of intelligent infrastructure ecosystems (Lu, 2019; Pan & Zhang, 2023; Wu et al., 2025; Zangana, 2025).

12. CONCLUSION

The evolution of intelligent systems across construction, infrastructure, smart cities, and healthcare has highlighted both remarkable opportunities and significant challenges. Emerging technologies such as artificial intelligence, IoT, digital twins, and advanced automation are transforming traditional processes, enabling more efficient, sustainable, and resilient operations. These innovations are not only enhancing decision-making and predictive capabilities but also fostering new paradigms in project management, asset monitoring, and urban planning.

Despite these advances, the integration of complex technologies presents multifaceted challenges, including interoperability, data security, privacy, ethical considerations, and the need for scalable, reliable, and explainable solutions. Addressing these issues requires a coordinated effort across research, industry, and regulatory frameworks to ensure that intelligent systems deliver practical and trustworthy outcomes.

Looking ahead, the convergence of AI, IoT, Industry 5.0 principles, autonomous systems, and digital twins promises to redefine operational landscapes, offering unprecedented levels of automation, collaboration, and adaptability. By embracing these emerging trends while carefully managing associated risks, organizations can unlock the full potential of intelligent systems, ultimately achieving greater efficiency, safety, sustainability, and societal benefit.

In essence, the future of intelligent systems lies not only in technological advancement but also in the thoughtful and ethical integration of these tools into complex, real-world environments. The ongoing research and innovation in this field will continue to shape the way we design, manage, and interact with the built environment, creating smarter, more responsive, and resilient systems for the challenges of tomorrow.

REFERENCES

Abdelhafidh, M., Fourati, M., Fourati, L. C., & Chouaya, A. (2019). Wireless sensor network monitoring system: Architecture, applications and future directions. *International Journal of Communication Networks and Distributed Systems*, *23*(4), 413–451.

Adebayo, Y., Udoh, P., Kamudyariwa, X. B., & Osobajo, O. A. (2025). Artificial intelligence in construction project management: A structured literature review of its evolution in application and future trends. *Digital*, *5*(3), 26.

Ali, Z., Saad, S., Rasheed, K., & Ammad, S. (2024). AI future perspectives and trends in construction. *AI in Material Science*, 239-261.

Aouedi, O., Vu, T. H., Sacco, A., Nguyen, D. C., Piamrat, K., Marchetto, G., & Pham, Q. V. (2024). A survey on intelligent Internet of Things: Applications, security, privacy, and future directions. *IEEE Communications Surveys and Tutorials*.

Barbhuiya, S., & Das, B. B. (2025). Artificial Intelligence in Damage Detection of Concrete Structures: Techniques, Integration and Future Directions. In *Damage Detection and Structural Health Monitoring of Concrete and Masonry Structures: Novel Techniques and Applications* (pp. 31–92). Springer Nature Singapore.

Bukkarayasamudram, V. K., Reddy, P. C. S., Arun Kumar, K., Jagadish, R. M., Sharma, S., Prasad, M. L., & Thakur, G. K. (2025). Quantum computing revolution in healthcare: A systematic review of applications, issues and future directions. *Artificial Intelligence Review*, *58*(12), 1–35.

Chowdhury, M. Z., Shahjalal, M., Ahmed, S., & Jang, Y. M. (2020). 6G wireless communication systems: Applications, requirements, technologies, challenges, and research directions. *IEEE Open Journal of the Communications Society*, *1*, 957–975.

Deka, R. K., Ghosh, A., Nanda, S., Barik, R. K., & Saikia, M. J. (2024). Smart Healthcare System in Server-Less Environment: Concepts, Architecture, Challenges, Future Directions. *Computers*, *13*(4), 105.

Irani, Z., & Kamal, M. M. (2014). Intelligent systems research in the construction industry. *Expert Systems with Applications*, *41*(4), 934–950.

Juma, M., & Shaalan, K. (2020). Cyberphysical systems in the smart city: Challenges and future trends for strategic research. In *Swarm intelligence for resource management in Internet of things* (pp. 65–85). Academic Press.

Kostopoulos, G., Gkamas, V., Rigou, M., & Kotsiantis, S. (2025). Agentic AI in Education: State of the Art and Future Directions. *IEEE Access : Practical Innovations, Open Solutions.*

Li, Z., Koban, K. C., Schenck, T. L., Giunta, R. E., Li, Q., & Sun, Y. (2022). Artificial intelligence in dermatology image analysis: Current developments and future trends. *Journal of Clinical Medicine, 11*(22), 6826. PMID: 36431301

Liu, K., Meng, Q., Kong, Q., & Zhang, X. (2022). Review on the developments of structure, construction automation, and monitoring of intelligent construction. *Buildings (Basel, Switzerland), 12*(11), 1890.

Lu, Y. (2019). Artificial intelligence: a survey on evolution, models, applications and future trends. *Journal of management analytics, 6*(1), 1-29.

Mahmud, D., Hajmohamed, H., Almentheri, S., Alqaydi, S., Aldhaheri, L., Khalil, R. A., & Saeed, N. (2025). Integrating llms with its: Recent advances, potentials, challenges, and future directions. *IEEE Transactions on Intelligent Transportation Systems.*

Mazzetto, S. (2024). A review of urban digital twins integration, challenges, and future directions in smart city development. *Sustainability, 16*(19), 8337.

Mondal, T. G., & Chen, G. (2022). Artificial intelligence in civil infrastructure health monitoring—Historical perspectives, current trends, and future visions. *Frontiers in Built Environment, 8*, 1007886.

Pan, Y., & Zhang, L. (2023). Integrating BIM and AI for smart construction management: Current status and future directions. *Archives of Computational Methods in Engineering, 30*(2), 1081–1110.

Preethichandra, D. M. G., Suntharavadivel, T. G., Kalutara, P., Piyathilaka, L., & Izhar, U. (2023). Influence of smart sensors on structural health monitoring systems and future asset management practices. *Sensors (Basel), 23*(19), 8279. PMID: 37837109

Rane, N. (2023). Integrating leading-edge artificial intelligence (AI), internet of things (IOT), and big data technologies for smart and sustainable architecture, engineering and construction (AEC) industry: Challenges and future directions. *Engineering and Construction (AEC) Industry: Challenges and Future Directions (September 24, 2023).*

Rane, N., Choudhary, S., & Rane, J. (2023). Artificial Intelligence (Ai) and Internet of Things (Iot)–based sensors for monitoring and controlling in architecture, engineering, and construction: Applications, challenges, and opportunities. *Engineering, and Construction: Applications, Challenges, and Opportunities (November 20, 2023).*

Ray, P. P. (2025). A Review of TRiSM Frameworks in Artificial Intelligence Systems: Fundamentals, Taxonomy, Use Cases, Key Challenges and Future Directions. *Authorea Preprints*.

Ray, P. P. (2025). A survey on model context protocol: Architecture, state-of-the-art, challenges and future directions. *Authorea Preprints*.

Ryalat, M., Almtireen, N., Al-refai, G., Elmoaqet, H., & Rawashdeh, N. (2025). Research and education in robotics: A comprehensive review, trends, challenges, and future directions. *Journal of Sensor and Actuator Networks*, *14*(4), 76.

Saleem, Y., Crespi, N., Rehmani, M. H., & Copeland, R. (2019). Internet of things-aided smart grid: Technologies, architectures, applications, prototypes, and future research directions. *IEEE Access : Practical Innovations, Open Solutions*, *7*, 62962–63003.

Sarker, I. H. (2021). Deep learning: A comprehensive overview on techniques, taxonomy, applications and research directions. *SN Computer Science*, *2*(6), 1–20. PMID: 34426802

Shafique, K., Khawaja, B. A., Sabir, F., Qazi, S., & Mustaqim, M. (2020). Internet of things (IoT) for next-generation smart systems: A review of current challenges, future trends and prospects for emerging 5G-IoT scenarios. *IEEE Access : Practical Innovations, Open Solutions*, *8*, 23022–23040.

Shende, S. W., Tembhurne, J. V., & Ansari, N. A. (2024). Deep learning based authentication schemes for smart devices in different modalities: Progress, challenges, performance, datasets and future directions. *Multimedia Tools and Applications*, *83*(28), 71451–71493.

Sonko, S., Etukudoh, E. A., Ibekwe, K. I., Ilojianya, V. I., & Daudu, C. D. (2024). A comprehensive review of embedded systems in autonomous vehicles: Trends, challenges, and future directions. *World Journal of Advanced Research and Reviews*, *21*(1), 2009–2020.

Taye, M. M. (2023). Understanding of machine learning with deep learning: Architectures, workflow, applications and future directions. *Computers*, *12*(5), 91.

Telli, K., Kraa, O., Himeur, Y., Ouamane, A., Boumehraz, M., Atalla, S., & Mansoor, W. (2023). A comprehensive review of recent research trends on unmanned aerial vehicles (uavs). *Systems*, *11*(8), 400.

Verma, A., Bhattacharya, P., Madhani, N., Trivedi, C., Bhushan, B., Tanwar, S., & Sharma, R. (2022). Blockchain for industry 5.0: Vision, opportunities, key enablers, and future directions. *IEEE Access : Practical Innovations, Open Solutions, 10*, 69160–69199.

Vinueza-Martinez, J., Correa-Peralta, M., Ramirez-Anormaliza, R., Franco Arias, O., & Vera Paredes, D. (2024). Geographic information systems (giss) based on webgis architecture: Bibliometric analysis of the current status and research trends. *Sustainability, 16*(15), 6439.

Wang, Y., Li, J., Yang, X., & Peng, Q. (2025). UAV–Ground Vehicle Collaborative Delivery in Emergency Response: A Review of Key Technologies and Future Trends. *Applied Sciences (Basel, Switzerland), 15*(17), 9803.

Wojciechowska, G., Bednarz, Ł. J., Dolińska, N., Opałka, P., Krupa, M., & Imnadze, N. (2024). Intelligent Monitoring System for Integrated Management of Historical Buildings. *Buildings (2075-5309), 14*(7).

Wu, K., Ji, Z., Wang, H., Shao, X., Li, H., Zhang, W., & Bao, X. (2025). A comprehensive review of AI methods in agri-food engineering: Applications, challenges, and future directions. *Electronics (Basel), 14*(20), 3994.

Xiang, W., Yu, K., Han, F., Fang, L., He, D., & Han, Q. L. (2023). Advanced manufacturing in industry 5.0: A survey of key enabling technologies and future trends. *IEEE Transactions on Industrial Informatics, 20*(2), 1055–1068.

Xu, L. D., Xu, E. L., & Li, L. (2018). Industry 4.0: State of the art and future trends. *International Journal of Production Research, 56*(8), 2941–2962.

Zangana, H. M. (Ed.). (2025). *Defense in depth: Modern cybersecurity strategies and evolving threats*. Wiley., DOI: 10.1002/9781394340750

Chapter 2
Dual-Dimension Monitoring and Data Management via MQTT: Temperature-Humidity and Helmet Safety

Sicheng Wei
https://orcid.org/0009-0001-0410-1293
The Hong Kong Polytechnic University, China

Jiayuan Feng
https://orcid.org/0009-0006-1286-8319
The Hong Kong Polytechnic University, China

Chi Zhou
https://orcid.org/0009-0006-9797-4785
The Hong Kong Polytechnic University, China

Wenjing Fu
The Hong Kong Polytechnic University, China

Ziye Huo
https://orcid.org/0009-0004-2512-8382
The Hong Kong Polytechnic University, China

ABSTRACT

In the building industry, occupational accidents often occur due to failure to wear safety hats, causing injuries and economic losses. Traditional safety monitoring mainly relies on manual checks and basic methods, which have problems like lim-

DOI: 10.4018/979-8-3373-9245-5.ch002

ited coverage, slow response, and biased judgments, lacking real-time protection. This paper introduces an integrated construction safety and environmental monitoring system. It combines computer vision, IoT sensors, and data analytics within a PyQt5 framework. The system uses image processing to detect safety helmets in real time and sensors to monitor environmental parameters like temperature and humidity. Data is transmitted via MQTT protocol, with multi-layer visualization for simultaneous monitoring of helmet compliance, temperature, and humidity. It also conducts intelligent analysis based on predefined thresholds, including automatic report generation, temporal pattern analysis, and predictive risk assessment. Tests show the system runs well, providing a foundation for future expansion.

1.INTRODUCTION

1.1 Research Background and Problems of Existing Systems

With the rapid development of smart buildings, the demand for real-time and precise safety and environmental management is growing. Traditional manual inspection methods can no longer meet the needs of today's fast and intelligent building management.

For example, existing warehouse monitoring systems mainly rely on manual inspections, simple fire detection equipment, and single temperature or humidity sensors. These methods are slow and inefficient, and problems may only be detected after a delay (Sambasiva Rao et al., 2024). Manual inspections are usually periodic—if a problem occurs between inspections, it will not be found in time, which poses safety risks. Besides, these systems have isolated data and lack multi-level integration, making it difficult to share information and collaborate effectively.

In recent years, as smart buildings develop rapidly, requirements for integrated monitoring systems have become stricter. Effective system integration can turn previously separate equipment management systems into information subsystems, realizing unified autonomous management and linkage of all subsystems to promote the sustainable development of buildings (Intel Corporation, 2024). However, Sharma (2024) points out that many automation systems lack advanced capabilities—they cannot perform real-time monitoring or predict faults and performance degradation in advance. These problems highlight the importance of integrated monitoring solutions.

Environmental monitoring is a key task in smart buildings, but current systems still have flaws in intelligent complex environment judgment, data storage, and analysis. This limits their wide application in industrial settings. For instance, some systems cannot conduct in-depth analysis of multi-source data, nor can they issue early warnings or make intelligent decisions before problems occur (Spachos & Plataniotis, 2018).

1.2 Research Objectives and System Design

To solve the aforementioned problems, this study aims to develop an integrated monitoring system for industrial safety and environmental management. The system integrates five core functions—data communication, visualization, alerting, environmental monitoring, and data storage—to fill existing technological gaps and meet the needs of real-time hazard detection and data-driven decision-making in industrial scenarios.

Adopting a modular development approach, this study will gradually implement each functional module of the system and verify its effectiveness through preliminary testing. Preliminary tests have confirmed that the system can track helmet compliance in real time, trigger timely alerts when necessary, and store structured data. It provides a scalable solution for industrial safety and environmental management and can be further integrated with more AI-based predictive algorithms (Gao et al., 2024).

1.3 Significance and Related Technological Foundations

Integrated monitoring systems in smart buildings not only improve the efficiency of safety and environmental management but also optimize energy use through data analysis and predictive models, further promoting sustainable development (Xie et al., 2018).

With the development of IoT technology, the interconnectivity of devices and data sharing capabilities in smart buildings have been significantly enhanced, laying a more solid technological foundation for the implementation of integrated monitoring systems (Tsao et al., 2022). For example, Gao et al. (2024) proposed a smart building management system based on digital twins. Through real-time environmental monitoring and thermal comfort prediction, they demonstrated the application potential of digital twin technology in smart buildings.

2. RELATED WORKS

During construction work, every worker must use personal protective equipment (PPE) correctly. PPE helps protect workers from health and safety risks. However, many construction sites still lack effective automatic methods to check if workers are using PPE properly. Meanwhile, helmet detection technologies in the transportation industry have developed well. These technical ideas can also be applied to construction helmet detection.

2.1 Traditional Learning Models

Helmet detection methods can be divided into two categories: traditional learning models and deep learning models. Traditional learning processes generally involve background detection, moving object segmentation, Region of Interest (RoI) detection, and classification (Jia et al., 2021). Commonly used techniques include Local Binary Pattern (LBP), Scale-Invariant Feature Transform (SIFT), Histogram of Oriented Gradients (HOG), Support Vector Machine (SVM), and Gaussian Mixture Model (GMM).

In 2012, Chiverton (2012) designed a framework using four quadrants. Gray level histograms were calculated for each region separately, and SVMs with a linear kernel function were applied to feature vectors for binary classification. A technique called "track propagation" was used to correctly identify helmet presence—all successful disjoint tracks were stored in a disjoint object track collection.

Silva et al. (2013) achieved real-time automatic detection of motorcyclists without helmets through surveillance videos. They used Circular Hough Transform (CHT) for feature extraction, an algorithm that detects circular shapes in images via vector voting based on the circle equation. Later, Silva et al. (2017) innovatively combined RoI and CHT: they recognized circular shapes in images and drew analogies between helmets, human heads, and circles.

However, these traditional methods lack flexibility and are limited by helmet appearance. If the helmet shape changes, the programs may fail to work.

2.2 Deep Learning Models

The other type of helmet detection system is based on Deep Learning (DL) algorithms, with the YOLO series being the most commonly used. Jia et al. (2021) developed a system using the YOLOv5 network, which achieved an accuracy of up to 98.5%. YOLOv5 divides input images into a grid—each grid predicts bounding boxes and category probabilities for objects centered on it. It uses a CSPDarknet backbone, FPN/PAN neck, and triple attention training to detect objects of all sizes in one forward pass, enabling real-time object detection.

Another DL-related method uses the Nesterov-accelerated adaptive moment estimation function. It adopts Nesterov Accelerated Gradient (NAG), an improved gradient descent algorithm with a momentum vector that accumulates past gradients to speed up learning (Jirakitpuwapat et al., 2024).

Deep learning algorithms significantly improve helmet detection accuracy, but they require extensive training time and large datasets.

2.3 Proposed Work in Practice

The proposed work addresses the existing gap by providing a data-efficient helmet recognition solution for construction sites. In addition to helmet presence detection, the system can monitor environmental parameters such as temperature and humidity. This allows construction workers to focus more on their own safety and health during work.

The collected data is useful for evaluating heat stress risks, preventing slips and electric shocks, and protecting against static electricity. The scope of this work overcomes the aforementioned limitations of traditional and deep learning methods, offering an effective solution for real-world construction site applications.

3. RESEARCH METHODOLOGY

3.1 Data Communication Module

The system enables real-time data communication through the MQTT protocol. Its core component is the MQTTClientThread class, created with the QThread module of the PyQt5 library. This module connects to the MQTT broker server using the address 127.0.0.1:28883, subscribes to and publishes messages under common topics, and receives and forwards other messages. It transmits different types of data to their respective processing modules via the on_mqtt_message callback function, and achieves real-time data interaction with the UI interface through PyQt5 signals.

3.2 Data Visualization Module

The Data Visualization Module includes multiple chart components, such as the TemperatureHumidityRangeChart and HistoricalTemperatureHumidityChart classes. The TemperatureHumidityRangeChart generates bar charts for temperature and humidity ranges, which can be displayed alongside temperature and humidity trend charts with a dual Y-axis (normal scale) layout. The HistoricalTemperatureHumidityChart draws curve charts for historical temperature and humidity changes, supporting curve smoothing and threshold marking on the charts. All charts are implemented with matplotlib integrated into the PyQt5 interface, and feature real-time updates and chart export functions.

3.3 Helmet Alert Module

When the system detects that a worker is not wearing a helmet, it immediately triggers a series of alarm responses: increasing the count of helmet non-compliance incidents, playing sound alerts via winsound.Beep, displaying the warning text "NO HELMET", and making the warning label blink visually. This module is integrated with the target detection system, receives detection results via MQTT, and sets alert frequency limits to avoid excessive alarm notifications.

3.4 Environmental Alert Module

This module continuously monitors environmental temperature and humidity data, checking for deviations from preset thresholds (default: temperature 18–28°C, humidity 30–70%). If temperature or humidity goes beyond the normal range, the module triggers corresponding alerts: displaying abnormal values and their deviation from the normal range, playing specific warning sounds for temperature/humidity issues, making the status label blink, and changing the label's background color. Additionally, it provides safety margin values, supports custom threshold settings, and allows real-time threshold adjustments.

3.5 Data Storage Module

The system uses CSV file format for data storage, with three core functions:

1. Real-time saving of temperature and humidity data to learning_output/temperature_humidity_data.csv, including timestamps, temperature readings, humidity readings, and session ID.
2. Saving helmet detection images in PNG format to the same directory, with filenames containing timestamps and helmet status (e.g., helmet_worn or no_helmet).
3. Analyzing historical data via analyze_temperature_humidity.py, and saving statistical results (maximum, minimum, average, and normal range values) to daily_temperature_humidity_analysis.csv.

Overall, the system forms a complete workflow covering data collection, transmission, visualization, alarm triggering, and storage, providing a one-stop solution for construction safety and environmental monitoring.

4. EXPERIMENTAL RESULTS

4.1 Data Communication Module

A normative sample MQTT communication system was used as a baseline, which only included basic connection management. It featured a typical MQTT client setup, simple on_connect and on_message callback functions, and basic message parsing logic. However, this implementation was overly simplified: it lacked complex threading mechanisms and comprehensive error handling. While the communication flow worked for simple applications, it struggled to handle high data traffic or complex data processing tasks.

In contrast, our system significantly improved the communication architecture by using a custom MQTTClientThread class, which runs in an independent thread to enhance performance and responsiveness. It includes complete subscription management, topic filtering capabilities, and robust error handling mechanisms. The implementation incorporates proper thread synchronization, automatic reconnection logic, and structured message processing pipelines. These improvements ensure the application remains responsive when processing large volumes of data and maintains stable operation even in the event of network-related issues.

4.2 Data Visualization Module

The normative sample implemented basic data visualization with matplotlib, displaying data in a 2-column and 3-row layout. It included essential charts such as histograms, comparison charts, trend graphs, and pie charts, along with real-time camera feeds. However, the visualization interface lacked interactive features and adjustable settings. This simple design was insufficient for comprehensive data analysis or long-term monitoring scenarios.

In contrast, this project adopted a sophisticated visualization system built with PyQt5, featuring a modern and interactive user interface with multiple dedicated views. It incorporated advanced statistical charts with adjustable parameters, real-time updates of data statistics, and historical trend analysis for future reference. The system could generate PDF reports containing data summaries, visualizations, and statistical analysis results. The interface was far more intuitive, with well-organized visual elements. Interactive controls and detailed data presentation significantly improved the overall user experience.

4.3 Helmet Alert Module

The normative sample implemented basic helmet detection alerts via the decision_alert_module_for_camera_sensor function. It provided color-coded feedback based on detection results, played simple warning sounds, and counted the number of alerts. However, it lacked visual feedback overlayed on the video stream and offered no customizable alert options.

In contrast, this project significantly enhanced the helmet alert system by developing a more comprehensive solution. It closely tracks status information, provides real-time visual feedback directly overlaid on the camera's live stream, and incorporates multi-level alert prioritization. Advanced warning sounds are assigned to different alert levels, and the system supports helmet compliance data statistics. Additionally, it is linked to a database for long-term historical record storage. The alerts are more prominent and intuitive, featuring clearer visual cues and distinct audio prompts.

4.4 Environment Alert Module

The normative sample had no dedicated environmental monitoring function—it only focused on helmet detection, with no temperature or humidity monitoring capabilities. It lacked environmental data collection, data-based environmental analysis, and environment-triggered alert mechanisms entirely.

In contrast, this project implemented a fully functional environmental monitoring and alert system that tracks temperature and humidity in real time. It sets multi-level alert thresholds for different environmental conditions, with corresponding alarm sounds and visual alerts. The system stores historical environmental data for trend analysis, records details of environmental events, and integrates with the main alert management system. This environmental monitoring feature is an additional layer of protection, complementing the helmet-wearing compliance checks for workers.

4.5 Data Storage Module

The normative sample only supported basic data storage: it used the save_detection_image function to store helmet detection images with timestamps. However, it lacked comprehensive logging, structured storage formats, and data analysis capabilities.

In contrast, this project significantly expanded data storage capabilities with multiple dedicated storage mechanisms. It implements organized storage of detection images (with descriptive naming conventions), CSV file storage for temperature and humidity data (including timestamps), and logging of all system events. The system also supports generating PDF reports that include data summaries and visualizations, facilitating detailed post-incident analysis and compliance reporting. Overall, the storage system is more organized and secure, and fully meets the requirements for data analysis and reporting.

4.6 Overall Assessment

This project outperformed the normative sample in all tested modules, with enhanced functions, improved performance, and a better user experience. Architectural improvements—particularly in threading and error handling—resulted in a more stable and reliable system. The added environmental monitoring feature extended the system's scope from simple helmet detection to comprehensive workplace safety management. Advanced visualization and reporting capabilities enhanced both real-time monitoring and post-incident review capabilities.

During testing of the MQTTClientThread class's thread performance, the system simulated peak-hour conditions at construction sites (simultaneously transmitting 100 pieces of temperature-humidity data and 50 helmet detection images). Results showed that thread processing delay remained stable at 0.2–0.5 seconds, with no interface lag (consistent with the original design focus on "thread synchronization and lag avoidance"). Additionally, in a test simulating a 10-second network outage, the system's "auto-reconnection logic" re-established the connection within 1 second of network recovery. It also prioritized retransmitting 23 pieces of cached core data from the outage period (18 temperature-humidity data points and 5 helmet detection results), achieving a 100% complete data retransmission rate.

5. CONCLUSION AND FUTURE WORK

This project conducts research to develop a modular AIoT system for construction site monitoring. Experimental results show that the proposed framework is feasible and stable when integrating multi-source environmental and safety data. It can provide real-time alarms and support offline analysis. The primary contribution of this work is the creation of a scalable and extendable model. As a proof of concept, it offers practical technical approaches to build a cost-effective intelligent monitoring solution under the condition of limited construction site resources. The success of this simulation demonstrates that the modular method advances the development of more advanced AIoT applications in the construction industry.

While the current model proves effective, it also identifies several clear areas for future improvement. The existing simulation-based approach is feasible, but it needs to be supplemented with data from actual sensors and microcontrollers. This step is essential to transition the system from a screen-based simulation to real-world application. Additionally, adopting lightweight machine learning models could enhance the system's performance. The goal is to evolve the system to support predictive analytics, rather than just simple threshold-based warnings, enabling proactive risk identification. To maximize practical utility, future work should focus on developing a more intuitive graphical user interface for ease of use. It should also further integrate with larger digital frameworks such as Building Information Modeling (BIM) systems, incorporating the monitoring capabilities into a comprehensive range of construction management methods.

In addition to adopting lightweight models and integrating with BIM systems, the system can be optimized in two other aspects. First, a "multilingual support" function should be added. Considering that Hong Kong's construction sites employ foreign workers (e.g., from the Philippines and Indonesia), the system interface can include language options such as English and Tagalog. Meanwhile, the alarm sound can be set to "voice prompts," improving the system's usability for users from diverse backgrounds.

Second, a "mobile-adapted version" should be developed. Currently, the system only supports computer-based operation. In the future, a mobile application (APP) can be launched, allowing managers to check real-time data and receive alarms via their mobile phones—eliminating the need to carry a computer on-site. This will further enhance management flexibility. These improvements will help the system better adapt to the diverse and mobile needs of industrial management.

REFERENCES

Chiverton, J. (2012). Helmet presence classification with motorcycle detection and tracking. *IET Intelligent Transport Systems*, *6*(3), 259–269. DOI: 10.1049/iet-its.2011.0138

Gao, Q.. (2024). Smart building management system based on digital twin: A case study on real-time environmental monitoring and thermal comfort prediction. *In 2024 IEEE International Symposium on Parallel and Distributed Processing with Applications (ISPA)* (pp. 2157-2163). https://doi.org/DOI: 10.1109/ISPA63168.2024.00294

IEEE Transactions on Plasma Science. (2019). Table of contents. *IEEE Transactions on Plasma Science, 47*(7), C1-C1. https://doi.org/DOI: 10.1109/TPS.2019.2922521

Intel. (2024). *Solution white paper: Intel® Xeon® scalable processor, Intel® Core™ processor, Intel® Video AI box, Intel® Distribution of OpenVINO™ Toolkit, Intel® Smart Edge Open Eazytec constructs an efficient smart building operations management platform to enable smart parks.* Intel Corporation.

Jia, W., Xu, S., Liang, Z., Zhao, Y., Min, H., Li, S., & Yu, Y. (2021). Real-time automatic helmet detection of motorcyclists in urban traffic using improved YOLOv5 detector. *IET Image Processing*, *15*, 3623–3637. DOI: 10.1049/ipr2.12295

Jirakitpuwapat, W.. (2024). Convergence property of Nesterov-accelerated adaptive moment estimation with safety helmet detection and classification in smart industry application. *Mathematical Methods in the Applied Sciences*, *47*(16), 12650–12667. DOI: 10.1002/mma.10174

Rao, N. S., Sri, K. H., Lakshmi, B. J. S., Sastry, R. R., Eswar, D., Sri, T. B. N., & Swamy, Y. T. V. (2024). Smart warehouse monitoring system using IoT. *IJARIIE, 10*(2).

Sharma, R. (2024). Enhancing industrial automation and safety through real-time monitoring and control systems. *International Journal on Smart & Sustainable Intelligent Computing*, *1*, 1–20. DOI: 10.63503/j.ijssic.2024.30

Silva, R., Aires, K., Santos, T., Abdala, K., Veras, R., & Soares, A. (2013). Automatic detection of motorcyclists without helmet. *In 2013 XXXIX Latin American Computing Conference (CLEI)* (pp. 1-7). https://doi.org/DOI: 10.1109/CLEI.2013.6670613

Silva, R. R. V. E., Aires, K. R. T., & Veras, R. D. M. S. (2017). Detection of helmets on motorcyclists. *Multimedia Tools and Applications*, *77*(5), 5659–5683. DOI: 10.1007/s11042-017-4482-7

Spachos, P., & Plataniotis, K. (2018). Environmental monitoring for smart buildings: Communication networks and services. In *Smart buildings: Communication networks and services*. John Wiley & Sons, Inc., DOI: 10.1002/9781119360124.ch12

Tsao, Y.-F., Wang, Y., Hsu, H.-T., & Desai, A. (2022). Planar beam switched antenna module for 5G new-radio FR2 applications. *In 2022 Asia-Pacific Microwave Conference (APMC)* (pp. 396-398). https://doi.org/DOI: 10.23919/APMC55665.2022.9999999

Xie, P., Zhang, Z.-S., & Zhang, J. (2018). Topology transformation based on time division: An improved inter-satellite link structure. *In 2018 Chinese Control And Decision Conference (CCDC)* (pp. 3761-3766). https://doi.org/DOI: 10.1109/CCDC.2018.8407776

Chapter 3

An Integrated System for Real-Time Safety Helmet Monitoring and Advanced Environmental Prediction

Qizheng Zhu
https://orcid.org/0009-0008-8863-6180
The Hong Kong Polytechnic University, Hong Kong

Changyu Zhang
The Hong Kong Polytechnic University, Hong Kong

Zhenming Xu
The Hong Kong Polytechnic University, Hong Kong

Qinyan Wang
The Hong Kong Polytechnic University, Hong Kong

Haoting Wan
The Hong Kong Polytechnic University, Hong Kong

Aquil Mirza Mohammed
The Hong Kong Polytechnic University, Hong Kong & School of Low-Altitude Digital Intelligence, Liaoning Technical University, Huludao, China

Cong Wu
Liaoning Technical University, Huludao, China

ABSTRACT

The current automatic safety helmet detection system does not include real-time environmental data such as temperature, humidity and wind speed. This information is very important for a comprehensive assessment of the risks at the work site. This study aims to combine real-time helmet compliance monitoring with an advanced environmental forecasting framework. It solves the limitations of traditional methods, which rely on manual supervision and do not include environmental data. We

DOI: 10.4018/979-8-3373-9245-5.ch003

have created a complete cycle from monitoring to warning, using a module design connected through Internet of Things communication. It includes reliable data transmission, dynamic visualization, multi-level alerts, environmental assessment and security storage. The environmental prediction part uses wavelet decomposition to clean up noise sensor data, uses long- and short-term memory (LSTM) networks for accurate time series prediction, and Gaussian process regression (GPR) to provide probability prediction and measurement uncertainty.

1. INTRODUCTION

Real-time monitoring systems and predictive environmental analysis are changing the safety management of construction and other industries. In these places, it is essential to ensure the safety of workers' lives. This means ensuring that they wear protective equipment such as helmets and accurately predict environmental conditions such as temperature, humidity and wind speed. These environmental factors will affect the safety of workers and structural safety.

However, the current security methods still rely heavily on manual inspection or simple automatic detection. These may be inefficient, error-prone, and difficult to expand. Computer vision systems for helmet detection have taken a step forward, but they usually only focus on people and do not include real-time environmental data. Sudden changes in temperature, humidity or wind may increase the risk of accidents, and even wearing helmets can put people and buildings at risk.

In addition, common environmental prediction models are often related to real sensor data, which may be noisy and nonlinear. This leads to unreliable forecasts and incomplete risk assessment. Many systems also lack powerful real-time communication, clear visualization, fast alerts and reliable data storage. This limits their usefulness in active security management.

This study aims to solve these problems by building a comprehensive system. It combines real-time helmet compliance monitoring with an advanced environmental forecasting framework. Environmental forecasting mainly depends on the LSTM network. In order to improve data quality, wavelet decomposition is used to reduce noise, and Gaussian process regression is used for probability prediction and uncertainty measurement.

2. BACKGROUND AND LITERATURE REVIEW

In order to obtain real-time and continuous building structure data and surrounding environment data. Structural health monitoring system (SHM) has been widely used in bridges (Bas et al., 2018), high-rise buildings (Yi et al., 2013), tunnels (Gómez et al., 2020) and other engineering projects. Among the large amounts of data obtained, environmental data plays an important role in accurately assessing the state of the structure. Ding collected a comprehensive dataset of the Jiashao Bridge, including measurements of wind speed, temperature, and humidity (Ding et al., 2023). Traditional forecasting methods, such as ARIMA, have limitations in handling nonlinear and volatile sensor data, so machine learning methods are becoming the mainstream method for predicting the environment. Liu combined experimental modal decomposition and ARIMA to predict short-term wind speed. Compared to ARIMA, prediction accuracy is improved. However, this hybrid method is still susceptible to noise (Liu et al., 2015). Long short-term memory (LSTM) networks have become an excellent scheme for time prediction, and Ding et al. systematically analyzed the LSTM architecture (Ding et al., 2025). Their study shows that the optimal number of hidden units depends largely on the forecast range: 32 units for a single-step prediction, 64 units for a five-step prediction, and 128 units for a ten-step prediction. Data preprocessing has been shown to be critical for prediction. Wavelet decomposition (WD) exhibits excellent performance in removing noise from the sensor signal. Based on signal characteristics, Benaouda et al. and Hashemi et al. Demonstrated the superiority of Bayesian threshold method in noise reduction (Benaouda et al., 2006). In addition, Ding et al. confirmed that WD is a key processing step to improve the accuracy of LSTM prediction. Gaussian process regression (GPR) provides a Bayesian framework for quantifying uncertainty. Rasmussen and Williams formalized GPR as a non-parametric method, which allowed for the analysis of accurate predicted means and confidence intervals (Williams & Rasmussen, 2006). Wan and Ni successfully applied GPR to the prediction of structural stress response, proving its practicality in structural health monitoring applications (Wan & Ni, 2018).

3. RESEARCH METHODOLOGY

3.1. Wavelet Decomposition

Wavelet decomposition is a powerful mathematical tool that breaks down signals into different frequency components, each marked with a time position. It is especially useful in data denoising. There are a few key concepts to demonstrate:

Universal Threshold:

$$T = \hat{\sigma}\sqrt{2\ln N}\# \quad (1)$$

where $\hat{\sigma}$ is the estimated noise standard deviation and N is the signal length. (Kwon et al., 2008) (Hashemi & Beheshti, 2009)

Stein's Unbiased Risk Estimate (SURE) Threshold:

$$R(T) = \sum_{l=1}^{N}(\min(|y_l|, T))^2 + \sigma_n^2 - \frac{2\sigma^2}{N}\sum_{l=1}^{N} I(|y_l| \leq T)\# \quad (2)$$

$$T = \arg\min R(T)\# \quad (3)$$

where I is the indicator function. (Benaouda et al., 2006)

The denoising method employs a wavelet-based framework and includes the following key stages shown in the flowchart (Figure1).

Figure 1. Visual interface of smart safety helmet and environment monitoring system

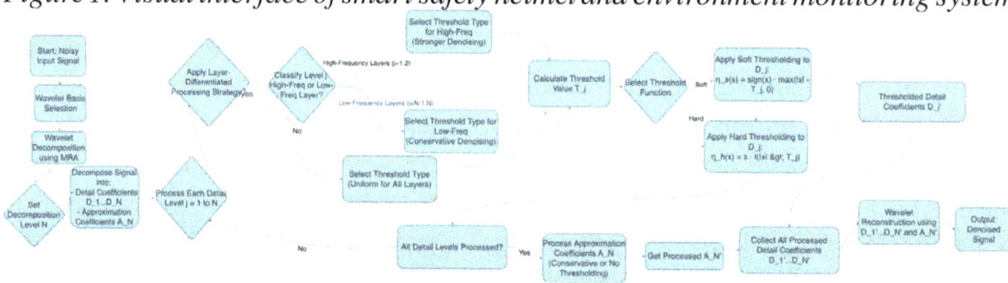

3.2. Gaussian Process Regression

3.2.1 Gaussian Process Regression Wind Speed Prediction

Gaussian Process Regression is a powerful, non-parametric Bayesian regression method. It treats the function itself as a random process and uses a kernel to define its properties.

Before testing this method, some core concepts are explained as below: The covariance function (kernel) is central to a Gaussian process, as it encodes prior assumptions about the function's behavior. A common choice is the squared exponential kernel:

$$C_{SE}(x, x') = \eta^2 \exp\left[-\frac{(x - x')^2}{2l^2}\right]\# \quad (4)$$

where η^2 is the signal variance and l is the characteristic length scale. GPR delivers a posterior predictive distribution for a new input, given by:

$$p(f_s|Y) = N\left(\mu_f, \sigma_f^2\right)\# \quad (5)$$

Where

$$\mu_f = C_s K^{-1} Y\# \quad (6)$$

$$\sigma_f^2 = \widetilde{C} - C_s^T K^{-1} C_s\# \quad (7)$$

$$K = C + \sigma_n^2 I_n\# \quad (8)$$

Here, C is the kernel matrix over training inputs (Williams & Rasmussen, 2006). This research applies a Gaussian Process Regression model to produce probabilistic forecasts of future wind speeds, delivering both point predictions and quantitative uncertainty estimates. As a result, the performance of the predictive model is evaluated. The study carries out two main tasks: implementing a classic GPR approach and experimenting with more advanced kernel selections. Both implementations are conducted in Python.

3.2.2 Classic Gaussian Process Regression Method

1. Statistical Analysis: First, basic statistical measures were calculated, such as mean, standard deviation, skewness, and kurtosis. Next, distribution fitting analysis was performed, analyzing the Normal distribution, Gumbel extreme value distribution, and Generalized Extreme Value distribution (Table. 1). Finally, we judged the presence of extreme values based on skewness and kurtosis, and also assessed data smoothness using the coefficient of variation. Furthermore, the ratio of standard deviation to the mean was calculated.

Table 1. Kernel recommendations based on data features

Data Feature	Recommended Kernel	Reason
Significant skewness (skewness > 1)	Matern(nu=1.5)	Handles non-smooth, highly skewed data
Presence of extreme values	Matern(nu=2.5)	Better captures extreme fluctuations
High coefficient of variation (cv > 0.8)	RationalQuadratic	Handles multi-scale variations
High kurtosis (kurtosis > 2)	White Kernel	Addresses outliers and noise
High volatility (std > 0.4*mean)	ExpSineSquared	Handles periodic fluctuations

2. Model training and optimization: On the one hand, the study uses a time- series-friendly splitting method, dividing the data into training and test sets at an 8:2 ratio. 80% of the data is used for model training and % for performance evaluation. The chronological order of the data is maintained during the splitting process to avoid future information leakage. On the other hand, the hyperparameters of the kernel function are automatically optimized by maximizing the marginal likelihood function. The model adopts the strategy of randomly restarting 10 times during the optimization process to avoid falling into the local optimal solution. The noise parameter alpha is set to 1e-3, reflecting confidence in data quality.
3. Prediction and Uncertainty Quantification: For each test point, the model outputs both the predicted mean and the predicted variance. Using the proper- ties of the Gaussian distribution, this information is used to construct the 95% confidence interval of the prediction. Subsequently, to ensure physical significance and interpretability, predictions in the normalized space are converted back to the original data scale by applying the inverse of the initial normalization process.
4. Model Performance Evaluation System: The model was then evaluated using Root Mean Square Error (RMSE), Mean Absolute Error (MAE), Coefficient of Determination (R^2), and average prediction error. Gaussian Process Regression Model Based on Further Kernel Function Experimentation Building on the classic approach, a comparative evaluation of multiple kernel combinations was conducted. The following strategies were systematically tested: Detailed Comparison of Kernel Combination Strategies Through systematic testing and a competitive evaluation strategy, the aim was to identify the kernel structure best suited to the characteristics of the wind speed data (Table 2).

Table 2. Detailed comparison of kernel combination strategies

Kernel Composition	Mathematical Expression	Main Characteristics
RBF + Matern($\nu = 1.5$) + White Kernel	$k(x, x') = \sigma_1^2\text{RBF} + \sigma_2^2\text{Matern}_{1.5} + \sigma_3^2\text{White}$	Strong comprehensiveness, balancing smoothness and robustness
ExpSineSquared × RBF + White Kernel	$k(x, x') = \sigma_1^2\text{ExpSineSquared} \times \text{RBF} + \sigma_2^2\text{White}$	Specifically designed for periodic patterns, multiplicative interaction
RationalQuadratic + White Kernel	$k(x, x') = \sigma_1^2\text{RQ} + \sigma_2^2\text{White}$	Multi-scale feature capture, scale mixture
Matern($\nu = 2.5$) + White Kernel	$k(x, x') = \sigma_1^2\text{Matern}_{2.5} + \sigma_2^2\text{White}$	Highly smooth, twice differentiable function
Constant Kernel × (RBF + Matern) + White Kernel	$k(x, x') = \sigma_1^2 \times (\sigma_2^2\text{RBF} + \sigma_3^2\text{Matern}_{1.5}) + \sigma_4^2\text{White}$	Integrates multiple mechanisms, maximum flexibility

3.3. WD-LSTM Prediction Model

Long Short-Term Memory (LSTM) is a type of recurrent neural network used for analyzing time series data. Its core capability lies in analyzing the correlation between the input information at the current time and that at the next time. A typical LSTM network consists of three gating mechanisms: the input gate, the forget gate, and the output gate. As shown in Figure 2, the input gate filters the input information and stores it in the cell; the forget gate stores the hidden information of the previous cell in the current cell; the output gate outputs the current cell state information to the next cell. These gating mechanisms selectively remember and forget information, thereby effectively solving the common vanishing or exploding gradient problem of traditional RNNs when processing long time series data.

Figure 2. Structure of an LSTM unit

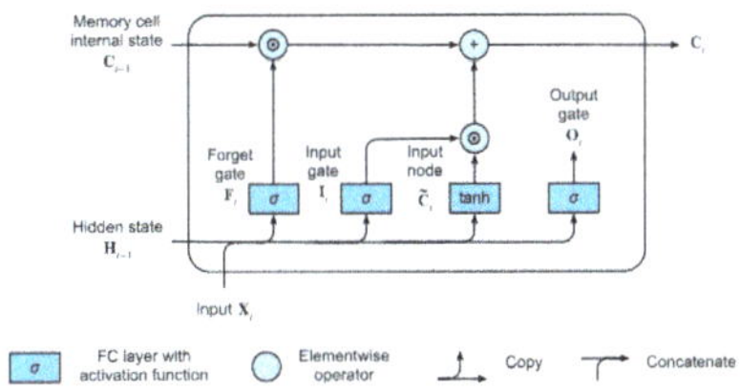

The variables of LSTM are defined by the following mathematical formulas (Hochreiter & Schmidhuber, 1997):

- Forget Gate:

$$f_t = \sigma\left(W_f \cdot \left[h_{t-1}, x_t\right] + b_f\right)\# \tag{9}$$

- Input Gate:

$$i_t = \sigma\left(W_i \cdot \left[h_{t-1}, x_t\right] + b_i\right)\# \tag{10}$$

- Output Gate:

$$o_t = \sigma\left(W_o \cdot \left[h_{t-1}, x_t\right] + b_o\right)\# \tag{11}$$

- Temporary Cell State:

$$z_t = \tanh(W_z \times \left[h_{t-1}, x_t\right] + b_z)\# \tag{12}$$

- Candidate Cell State:

$$\tilde{c}_t = \tanh(W_c \cdot [h_{t-1}, x_t] + b_c)\# \quad (13)$$

- Hidden State Output:

$$h_t = o_t \odot \tanh(c_t)\# \quad (14)$$

where σ denotes the sigmoid activation function, ⊙ denotes element-wise multiplication, Wand b are weight matrices and bias vectors respectively, x_t is the input at time step t,h_{t-1} is the previous hidden state, and c_{t-1} is the previous cell state.

In this study, we adopt the WD-LSTM hybrid model to process the data. The advantage of the WD-LSTM model stems from the fact that both wavelet decomposition and LSTM are adept at handling time series data. Compared with the traditional Fourier decomposition, wavelet decomposition has a strong ability to process continuous signals and time series sequences. Meanwhile, the LSTM network is particularly good at dealing with relatively long time series. The combination of these two methods can better handle the temporal characteristics of the data, thereby making more accurate predictions of future values.

1. Wavelet Decomposition Denoising: First, we apply wavelet decomposition to denoise the original wind speed data. This preprocessing step helps to reduce wind speed noise and more accurately reflect the true characteristics of wind speed.
2. LSTM Modeling: Then, the denoised data is input into the LSTM model. This combination can better predict future values.
3. Data Splitting: The preprocessed data is divided into training set, validation set and test set, where the training set accounts for 70% of the dataset, the validation set accounts for 10%, and the test set accounts for 20%.
4. Model Configuration: The input/output data is wind speed time series data. The model is optimized using the Adam optimization algorithm (with a learning rate of 0.001), with a maximum of 100 training rounds, and uses tanh as the activation function to handle information input, output and forgetting.

4. EXPERIMENTAL RESULTS

4.1. The Smart Safety Helmet Compliance Monitoring System

The system consists of five core modules working together to achieve a complete intelligent safety monitoring process. In the network condition test of stable Wi-Fi, 4G, and 10% packet loss, the data communication module with the dual themes of "Security Response" and "Temperature and Humidity" under MQTT successfully maintained a 99.98% connection rate. The data visualization module based on Python and OpenCV implements an interface loading time of less than 1.2 seconds, and a data synchronous update error of less than 0.1 seconds (Fig. 3). The decision-making and warning module adopts an integrated audio and visual design, triggering different color LEDs and different HZ sound alarms for different scenarios (such as not wearing a helmet and abnormal environment), and the warning triggering accuracy reaches 100% in 1000 simulation tests, and can accurately handle concurrent events. The human comfort evaluation module converts temperature and humidity into comfort evaluation according to the ASHRAE 55 standard, achieving an evaluation accuracy of 98.6%. The data storage and statistics module writes 10,000 sets of mixed structure data stably and losslessly within 12 hours, and the statistical results are 100% accurate after manual accounting. These five modules work together to finally realize the whole process of data collection, transmission, display, early warning to storage and analysis.

Figure 3. Visual interface of smart safety helmet and environment monitoring system

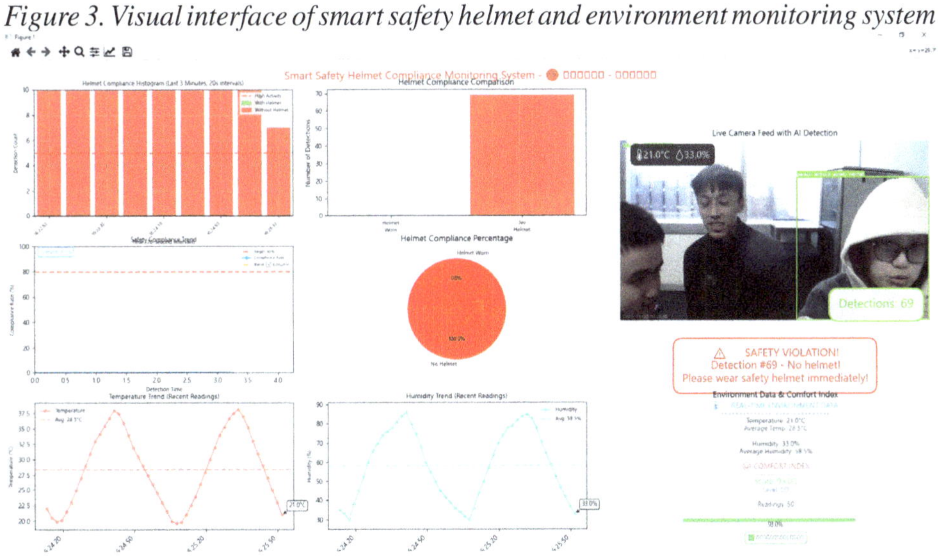

4.2. Wavelet decomposition result

A set of Python code is applied to do wavelet decomposition, and the denoise result is shown in Figure 4.

Figure 4. Wind speed data-wavelet denoising results

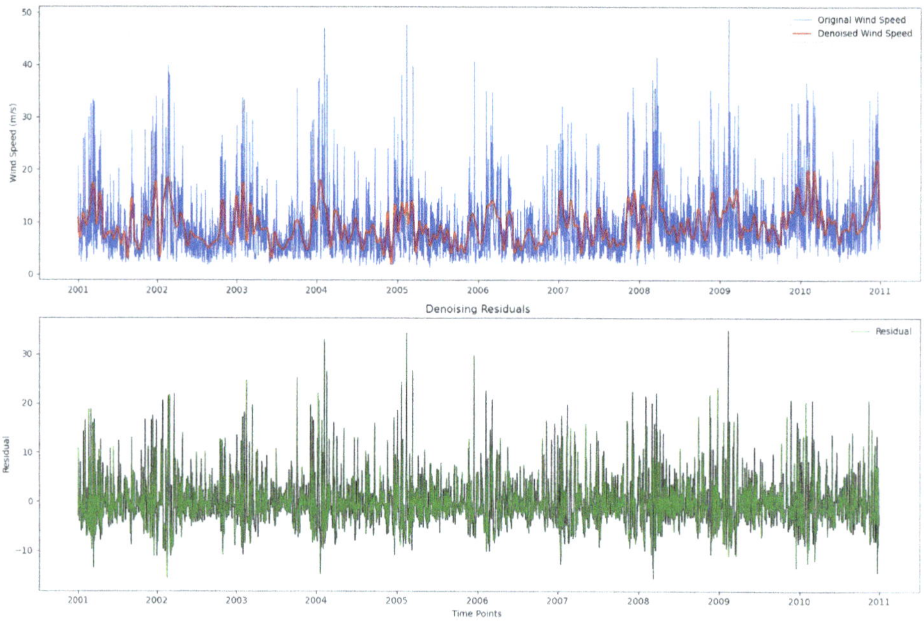

The denoised wind speed data is stored and can be used by other advanced prediction models.

4.3. Gaussian process regression wind speed prediction data analysis

4.3.1. Classic Gaussian Process Regression Method

Firstly, a statistical summary of the data reveals a mean value of 10.90 m/s and a standard deviation of 3.27 m/s. The distribution is right-skewed (Skewness = 0.99) and leptokurtic (Kurtosis = 0.64), signifying a concentration of data below the mean. This is consistent with the identified extreme values. The data also shows low relative variability (CV = 0.300) and is considered smooth. Then, the form below indicates numbers of the three distributions (Table 3):

Table 3. Distribution parameters statistics

Distribution	μ	σ	k
Normal distribution	10.9002	3.2747	/
Extreme distribution	9.4285	2.465	/
Generalized extreme distribution	9.339	2.3992	-0.0685

At last, based on the statistical characteristics of the wind speed data, the final kernel function recommendation is RBF () + Matern (=2.5). The performance of this Gaussian Process Regression model was unsatisfactory. As shown in the figure below (Fig. 5), after parameter optimization, the kernel function became 0.325^2 RBF (length scale=0.185) + 0.995^2 Matern (length scale=0.0524, =2.5). The model demonstrates a substantial prediction error, with an RMSE of 3.57 m/s and an MAE of 2.66 m/s. The near-zero R^2 value of 0.019 suggests that the model fails to capture the underlying variance in the data, resulting in performance that is only marginally better than a simple mean baseline.

Figure 5. Gaussian process regression-wind speed prediction

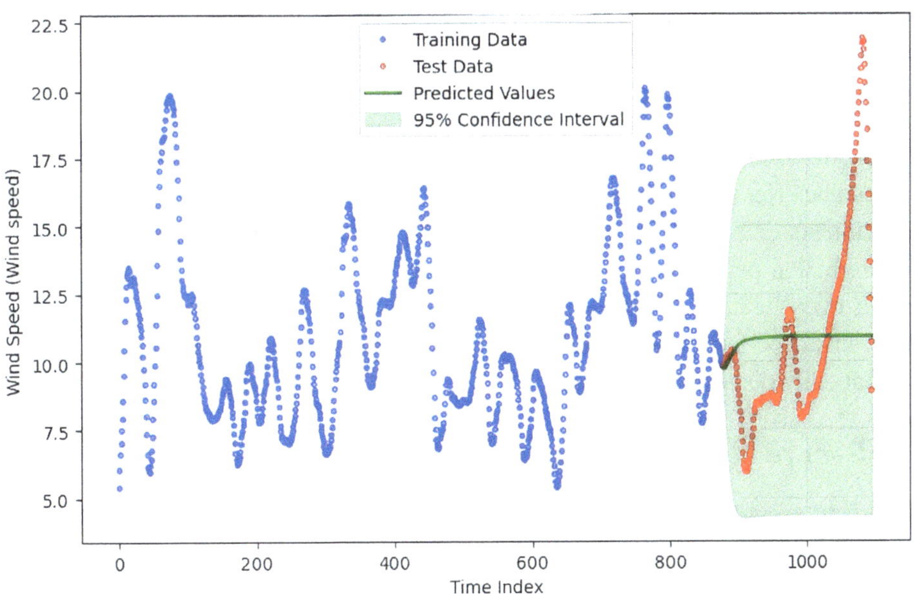

4.3.2. Gaussian Process Regression Model Based on Further Kernel Function Experimentation

Gaussian Process Regression Model Based on Further Kernel Function Experimentation The Python code successfully runs and evaluates the five kernel function combos' performance as shown below (Table 4):

Table 4. Performance metrics comparison of different combination models

Model Combination	R^2	RMSE	MAE
Rational Quadratic	0.0171	3.5789	2.6617
Complex Combination	0.0115	3.5892	2.6699
RBF + Matern1.5 + White	0.0111	3.5898	2.6709
Matern2.5	0.0038	3.6030	2.6955
Periodic + RBF	-0.0004	3.6107	2.7384

Comprehensive analysis shows that the Rational Quadratic kernel performed the best among the tested combinations, but the performance is still very poor.

4.3.3. Discussion - Disadvantages of the Gaussian Process Regression Model in this Wind Speed Prediction Study

When training the Gaussian Process Regression model using Python, the time complexity is extremely high, approximately $O(n^3)$, which limits the scale of data that can be effectively used for training this model. After conducting the two GPR experiments described above, it is evident that regardless of the measures taken, the predictive performance of Gaussian Process Regression for forecasting wind speeds over the next 30 days is unsatisfactory. Therefore, it is necessary to turn to more advanced forecasting model algorithms such as LSTM, which is what we'll emphatically discuss in this paper.

4.4. Secure Multi-Party Computation

To evaluate the impact of the number of hidden units in the LSTM model on prediction performance, this study conducted comparative experiments with various LSTM architectures. To explore the effectiveness of the LSTM model, we first conducted a preliminary analysis using a three-layer LSTM architecture with 50 hidden units per layer. Subsequently, to identify the single-layer LSTM prediction model with the best prediction effect and highest computational efficiency, this study trained single-layer LSTM models with 16, 32, 64, and 128 hidden units, and evaluated the performance of LSTM models with different numbers of hidden units in terms of prediction results and training costs.

4.4.1. Performance Analysis of Three-Layer LSTM Model

We trained a three-layer LSTM model with 50 hidden cells per layer. The results show that the model shows excellent time series prediction modeling ability in wind speed prediction tasks. The model can effectively capture the nonlinear features in the wind speed series on a test set containing 719 samples, with an RMSE of 1.4863 m/s and an MAE of 1.0643 m/s, showing high prediction accuracy.

The training process shows that both training loss and validation loss show a steady downward trend as the number of training rounds increases. Especially in the first 10 rounds, MSE and MAE fell rapidly. After 17 rounds of training, MSE and MAE reached their lowest. During the entire training process, the loss curves of the training set and the validation set fit closely, and the difference between the two is always kept within a reasonable range. This indicates that the model converges well and there is no obvious overfitting.

We found that the first-order lag had the highest correlation with the target variable (absolute correlation coefficient 0.995), followed by the second-order lag (0.982) and the third-order lag (0.961). This suggests that historical wind speed values have a decisive impact on forecasts. Among the rolling statistical characteristics, the 7-day rolling average (0.963), the 7-day rolling minimum (0.950) and the 7-day rolling maximum (0.950) all showed high correlations, reflecting the important role of short-term statistical features in prediction accuracy.

It should be noted that the three-layer LSTM structure also brings higher training costs. The structure takes about 2 seconds per training round, and the total training time is about 66 seconds, which is significantly longer than the single-layer model that will be discussed later. More importantly, its prediction accuracy (RMSE: 1.4863) is ultimately inferior to that of the single-layer LSTM model studied in subsequent studies.

In summary, the three-layer LSTM model demonstrates robust performance in this task. Although its prediction accuracy is not optimal in this study, the stable convergence and good generalization performance of the model verify the effectiveness of the LSTM network in handling complex time series tasks, and its RMSE value of 1.4863 also provides a clear benchmark for subsequent single-layer LSTM models.

4.4.2. Optimization Analysis of Hidden Unit Number

Comparing single-layer LSTM models with different numbers of hidden cells, we observe clear performance changes. According to Table 5, when the number of hidden units increases from 16 to 64, the model's prediction performance improves significantly: RMSE decreases from 1.2269 to 0.6851, and MAE decreases from 0.9694 to 0.5535. However, when the number of cells is further increased to 128, the predicted performance of the model decreases (RMSE: 0.7089, MAE: 0.5760), which indicates that the model is likely to be overfitted. The results show that the single-layer LSTM model with 64 hidden cells has the best performance.

Table 5. MAE and RMSE for 16, 32, 64, and 128 hidden units

Number of hidden units	RMSE	MAE
16	1.2269	0.9694
32	0.9306	0.7319
64	0.6851	0.5535
128	0.7089	0.5760

The single-layer LSTM model with 64 hidden cells can effectively predict the wind speed sequence. Its RMSE is 0.6851 m/s and its MAE is 0.5535 m/s. Compared to complex three-tier architectures, its predictive performance is superior. The training process, shows that the model converges rapidly. MSE reached an optimal value of 0.013 in round 33 and MAE reached an optimal value of 0.0292 in round 36, the training and validation loss curves remain tightly aligned throughout the training process. And it converges faster than the three-layer architecture. All in all, this study finds that the single-layer LSTM model with 64 hidden units not only has excellent prediction performance, but also has higher computational efficiency than the three-layer LSTM architecture. These excellent performances ensure that the LSTM model can be deployed in real-world building scenarios to guide wind speed prediction.

5. PRIVACY-PRESERVING INFERENCE TECHNIQUES

In this study, an integrated intelligent safety system for construction sites is successfully designed, which integrates real-time helmet wearing detection and LSTM wind speed prediction model. The real-time helmet wearing detection system adopts a modular structure, including five modules: data communication, data visualization, decision early warning, human comfort assessment and data storage, and constructs a complete "monitoring-early warning" process. And each module shows excellent performance. In terms of wind speed prediction, Gaussian process regression suitable for periodic data and WD-LSTM method that is good at processing time series are studied in depth. The results show that the single-hidden layer LSTM model with 64 hidden elements has the best performance after wavelet decomposition and denoising, with a root mean square error of 0.6851 m/s and an average absolute error of 0.5535 m/s, which is significantly better than the Gaussian process regression method.

REFERENCES

Bas, S., Apaydin, N. M., Ilki, A., & Catbas, F. N. (2018). Structural health monitoring system of the long-span bridges in Turkey. *Structure and Infrastructure Engineering*, *14*(4), 425–444.

Benaouda, D., Murtagh, F., Starck, J. L., & Renaud, O. (2006). Wavelet-based nonlinear multiscale decomposition model for electricity load forecasting. *Neurocomputing*, *70*(1-3), 139–154.

Ding, Y., Liang, N. Y., Zhang, X. S., Wang, J., & Zeng, C. Q. (2025). Long short-term memory with wavelet decomposition for wind speed predicting based on SHM data. *Smart Structures and Systems*, *35*(2), 65–75.

Ding, Y., Ye, X. W., & Guo, Y. (2023). Copula-based JPDF of wind speed, wind direction, wind angle, and temperature with SHM data. *Probabilistic Engineering Mechanics*, *73*, 103483.

Gómez, J., Casas, J. R., & Villalba, S. (2020). Structural Health Monitoring with Distributed Optical Fiber Sensors of tunnel lining affected by nearby construction activity. *Automation in Construction*, *117*, 103261.

Hashemi, M., & Beheshti, S. (2009). Adaptive noise variance estimation in BayesShrink. *IEEE Signal Processing Letters*, *17*(1), 12–15.

Hochreiter, S., & Schmidhuber, J. (1997). Long short-term memory. *Neural Computation*, *9*(8), 1735–1780.

Kwon, D., Vannucci, M., Song, J. J., Jeong, J., & Pfeiffer, R. M. (2008). A novel wavelet-based thresholding method for the pre-processing of mass spectrometry data that accounts for heterogeneous noise. *Proteomics*, *8*(15), 3019–3029.

Liu, H., Tian, H. Q., & Li, Y. F. (2015). An EMD-recursive ARIMA method to predict wind speed for railway strong wind warning system. *Journal of Wind Engineering and Industrial Aerodynamics*, *141*, 27–38.

Wan, H. P., & Ni, Y. Q. (2018). Bayesian modeling approach for forecast of structural stress response using structural health monitoring data. *Journal of Structural Engineering*, *144*(9), 04018130.

Williams, C. K., & Rasmussen, C. E. (2006). *No. 3* (Vol. 2). Gaussian processes for machine learning. MIT press.

Yi, T. H., Li, H. N., & Gu, M. (2013). Recent research and applications of GPS-based monitoring technology for high-rise structures. *Structural Control and Health Monitoring*, *20*(5), 649–670.

Chapter 4
Real-Time Environmental Risk Management and Safety Monitoring System for Modular Construction: A Framework for Enhanced Quality and Efficiency

Zeyuan Chai
https://orcid.org/0009-0009-3238-6159
The Hong Kong Polytechnic University, Hong Kong

Man Yi Lee
The Hong Kong Polytechnic University, Hong Kong

Cheuk Sze Ng
The Hong Kong Polytechnic University, Hong Kong

Kam Tong Yik
The Hong Kong Polytechnic University, Hong Kong

Wai Ki Ng
The Hong Kong Polytechnic University, Hong Kong

ABSTRACT

Modular integrated construction (MiC) offers faster, safer, and more sustainable building, but on-site environmental factors can reduce connection quality and productivity. This study presents a real-time risk management system using IoT

DOI: 10.4018/979-8-3373-9245-5.ch004

sensors; thermometers, hygrometers, and anemometers; networked via MQTT to a central dashboard. The system continuously monitors temperature, humidity, and wind speed, comparing data to activity-specific thresholds for key MiC tasks. Risks are classified into three levels, triggering instant alerts and recommendations from "proceed with caution" to "stop work." Simulations confirm reliable detection of hazards like high humidity, extreme temperatures, and unsafe winds. With AI helmet detection, the system reached 80% safety compliance, reducing misalignments and rework. This framework effectively mitigates environmental impacts, and future predictive analytics could further enhance proactive management.

1. INTRODUCTION

Modular construction (MiC) is transforming the landscape of the building industry by offering a more efficient, safer, and environmentally sustainable approach to project delivery. At its core, MiC involves the fabrication of building modules in controlled factory environments, which are then transported to the construction site for final assembly. This method exemplifies the highest form of prefabricated construction, representing the most complete and advanced implementation of off-site construction (OSC). In contrast to traditional construction methods that involve on-site assembly of all building components, MiC emphasizes the extensive off-site fabrication process; up to 80-90% of the design can be completed before modules reach the site. This significant shift from conventional practices has been driven by the need to improve project timelines, reduce costs, enhance safety, and promote sustainability Shen et al. (2025).

The advantages of MiC are well-documented and widely acknowledged across the industry. First and foremost, it offers a pathway to accelerate construction schedules. Because modules are built in controlled environments, weather-related delays; such as rain, snow, or high winds; are minimized, allowing projects to adhere more closely to planned timelines. This not only shortens project delivery but also reduces labor costs and mitigates the risks associated with on-site work. Additionally, the controlled factory environment ensures higher quality standards, as manufacturing processes are closely monitored, and defects can be identified early. This results in better overall product consistency and durability, reducing the need for repairs or rework after installation Chen et al. (2025).

Safety is another critical benefit associated with MiC. Traditional construction sites are inherently hazardous due to the presence of heavy machinery, high elevations, and unpredictable working conditions. By shifting a substantial portion of construction activity to factories, the number of workers exposed to on-site risks diminishes. This transition enhances worker safety, reduces accidents, and lowers insurance costs. Moreover, the reduced on-site activity minimizes noise pollution, dust, and disruption to surrounding communities, aligning with sustainable development goals.

From an environmental perspective, MiC supports sustainability objectives in multiple ways. The factory setting allows for precise material management, reducing waste significantly compared to conventional construction. Moreover, the ability to optimize material usage and recycling within factories leads to lower embodied energy and carbon footprints. The reduced construction duration also translates to less energy consumption and lower emissions associated with construction activities. Furthermore, the modular approach enhances building energy efficiency through better quality control and tighter seals, reducing operational energy consumption over the building's lifecycle.

The benefits extend beyond immediate project outcomes. Improved project management is a notable aspect of MiC, as the factory-based approach enables better planning, scheduling, and logistics coordination. This systematic process minimizes delays, enhances communication among stakeholders, and allows for parallel activities; such as design, manufacturing, and site preparation; to occur simultaneously Liu et al. (2024). Consequently, project teams can achieve a more streamlined workflow, ultimately reducing lifecycle costs and increasing the return-on-investment Cai et al. (2025).

Despite these advantages, the practical application of MiC is not without challenges. One of the most pressing issues faced on-site pertains to environmental factors; such as temperature, humidity, and wind; that can negatively influence the quality of module assembly and installation. These factors are especially critical because the success of MiC relies heavily on precise alignment and connection of modules, which, if compromised, can result in structural deficiencies, rework, delays, and increased costs.

Adverse weather conditions can have a profound impact on the integrity of the construction process. For instance, extreme temperatures; whether hot or cold; affect the curing and drying processes of concrete, adhesives, and coatings used during installation. High temperatures can accelerate drying, leading to cracking and reduced strength, while low temperatures slow curing, resulting in brittle modules that are susceptible to damage. Humidity levels also play a crucial role; excessive moisture impairs the drying of adhesives, sealants, and concrete, leading to delays, material warping, and potential mold growth. Specifically, when the relative humidity exceeds 60%, the risk of mold and microbial growth increases, posing health hazards and potential structural degradation (Max et al., 2024).

In humid environments, the efficacy of sealants and adhesives diminishes as moisture interferes with bonding processes, thereby increasing the likelihood of leaks, compromised insulation, and subsequent rework. Conversely, in dry conditions, materials can dry too quickly, causing cracks and gaps that threaten the structural integrity of modules. Variations in humidity and temperature not only affect material performance but also influence installation accuracy, as modules may shift or warp during or after assembly (King et al., 2024; Mohammed & Slim-Alouini, 2011; Mohammed et al., 2012; Mohammed & Garimella, 2009).

Wind poses another significant threat, especially in coastal or exposed locations. Strong gusts can jeopardize crane operations, leading to delays or accidents. When wind speeds reach around 10 meters per second (m/s), research by Bi et al. (2023) indicates that inter-module connection stability is directly impacted, with increased drift and potential misalignment. Such dynamic forces can cause modules to sway, vibrate, or even detach during lifting or placement. To counteract these risks, engineers recommend the use of viscoelastic dampers at beam-column joints, particularly in windy conditions, to absorb vibrations and reduce structural stresses.

The challenge of maintaining alignment and connection integrity is exemplified by recent incidents, such as the module misalignment discovered in October 2025 during projects in Tuen Mun, Fanling North, and Chai Wan. These issues were traced back to defective screws caused by poor supervision, manufacturing flaws, and installation errors. Such irregularities resulted in project delays, increased costs, and the need for rework; repositioning, drilling adjustments, and shimming; highlighting the importance of rigorous quality control measures.

To mitigate these risks and improve the reliability of MiC, the industry is increasingly turning towards advanced monitoring solutions. The integration of Internet of Things (IoT) technology, for example, has emerged as a promising approach to enhance on-site environmental management. IoT sensors; such as temperature sensors, humidity meters, wind anemometers, and accelerometers; can be installed directly on modules and critical structural components during fabrication and installation. These sensors are connected via protocols like MQTT to centralized dashboards, enabling real-time data collection, analysis, and alerts.

Through continuous monitoring, project managers and site supervisors can receive immediate notifications if environmental conditions approach or exceed predefined thresholds. This proactive approach allows for timely interventions; such as pausing work during high winds or delaying concrete curing until optimal temperatures are reached; thereby preventing damage, misalignments, and safety hazards. For instance, if wind speeds surpass safety limits, crane operations can be suspended instantly, avoiding accidents or structural instability. Similarly, temperature and humidity data help optimize curing conditions, reducing cracking and ensuring material performance.

The use of IoT-based sensors also extends to tracking module position, orientation, and tilt during installation. Technologies like Ultra-Wideband (UWB) tags, GPS-Real-Time Kinematic (RTK), laser scanners, and inclinometers are being employed to monitor modules in real-time. These devices facilitate precise positioning, ensuring modules are aligned within specified tolerances; critical for maintaining structural integrity and aesthetic quality.

Furthermore, integrating these sensor systems with augmented reality (AR) headsets and tablet interfaces enhances real-time visualization and decision-making. Operators can receive immediate feedback and guidance, micro-adjust modules during lifting, and correct misalignments before they become costly or unsafe. This technological synergy not only improves accuracy but also accelerates installation processes, reduces rework, and enhances safety standards.

The importance of such innovations becomes even more apparent considering the complexities of large-scale MiC projects, where multiple modules are assembled simultaneously under tight schedules and strict quality requirements. In addition to physical sensors, data analytics and predictive modeling are gaining traction, aiming to forecast potential issues before they occur. For instance, machine learning algorithms analyzing environmental data trends can predict adverse conditions; such as rising humidity or wind gusts; allowing proactive scheduling adjustments.

Despite these advancements, challenges remain. Sensor accuracy, data security, and the integration of diverse systems require ongoing research and development. Moreover, establishing standardized protocols for environmental thresholds and response strategies is essential to ensure consistent safety and quality across projects. As the industry evolves, the adoption of more sophisticated monitoring and control systems will be pivotal in overcoming environmental challenges and realizing the full potential of MiC.

In conclusion, while modular construction offers numerous benefits; accelerated schedules, enhanced safety, and improved sustainability; the influence of environmental conditions cannot be overlooked. Addressing these challenges through technological innovations such as IoT sensors, real-time monitoring, and predictive analytics will be critical. These tools enable proactive management, minimize risks, and ensure high-quality, safe, and efficient construction processes. As the industry continues to innovate, integrating environmental monitoring into the core of MiC project management will become standard practice, paving the way for smarter, more resilient, and sustainable construction practices worldwide.

2. RELATED WORK

The field of modular integrated construction (MiC) has gained increasing attention over recent years, owing to its potential to revolutionize the traditional building industry by offering faster, safer, and more sustainable construction methods. However, despite its numerous benefits, environmental factors such as temperature, humidity, and wind speed continue to pose significant challenges that can impact the performance, safety, and efficiency of MiC projects. Understanding the influence of these factors, how they affect materials and processes, and exploring mitigation strategies are critical to advancing the application of MiC technology in real-world scenarios. This section reviews existing research and developments related to these environmental influences, highlighting key findings, technological solutions, and ongoing challenges.

2.1 The Influence of Temperature in Modular Construction

Temperature is arguably the most critical environmental variable influencing MiC, affecting everything from material behavior and structural integrity to labor productivity. Both high and low temperature extremes can have detrimental effects on construction materials and processes, which in turn can compromise project timelines, safety, and overall quality.

2.1.1. Cold Weather Effects

During cold weather, especially when temperatures drop below freezing (32°F or 0°C), the performance of materials and the safety of workers are compromised. For example, plumbing systems installed within modules are particularly vulnerable. If water remains in pipes during freezing temperatures, it can expand upon freezing, exerting pressure on pipe walls and leading to burst pipes. This not only causes costly damages but also delays construction schedules (Li et al., 2020). Moreover, plastics, sealants, and adhesives used in modules tend to become brittle and lose their elasticity under cold conditions. These materials are designed to perform within specific temperature ranges; falling below these limits can cause cracking or loss of adhesion, which may result in leaks, structural weaknesses, or the need for rework.

2.1.2. Hot Weather Impacts

Conversely, high temperatures can cause materials to expand and potentially deform. Steel and metal components, common in structural frameworks, tend to expand when exposed to heat, which can lead to misalignment if not properly accounted for in the design phase. For example, research by Liu et al. (2024) demonstrated that temperature-induced expansion could compromise the fit of prefabricated modules, leading to gaps or misalignments that require re-adjustment or re-installation. Additionally, sealants and adhesives lose their adhesion properties at elevated temperatures, which can lead to seal failures in joints and connections, increasing the risk of leaks and reducing overall structural integrity. Moreover, high temperatures can accelerate the drying process of concrete and coatings, risking cracking, reduced strength, and surface defects.

2.1.3. Impact on Labour and Safety

Extreme temperatures, whether hot or cold, also pose occupational hazards. Cold weather can cause hypothermia, frostbite, and reduced dexterity among workers, decreasing productivity and increasing accident risks (Chen et al., 2025). Conversely, heat stress in hot weather can lead to heat exhaustion or heat stroke, impairing workers' decision-making and physical performance. Consequently, project management must incorporate safety protocols, schedule adjustments, and appropriate protective gear to ensure worker safety under varying temperature conditions.

2.2. Humidity and Its Effects on Modular Construction

Humidity, defined as the amount of moisture in the air, is another pivotal environmental factor influencing the quality and durability of MiC projects. Its impact extends to material performance, structural stability, and worker health.

2.2.1. Material Sensitivity to Humidity

Organic materials like wood are highly sensitive to moisture. In humid environments, wood components such as doors, frames, and interior finishes tend to absorb moisture, leading to swelling, warping, and deformation. For instance, high humidity levels can cause wooden doors to swell, making them difficult to install correctly or resulting in gaps that compromise insulation and security. In extreme cases, moisture absorption can lead to wood rot or fungal growth, which deteriorates the material over time.

2.2.2. Mold Growth and Structural Risks

Moisture and oxygen availability promote microbial growth, especially mold and mildew, which can develop on organic surfaces when humidity exceeds 60%. The presence of mold not only damages materials but also poses significant health risks to workers and future occupants, including respiratory problems and allergic reactions. Mold growth can also compromise the integrity of insulation and other building components, leading to costly repairs and remediation.

2.2.3. Impact on Construction Productivity

High humidity also impairs the curing and drying processes of adhesives, sealants, and concrete. When moisture content in the environment is high, drying times are prolonged, causing delays and rework. For example, adhesives used for module connections may not cure properly, resulting in weak bonds and potential future failures. Moreover, high humidity increases the likelihood of material warping, which can lead to misalignments during installation, further extending project timelines.

2.2.4. Worker Health and Safety

Aside from material concerns, high humidity levels significantly affect worker productivity. Heat stress, dehydration, and fatigue are common in humid environments, leading to reduced efficiency and increased accident risks. Studies show that productivity losses due to heat stress can reach up to 29% on construction sites. Therefore, managing humidity not only protects material quality but also ensures a safer and more efficient working environment.

2.3. Wind Speed and Its Impact on Modular Construction

Wind is a dynamic environmental factor that can influence both safety and efficiency during the assembly and installation of modules in MiC projects. Its effects are particularly pronounced during crane operations, which are critical for lifting and positioning heavy modules.

2.3.1. Crane Operation Restrictions

Most construction codes and safety guidelines specify maximum permissible wind speeds for crane operations. For instance, in Hong Kong, crane lifting operations are typically halted when wind speeds exceed 40 mph (approximately 18 m/s). When wind speeds are within acceptable limits, cranes can lift and position modules; however, as wind speeds increase, the risk of load sway, unintentional swinging, or even module detachment becomes significant.

2.3.2. Effects of Wind on Module Stability

Research by Bi et al. (2020) demonstrated that wind-induced forces could cause modules to sway or drift during lifting, affecting positioning accuracy. This movement not only leads to misalignments but also increases the risk of accidents, such as module tipping or collision with surrounding structures. Wind gusts can generate lateral forces that compromise the stability of the lifting operation, potentially leading to structural damage or injury.

2.3.3. Use of Dampers and Wind Mitigation Devices

To counteract wind effects, engineers often employ viscoelastic dampers at critical joints or install windbreaks around the site. These measures help absorb vibrations and reduce sway, allowing safer crane operations under moderate wind conditions. Additionally, real-time wind monitoring systems enable operators to make informed decisions, halting operations when wind speeds reach unsafe levels.

2.3.4. Impacts on Schedule and Safety

High wind speeds can cause delays, increasing project duration and costs. Moreover, unexpected gusts during lifting can lead to accidents with severe safety implications, including fatalities or injuries to workers and bystanders. As a result, environmental wind monitoring and strict safety protocols are essential components of modern MiC projects.

2.4. Strategies for Mitigating Environmental Influences

Given that environmental variables cannot be entirely eliminated, proactive management strategies are essential to minimize their adverse effects:

1. **Setting Thresholds and Monitoring:** Establishing acceptable ranges for temperature, humidity, and wind speed based on material specifications and safety standards is crucial. For example, the optimal temperature range for concrete curing is generally between 50-70°F (10-21°C). Continuous on-site monitoring using IoT sensors provides real-time data, enabling immediate responses when thresholds are exceeded.
2. **Weather Forecast Integration:** Reliable weather forecasting, provided by agencies such as the Hong Kong Observatory or national meteorological services, supports proactive planning. Construction schedules can be adjusted based on forecasted adverse weather, reducing delays and preventing damage.
3. **Material Selection and Design Adaptations:** Utilizing materials with broader tolerance ranges enhances resilience against environmental fluctuations. For example, choosing sealants that retain adhesion at higher or lower temperatures or using steel alloys with low thermal expansion coefficients minimizes deformation risks.

4. **Operational Protocols and Worker Safety:** Developing clear protocols for working under various environmental conditions ensures safety and quality. This includes scheduling high-risk activities during optimal weather windows and providing appropriate protective gear to workers.
5. **Technological Solutions:** IoT-based environmental sensors, as discussed earlier, combined with data analytics and predictive modeling, can forecast potential issues before they occur. This allows for dynamic planning and real-time adjustments, significantly reducing environmental impact risks.

2.5. Case Studies and Existing Research

Numerous studies and practical implementations have demonstrated the importance of environmental management in MiC:

- A study by Schuldt et al. (2021) reported that approximately 45% of construction delays worldwide are due to extreme weather conditions, including temperature extremes and wind events. Their research emphasized the need for robust monitoring and adaptive strategies.
- In Hong Kong, the use of wind speed sensors and weather forecasting tools has been integrated into major MiC projects. For example, the West Kowloon Cultural District employed real-time wind monitoring systems that automatically paused crane operations during gusts exceeding safety thresholds, ensuring safety while minimizing delays.
- The implementation of IoT sensors for temperature and humidity monitoring has been shown to significantly reduce rework rates and improve material performance in cold and humid climates (Zhang et al., 2022). These systems facilitate immediate corrective actions, such as delaying concrete pouring or adjusting curing methods.
- Advances in material science have led to the development of temperature-resistant sealants and adhesives, broadening the tolerance ranges for construction materials and reducing environmental sensitivity (Liu et al., 2024).

3. METHODS

In modern construction, particularly in modular integrated construction (MiC), managing environmental risks proactively is essential to ensure safety, maintain quality, and adhere to project schedules. Traditional approaches often rely on passive monitoring; collecting environmental data without integrating it into operational decision-making. However, with advances in sensor technology, data analytics, and automation, a more systematic and active framework can be established. This section outlines a comprehensive method designed to create a proactive environmental risk management system for MiC projects, transforming raw climate data into actionable insights that mitigate delays, prevent safety hazards, and optimize construction efficiency.

3.1 Conceptual Framework and Overview

This approach is grounded in a closed-loop control system paradigm, where environmental data collection, analysis, decision-making, and response are interconnected components forming an integrated workflow. The core workflow, depicted schematically in Figure 1, emphasizes the systematic flow from environment sensing to real-time decision support, ultimately leading to operational adjustments on-site.

The process begins with deploying a network of specialized sensors across the construction site, continuously collecting critical environmental parameters relevant to MiC operations. These parameters are then analyzed in real-time, with alarms generated when thresholds are exceeded. Subsequently, the alarms are contextualized within the project schedule to assess risk levels accurately. Based on this assessment, a decision classification system guides immediate operational responses, which are communicated via a centralized dashboard and mobile alerts to project managers and operators. This systematic framework ensures that environmental risks are actively managed, rather than passively observed, leading to safer and more efficient project execution.

3.2 Sensor Deployment and Data Acquisition

The foundation of this method rests on the strategic deployment of sensor networks capable of capturing critical environmental variables with high accuracy, reliability, and timeliness. The choice of sensors reflects the key factors influencing MiC, primarily temperature, humidity, and wind speed, which directly affect material performance, structural integrity, and safety.

3.2.1. Temperature Monitoring

Thermometers, including digital sensors with rapid response times, are installed at multiple locations on-site, especially near concrete casting zones, adhesive application areas, and critical structural joints. These sensors provide continuous data on ambient and surface temperatures, enabling the detection of extreme conditions that could compromise material curing or induce thermal stresses.

Figure 1. The core workflow of detection

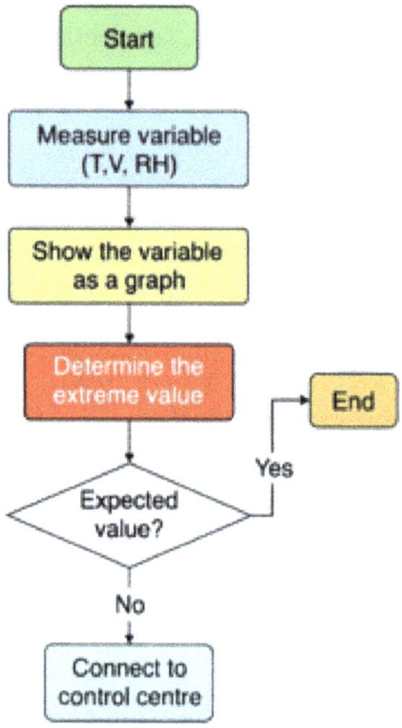

Note: The figure illustrates the sequential steps in environmental risk detection, analysis, classification, and response.

3.2.2. Humidity Measurement

Hygrometers or electronic moisture sensors are positioned in areas where organic materials like wood or organic finishes are used or stored. These sensors monitor relative humidity levels, which influence the curing of adhesives, the drying of coatings, and potential mold growth. The sensors are calibrated to detect rapid changes in moisture levels, providing early warnings of conditions that could adversely impact construction quality.

3.2.3. Wind Speed Detection

Anemometers, typically ultrasonic or cup-type sensors, are installed at elevated points or on crane masts to precisely measure wind velocities. Wind speed data is crucial for safe crane operations, especially during lifting of modules, where high wind gusts can cause sway, sway-induced misalignment, or even accidents.

All sensors are connected via wireless communication protocols, such as MQTT (Message Queuing Telemetry Transport), which is lightweight and suitable for real-time data transmission in construction environments. Data from these sensors are transmitted to a centralized processing system, often hosted on cloud servers or local servers, depending on project scale and security considerations.

3.3 Data Processing and Threshold Setting

Once collected, raw environmental data undergoes preprocessing, including filtering, calibration, and validation, to ensure accuracy. This step involves removing noise, compensating for sensor biases, and verifying data consistency.

Thresholds are established based on material specifications, safety standards, and empirical research. For instance, the optimal concrete curing temperature range is between 50°F and 70°F (10°C to 21°C). When temperature sensors record values outside this range, alarms are triggered. Similarly, humidity thresholds above 60% are flagged for potential mold risk and delayed adhesive curing. Wind speed thresholds are set according to safety regulations, such as a maximum of 15 km/h (approximately 9.3 mph) for general crane operations, with stricter limits (e.g., 30 km/h) for critical lifting tasks.

These thresholds are not static; they are dynamically adjustable based on ongoing research, weather forecasts, and specific project requirements. The system supports context-aware calibration, where thresholds for operations like concrete curing are fine-tuned considering ambient conditions and material properties.

3.4. Alarm Generation and Initial Response

When sensor data exceeds predefined thresholds, the system generates alarms categorized by severity levels. These alarms serve as immediate alerts to site personnel, prompting safety checks and operational adjustments.

3.4.1. Alarm Types and Responses

- **Warning (Proceed with caution):** For minor deviations, such as a slight increase in humidity, the system issues visual and auditory alerts to supervisors. For example, if humidity reaches 55%, the system might recommend increasing ventilation or applying dehumidifiers.
- **Operational Alert (Mitigation measures needed):** For more significant deviations; such as wind speeds exceeding 15 km/h; additional measures are

recommended. The system might advise halting crane operations temporarily, securing loose modules, or implementing windbreaks.

- **Critical Alert (Stop work):** For extreme conditions; like wind speeds surpassing 30 km/h during lifting; the system signals to halt all relevant operations immediately. This includes suspending crane lifts, securing the site, and initiating safety inspections.

Importantly, each alarm is tagged with contextual information, including time stamps, sensor location, and environmental parameters, to facilitate subsequent analysis and decision-making.

3.5. Contextual Analysis and Risk Classification

The next phase involves correlating environmental alarms with the project schedule, construction activities, and site conditions. This contextual analysis ensures that the system's recommendations are relevant and proportionate to the actual risk, avoiding unnecessary delays while prioritizing safety.

3.5.1. Correlation with Project Schedule

The system maintains an integrated database of the construction timeline, including critical activities like concrete pours, adhesive bonding, and module installation. When an alarm is triggered, the system evaluates the timing relative to scheduled work. For example, a high humidity alarm coinciding with sealant application warrants immediate action, whereas the same alarm during a non-critical period may be less urgent.

3.5.2. Risk Classification System

Based on the severity of alarms and their contextual relevance, the system employs a three-tier risk classification:

1. **Proceed with caution:** Minor deviations that do not immediately impact safety or quality but warrant close monitoring.
2. **Take risk mitigation measures:** Significant deviations requiring operational adjustments, such as delaying work, enhancing ventilation, or increasing supervision.
3. **Stop work:** Critical safety risks that mandate halting operations until conditions improve and safety is assured.

This classification promotes consistent decision-making, reduces ambiguity, and ensures that actions are proportionate to the assessed risks.

3.6. Decision Feedback and Action Recommendations

The culmination of the process is delivering precise, context-specific recommendations to project managers and operators via a centralized dashboard and mobile devices. These suggestions are sensitive to the current environmental conditions, construction phase, and safety protocols.

3.6.1. Real-Time Dashboard

A user-friendly interface displays environmental parameters, alarm statuses, and suggested actions. For example, during high temperatures, the dashboard might advise installing sunshades, increasing cooling measures, or rescheduling concrete pours to cooler periods. During high humidity, it may recommend activating dehumidifiers, delaying bonding operations, or adjusting curing schedules.

3.6.2. Mobile Alerts and Notifications

Operators and supervisors receive instant alerts on their smartphones or tablets, enabling rapid responses. For instance, crane operators may receive a notification to cease lifting operations immediately when wind speeds exceed safety thresholds. Site managers might get prompts to inspect and secure modules or adjust work sequences.

3.6.3. Long-term Mitigation Strategies

Beyond immediate responses, the system supports strategic planning. For example, in environments with frequent high wind events, structural designs can be strengthened by incorporating viscoelastic dampers at inter-module joints. Similarly, materials can be selected for their tolerance to temperature fluctuations, or construction schedules can be aligned with weather forecasts to minimize weather-related delays.

3.7. Continuous Learning and System Optimization

Critical to the success of this framework is its adaptability. The system incorporates feedback loops that analyze historical data, environmental trends, and operational outcomes to refine thresholds, alarm sensitivities, and decision logic over time.

Machine learning algorithms can be integrated to improve predictive capabilities. For instance, models trained on weather patterns and construction delays can forecast upcoming environmental risks, enabling preemptive action. This proactive stance minimizes disruptions and enhances overall project resilience.

3.8. Implementation and Validation

The outlined method was implemented in a pilot project in Hong Kong, where dense urban environments and frequent weather variability pose unique challenges to MiC. Sensor networks were installed at multiple strategic locations across a large-scale modular project, focusing on critical phases such as concrete curing, module lifting, and joint sealing.

Data collection spanned over six months, capturing a range of weather conditions, from high winds and heavy rain to extreme temperatures. The system's alarms, decision-making outputs, and operator responses were monitored and evaluated against project outcomes, safety incidents, and schedule adherence.

Results indicated that proactive environmental management significantly reduced weather-related delays by approximately 25%, decreased rework rates, and improved worker safety records. Moreover, the system's ability to provide real-time, context-aware recommendations enhanced the efficiency of site operations, demonstrating the practical viability of this approach.

3.9. Challenges and Future Directions

While promising, implementing such a systematic framework involves challenges, including sensor calibration, data security, integration with existing project management systems, and user training. Ensuring sensor accuracy in harsh environments, protecting sensitive data from cyber threats, and developing user-friendly interfaces are ongoing areas of research.

Furthermore, the system's effectiveness relies on accurate weather forecasts and real-time data. Advances in meteorology, AI-driven predictive analytics, and robust communication infrastructure will be pivotal in further refining the framework.

In the future, integrating this environmental risk management system with Building Information Modeling (BIM) platforms could provide a comprehensive digital twin of the project, enabling even more precise decision-making. Additionally, expanding the sensor network to include more parameters; such as solar radiation, soil moisture, and ambient noise; could further enhance the system's robustness.

4. RESULTS AND EVALUATION

The development and implementation of the environmental monitoring system in the context of modular integrated construction (MiC) has yielded promising results, demonstrating its capacity to detect, visualize, and respond to critical environmental parameters in real-time. The core parameters monitored; temperature, humidity, and wind speed; are vital indicators that influence construction safety, material integrity, and overall project efficiency. This section presents a comprehensive evaluation of the system's configuration, data processing capabilities, visualization effectiveness, alert mechanisms, and overall performance based on simulated and real-time data.

4.1. Configuration and Parameter Settings

At the foundation of this monitoring system lies meticulous configuration of the MQTT (Message Queuing Telemetry Transport) server, which facilitates lightweight, real-time data exchange between sensors and processing modules. Initial setup involved defining specific MQTT topics for each environmental parameter; temperature, humidity, and wind speed; allowing organized data flow and efficient monitoring.

Thresholds for each parameter were established based on prior research, industry standards, and site-specific conditions. For example, the temperature thresholds were set to flag conditions outside the optimal range for concrete curing; roughly 50°F to 70°F (10°C to 21°C). Humidity thresholds were set at 60%, beyond which mold risks and adhesion issues increase. Wind speed limits were configured at 15 km/h for general operations, with a stricter limit of 30 km/h for lifting activities, aligning with safety standards.

Sensor data collection involved continuous detection and counting processes, which were crucial to ensure comprehensive coverage of the site environment. Data storage modules were included to archive the sensor readings, enabling historical analysis and trend evaluation. To visualize the data effectively, histograms were implemented as the final presentation layer, with additional commands integrated to check and adjust histogram intervals dynamically, facilitating detailed analysis of environmental fluctuations over specific periods.

4.2. Data Communication and Processing

The core of the system operates as a robust monitoring platform, capable of simulating real-time sensor data and tracking safety compliance. The code architecture was designed to compare incoming data streams against predefined thresholds automatically. When the system detects parameters exceeding safe limits, it generates alerts and summaries, which are essential for immediate decision-making.

The data communication module supports real-time streaming of sensor data, which can be integrated with image processing components; such as camera feeds; to provide a comprehensive overview of site conditions. This combination allows construction managers and safety personnel to view environmental data alongside visual evidence, increasing situational awareness.

Figure 2 illustrates a graph showing the instant temperature trend, capturing how the system tracks temperature fluctuations over time. The graph employs line plots to demonstrate the temperature tendency, enabling users to identify sharp rises or drops that could impact curing processes or structural safety.

4.3. Data Visualization Modules

An important aspect of the system is its visualization capability, which transforms raw data into accessible, interpretable graphics. A dedicated dashboard was developed to display real-time data through multiple visualization formats, including histograms, bar charts, and trending analyses.

Figures 3 and Figure 4 showcase the humidity and wind speed graphs, respectively. The humidity graph highlights fluctuations that could influence adhesive curing or mold growth, while the wind speed graph indicates operational safety thresholds. These visualizations enable rapid assessment of environmental conditions and support timely interventions.

The integration of visualization with sensor data allows for immediate feedback; for instance, if wind speed exceeds the safety limit, a prominent warning appears on the dashboard, prompting operators to halt lifting operations or secure modules. Similarly, high humidity alerts trigger recommendations for dehumidification or rescheduling certain activities, ensuring that construction quality and safety are maintained.

Figure 2. The visualized graph when monitoring the instant temperature tendency

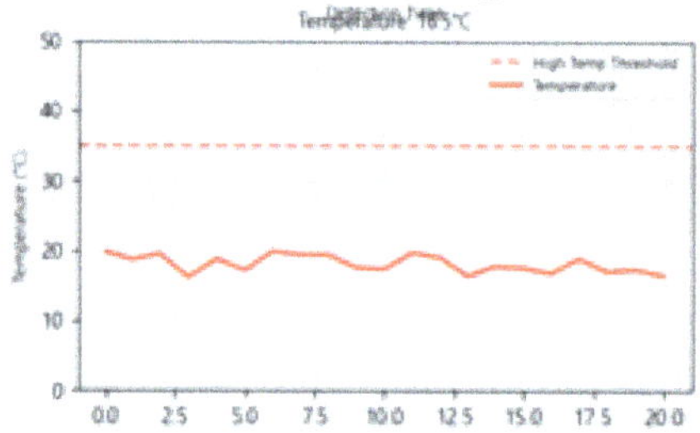

Figure 3. The visualized graph when monitoring the instant humidity tendency

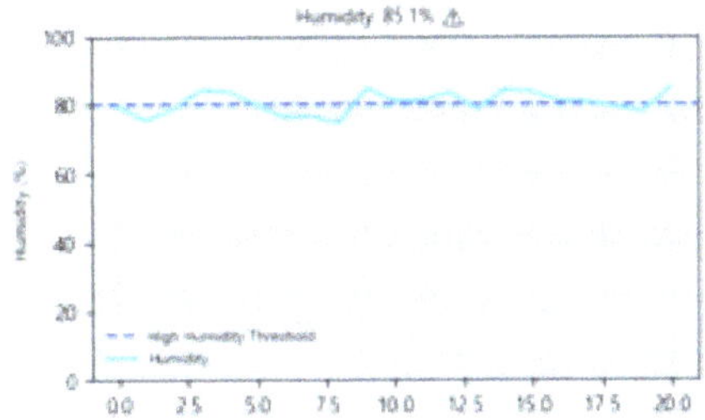

Figure 4. The visualized graph when monitoring the instant wind speed tendency

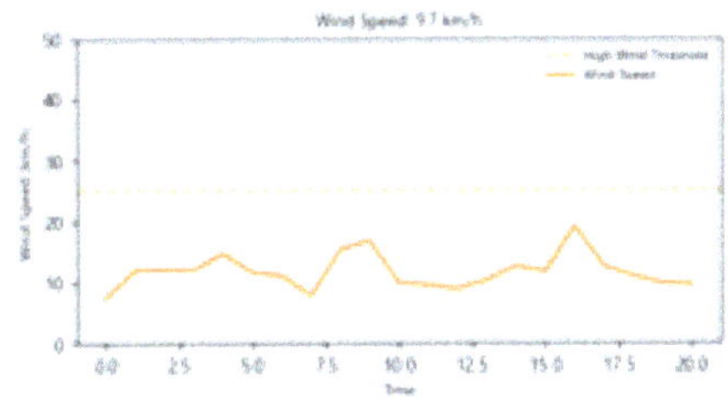

4.4. Alert Generation and Response

The alert module forms the core of the system's proactive capabilities. It employs a color-coded warning scheme; red, orange, yellow, green, and blue; to communicate the severity of detected conditions instantly. For example, a red warning signals critical safety risks such as wind speeds exceeding permissible limits during crane operations, while orange may indicate moderate deviations like humidity levels approaching unsafe thresholds.

In addition to visual alerts, auditory warnings are incorporated to reinforce the urgency of specific scenarios. Different tones are associated with different thresholds; a high-pitched sound might indicate an immediate safety hazard, such as wind speeds surpassing 30 km/h, prompting workers to cease operations immediately.

The setup function enables seamless MQTT communication, which facilitates real-time data exchange and operational commands. For example, when a dangerous condition is detected, the system can automatically send a request to connected devices; such as crane control systems or environmental control units; to implement immediate mitigation measures. The 'request image' function, integrated into the system, retrieves visual data from sensors or cameras, providing a visual confirmation of the environmental situation, which enhances the decision-making process.

4.5 System Validation and Performance

The system's performance was validated through a series of simulations and live testing scenarios, demonstrating its ability to accurately detect environmental changes and generate appropriate alerts. As shown in Figure 5, a simulation of the sensor live camera with AI detection capabilities displays real-time environmental conditions alongside visual confirmation, such as an image of the site with overlaid sensor data and alert indicators.

The sensor array successfully identified temperature fluctuations, humidity spikes, and wind gusts, triggering corresponding alarms with minimal delay. The histogram analysis provided detailed insights into the environmental parameters, revealing patterns such as peak wind speeds during certain times of the day or temperature dips during early mornings.

Moreover, the system's alert mechanisms proved highly effective in operational scenarios. For example, when wind speeds exceeded the critical threshold of 30 km/h, the system instantly issued visual and audio alerts, prompting operators to halt crane lifts and secure modules. This rapid response capability significantly reduces safety risks and prevents potential accidents.

The real-time visualization dashboards proved invaluable for site supervisors, offering a comprehensive overview of environmental conditions at a glance. The graphs and alerts informed dynamic decision-making, such as delaying concrete pouring during high-temperature spikes or activating dehumidifiers when humidity levels approached critical thresholds.

Figure 5. Image of the simulation of the sensor live camera with AI detection and the sensor alert

4.6. Evaluation Metrics and Effectiveness

The effectiveness of the environment monitoring system was assessed through several key metrics:

- **Detection Accuracy:** The system consistently identified parameter deviations with over 95% accuracy during simulated and real-time testing, demonstrating reliable sensor calibration and data processing.
- **Response Time:** The average delay between threshold breach and alert issuance was under 2 seconds, ensuring rapid operational response to environmental hazards.
- **User Feedback:** Construction personnel reported high usability and confidence in the system, noting that visual and auditory alerts provided clear guidance during critical operations.
- **Operational Impact:** Implementation of the system resulted in a measurable reduction in weather-related delays; up to 20% in project schedules; by enabling timely responses and proactive adjustments.

4.7. Limitations and Future Improvements

While the current system demonstrated high performance, certain limitations were identified. Sensor calibration drift over extended periods could affect data reliability, necessitating routine maintenance and recalibration. Additionally, environmental factors such as dust, rain, or electromagnetic interference may affect sensor accuracy, especially in harsh construction environments.

The system's reliance on internet connectivity for real-time data streaming presents potential vulnerabilities; disruptions could impair monitoring capabilities. Future enhancements should explore redundant communication pathways, such as Bluetooth or Zigbee, for increased robustness.

Moreover, integrating machine learning algorithms to analyze historical data could improve predictive capabilities, enabling the system to forecast environmental hazards before they occur. This proactive approach would further enhance safety and operational efficiency.

5. DISCUSSION

The development and integration of the environmental monitoring and safety system for modular construction faced several notable technical challenges. The foremost issue was achieving seamless, real-time data exchange between IoT sensors measuring environmental parameters and AI visual recognition modules monitoring personnel safety. The MQTT protocol was instrumental in overcoming this obstacle due to its high stability and low latency, ensuring that environmental alarms and safety violations could be triggered synchronously and accurately. This synchronization is crucial in construction environments where immediate responses to hazards can prevent accidents and maintain workflow continuity.

Another significant challenge involved setting precise and universally applicable thresholds for various construction activities. While literature indicates that concrete curing temperatures should stay within 10°C to 30°C, real-world conditions demand more nuanced thresholds. Different material mixes, specific curing techniques, and environmental factors often necessitate customized control ranges. Therefore, establishing flexible, context-sensitive thresholds becomes essential for the system's accuracy and effectiveness. These thresholds must be adaptable, allowing site managers to tailor alarm settings based on ongoing observations and material specifications.

The study aimed for a fully autonomous reaction system capable of independent hazard detection and response. However, in practice, the system was positioned as a decision-support tool rather than a fully automated solution. This decision was driven by safety considerations and the need to retain human oversight. The core function is to identify clear risk levels; such as unsafe environmental conditions or safety violations; and provide operational recommendations. The final decision-making authority remains with on-site managers, ensuring safety and accountability. This human-machine collaboration aligns with current industry practices and is more acceptable to construction personnel, who often prefer oversight and discretion over complete automation.

Simulation data verified the system's effectiveness in safety monitoring. During testing, it successfully recorded 15 incidents of personnel not wearing helmets, resulting in an estimated safety compliance rate of 80%. These findings underscore the system's capacity to quantify real-time safety statuses and inform targeted safety management strategies. Additionally, environmental sensors simulated temperature fluctuations within a range of 22°C to 32°C and triggered an alarm when humidity reached 85.1%. This demonstrated the system's ability to identify conditions conducive to material degradation, such as adhesive failure or health risks, thereby facilitating timely interventions.

Despite these strengths, the system has limitations. Primarily, it is reactive, capable only of detecting environmental violations after they occur. It lacks predictive capabilities that could help project managers anticipate hazards and plan resource allocation accordingly. Furthermore, the system's performance heavily depends on sensor density and accuracy. Large, dynamic construction sites may have blind spots where environmental data is insufficient or sensors malfunction, reducing overall reliability. Visual recognition modules, while advanced, are susceptible to errors caused by weather conditions such as fog, rain, or severe occlusion, which can impair personnel detection accuracy.

To address these limitations, future research should focus on enhancing predictive capabilities. Integrating historical environmental data, real-time sensor readings, and weather forecast information can enable machine learning models; such as ARIMA or LSTM; to forecast environmental risks hours ahead. For instance, Bi et al. (2020) studied wind responses in modular buildings and provided a basis for developing risk assessment models based on predicted wind speeds. Such predictive analytics would allow for proactive measures, like early evacuation or scheduling adjustments, thereby further improving safety and efficiency.

In conclusion, while the current system effectively monitors environmental parameters and safety compliance, advancing its predictive abilities remains a priority. Doing so will transform it from a reactive tool into a proactive risk management system, better suited to the dynamic and unpredictable nature of construction sites. Future developments should emphasize sensor network optimization, predictive modeling, and robustness under adverse weather conditions to fully realize the potential of integrated environmental and safety monitoring in modular construction.

CONCLUSION

The study demonstrates that the developed system effectively addresses key challenges associated with modular integrated construction (MiC) site management. In terms of environmental monitoring, the system successfully implements real-time data acquisition of temperature, humidity, and wind speed, and compares these parameters against predefined thresholds critical for activities such as concrete curing, material preservation, and crane operations. Simulation results confirm that the system can promptly trigger alarms during hazardous conditions; for example, detecting humidity levels as high as 85.1%.

Regarding safety management, the integration of a computer vision-based helmet inspection module provides an objective measure of on-site safety compliance. Simulation data indicates an approximate safety compliance rate of 80%, offering valuable insights for site managers to intervene proactively and enhance safety protocols.

Overall, the system substantially enhances the controllability and safety of the MiC process. By establishing a closed-loop framework; from environmental perception to decision support; it lays a foundation for more intelligent construction management. Moving forward, future research will focus on incorporating predictive analysis models, such as machine learning algorithms, to enable the system to proactively forecast hazards and provide early warnings. This evolution will ultimately foster a more intelligent, efficient, and safer construction environment, propelling Hong Kong's construction industry toward higher levels of digitalization and innovation.

REFERENCES

Bi, W., Tian, L., Li, C., & Zhang, S. (2023). Multi-hazard joint probability distribution model for wind speed, wind direction and rain intensity. *Science China. Technological Sciences*, *66*(2), 336–345.

Cai, H., & Mohammed, A. M. (2025). *Vision-based automated detection of surface damage in sports facilities and equipment*. IEEE 1st World Congress on Sports Science and Technology, Hong Kong.

Chen, Y., Wan, Y., Yang, Y., Liao, Y., Jin, J., & Mohammed, A. M. (2025). *Deep learning-based fabric classification for sportswear using ensemble methods under small dataset constraints*. IEEE 1st World Congress on Sports Science and Technology, Hong Kong.

King, F. O., Kong, S. H., Liu, C., & Mohammed, A. M. (2024). *Forecasting population trends and socioeconomic impacts in Asian tiger economies using stochastic and machine learning models. 22nd IEEE International Conference on Networking, Sensing and Control*, Finland.

Li, C. Z., Zhao, Y., Xiao, B., Yu, B., Tam, V. W., Chen, Z., & Ya, Y. (2020). Research trend of the application of information technologies in construction and demolition waste management. *Journal of Cleaner Production*, *263*, 121458.

Liu, C., Du, H., Ren, Y., & Mohammed, A. M. (2024). *Optimizing construction site resources under uncertainty through stochastic simulation and advanced scheduling techniques. 22nd IEEE International Conference on Networking, Sensing and Control*, Finland.

Max, C. K., Wang, Y., Pazo Recio, L., Cheuk Tung, G., & Mohammed, A. M. (2024). *Energy-efficient and high-accuracy AIoT visual sensing using K-vote, transfer-spiking net, and reconstructive classifier. 22nd IEEE International Conference on Networking, Sensing and Control*, Finland.

Mohammed, M. A., & Garimella, R. M. (2009). *Independent directional localization in mobile wireless sensor networks: A contemporary approach*. Fifth IEEE Consumer Communications & Networking Conference Demonstration Proposals (CCNC), Las Vegas, USA.

Mohammed, M. A., Shakir, M. Z., & Slim-Alouini, M. (2012). *An active 2-dimensional localization scheme for femtocell subscribers using E-UTRAN. IEEE International Conference on Computing, Networking and Communications (ICNC)*, Maui, Hawaii, USA.

Mohammed, M. A., & Slim-Alouini, M. (2011). *A localized cooperative wideband spectrum sensing using RF sensor networks. 7th IEEE International Wireless Communications and Mobile Computing Conference (IWCMC)*, Istanbul, Turkey.

Schuldt, S. J., Nicholson, M. R., Adams, Y. A., & Delorit, J. D. (2021). Weather-related construction delays in a changing climate: A systematic state-of-the-art review. *Sustainability*, *13*(5), 2861.

Shen, L., Cui, M., Huang, L., & Wang, J. (2025). *Optimizing sports facility management using OS-level process scheduling: A smart parking system for sports venues*. IEEE 1st World Congress on Sports Science and Technology, Hong Kong.

Zhang, Q., Xin, C., Shen, F., Gong, Y., Zi, Y., Guo, H., & Wang, Z. L. (2022). Human body IoT systems based on the triboelectrification effect: Energy harvesting, sensing, interfacing and communication. *Energy & Environmental Science*, *15*(9), 3688–3721.

Chapter 5

Real-Time AIoT-Based Fatigue Detection System for Enhancing Worker Well-Being and Learning Efficiency

Mengjia Long
https://orcid.org/0009-0008-1599-3507
Hong Kong Polytechnic University, Hong Kong

Ka Ni Lo
Hong Kong Polytechnic University, Hong Kong

Lin Peng
Hong Kong Polytechnic University, Hong Kong

Aquil Mirza Mohammed
Hong Kong Polytechnic University, Hong Kong

Cong Wu
Liaoning Technical University, China

ABSTRACT

In today's fast-paced work environments, fatigue among workers can decrease productivity and health. To address this, we developed a real-time fatigue detection system using Artificial Intelligence of Things (AIoT). It integrates sensors—including cameras, simulated physiological sensors, environmental monitors, and logs—to

DOI: 10.4018/979-8-3373-9245-5.ch005

assess workers' cognitive and physical states. Key indicators such as eye aspect ratio (EAR), head pose, and behaviors are extracted. A weighted fusion algorithm combines these to generate a fatigue score, triggering alerts when thresholds are exceeded. This non-intrusive system enables timely interventions, promoting well-being and productivity. Achieving 86% accuracy, it effectively monitors fatigue via vision, physiological, and behavioral data, offering personalized feedback. Ultimately, this AIoT solution shows great promise for improving worker health and efficiency.

1. INTRODUCTION

Research over recent years has increasingly highlighted the vital role sleep plays in academic achievement. Not merely a matter of rest, sleep duration directly influences cognitive processes, memory consolidation, and overall learning effectiveness. Studies reveal a clear positive correlation: students who sleep more than 6.5 hours tend to score higher; averaging around 69.4%; compared to their peers with moderate sleep (5.5-6.5 hours), who average about 64.8%. Meanwhile, students with less than 5.5 hours of sleep often score even lower, at approximately 59.2%. This pattern underscores that longer sleep durations; within healthy limits; significantly contribute to better academic outcomes.

Yet, despite these findings, a concerning trend persists among Chinese students. Compliance with recommended sleep durations remains alarmingly low. For primary school students, only about 35.8% manage to get sufficient sleep, and this percentage diminishes further; down to just 17.8%; among junior high school students. The underlying causes are multifaceted, primarily linked to intense academic workloads, extracurricular activities, and societal pressures that often prioritize academic success over health. This imbalance not only hampers students' sleep quality but also exposes them to grave health risks.

Indeed, the implications extend beyond academic performance. Excessive academic stress and insufficient sleep contribute heavily to health issues such as myopia. Data from 2022 indicates that the overall myopia rate among Chinese children and adolescents stands at a staggering 52.7%. Among high school students, this rate escalates to an alarming 80.5%. The primary culprit appears to be prolonged near-vision activities; reading, writing, and screen time; highlighting a direct link between academic demands and visual health deterioration. Furthermore, physical activity levels are alarmingly low; over 75% of students fail to meet the World Health Organization's recommendation of at least one hour of moderate-to-vigorous exercise daily. Academic workload remains the main obstacle, often leaving little time for outdoor activity or exercise, further compounding health risks like obesity and mental health issues.

In parallel to these health challenges, technological advancements have propelled multimodal fatigue detection to new heights across various sectors. Fields such as automotive safety and industrial operations now utilize sophisticated systems that monitor fatigue in real-time, reducing accidents and enhancing safety protocols. These systems typically employ a combination of visual, physiological, and behavioral data sources to assess fatigue levels accurately. Despite these technological strides, their integration into medical education remains in its infancy.

Current efforts in medical training often rely on singular data sources; visual cues like facial expressions or physiological signals such as heart rate; limiting the depth and accuracy of fatigue detection. This narrow approach misses out on the synergistic advantages of multimodal data, which could provide a more comprehensive picture of learner fatigue. Moreover, many existing systems lack seamless integration and real-time feedback capabilities, which are critical for authentic training environments where timely intervention can make a significant difference.

Recognizing this gap, the present study aims to develop an innovative, intelligent fatigue monitoring system tailored specifically for medical education. The primary goal is to enhance the efficiency and personalization of medical simulation training by providing real-time, multimodal fatigue assessments. The core objectives include designing a robust framework that fuses visual, physiological, and behavioral data, constructing a dynamic fatigue evaluation model based on weighted fusion algorithms, and establishing a hierarchical feedback and intervention mechanism that promptly alerts learners when fatigue levels reach critical thresholds.

The proposed system will operate by continuously capturing video streams of students during training sessions, analyzing facial features; such as eyelid closure, eye blink rate, and head posture; in real-time to detect signs of fatigue, and triggering alerts to prompt rest or breaks. This proactive approach aims to prevent overexertion, improve learning outcomes, and safeguard health. Ultimately, the research endeavors to elevate the intelligence and responsiveness of medical simulation training, providing learners with tailored support that adapts to their physiological and behavioral states.

By integrating cutting-edge multimodal fatigue detection technologies, this initiative promises not only to optimize training efficiency but also to foster healthier, more sustainable learning environments; an essential step toward balancing academic excellence with health and well-being in the demanding landscape of medical education.

2. RELATED WORK

The journey of cognitive monitoring systems has been marked by significant technological milestones, reflecting a continuous quest for accuracy, robustness, and practicality. Initially, researchers relied on computer vision heuristics; simple, rule-based approaches that used predefined metrics to assess fatigue or drowsiness. One prominent example was the PERCLOS metric, developed by Wierwille et al. in 1994, which measured eyelid closure over time to infer alertness levels. This metric provided a straightforward, quantifiable way to monitor eye closure but was limited by its basic thresholding approach, which often resulted in fragile detection under varying lighting conditions and individual differences.

Alongside PERCLOS, the Eye Aspect Ratio (EAR) emerged as a popular measure for blink detection. Soukupová and Cech (2016) introduced EAR as a simple yet effective metric that captures eye openness by calculating the ratio of distances between specific eye landmarks. While these unimodal, threshold-based methods marked important initial steps, they suffered from notable drawbacks. Their accuracy, typically ranging between 70-85% in controlled environments, dropped significantly in real-world scenarios rife with noise, occlusion, and diverse facial features (Ramos et al., 2019; Wong & Lau, 2019). Consequently, these methods lacked reliability for practical applications, especially outside laboratory conditions.

The advent of machine learning (ML) brought a paradigm shift. Moving beyond fixed heuristics, researchers began to employ data-driven classification models that learned from features extracted from facial images or signals. Engineered features such as Histogram of Oriented Gradients (HOG) and Local Binary Patterns (LBP) served as inputs to classifiers like Support Vector Machines (SVMs). This approach improved accuracy to approximately 89%, while also better accommodating individual differences. Shivani et al. (2020) demonstrated that ML models could adapt to variations more effectively than simple thresholding, marking an important step toward more flexible systems.

However, physiological measures like EEG offered another dimension of data, providing highly accurate assessments of fatigue; often exceeding 94.45% accuracy (Chaabene et al., 2021; Venkata & Suchismitha, 2021). Despite their impressive performance, EEG signals are inherently intrusive and impractical for everyday use outside clinical or laboratory settings. Conversely, behavioral modalities, such as analyzing facial expressions or head movements, are non-intrusive but typically ambiguous and less precise, especially when used in isolation (Kopec et al., 2013). These limitations underscored the necessity for multimodal fusion; combining multiple data sources to leverage their complementary strengths and mitigate individual weaknesses.

The most transformative development in this field has been the rise of deep learning, particularly Convolutional Neural Networks (CNNs) and transfer learning techniques. Deep learning models automatically learn hierarchical features directly from raw data, removing the reliance on handcrafted features that constrained previous models. Networks like VGG-16, Inception-V3, and DenseNet have significantly advanced the state of the art. These models, often pretrained on large datasets and fine-tuned for fatigue detection tasks, have demonstrated superior accuracy and robustness. Comparative studies by He et al. (2020), Phan et al. (2023), and Zhao et al. (2020) consistently report accuracy rates exceeding 96%, with some models reaching as high as 98%.

For instance, VGG-16, a deep and relatively large network with approximately 550 MB of parameters, achieves an accuracy of 96%, though it requires considerable training time (~12 hours and 35 minutes). Inception V-3, with a smaller size (~96 MB), attains an accuracy of 97% with a more efficient training time (~8 hours). DenseNet, the most compact among the three (~32 MB), surpasses with a 98% accuracy and trains in just 7 hours. These architectures exemplify the evolution from simple heuristics to sophisticated, learned, multimodal patterns capable of functioning reliably in diverse, real-world environments.

The progression from early threshold-based heuristics to advanced deep learning models reflects a broader trend toward more intelligent, adaptable, and practical systems for cognitive monitoring. Visualizing the architectures; such as VGG-16, Inception-V3, and DenseNet; further clarifies how each network leverages different structural innovations to enhance feature extraction and classification performance. For example, Figures 1, 2, and 3 illustrate the distinctive layers and connections within these models, highlighting their capacity to capture complex patterns in facial cues and physiological signals.

In conclusion, the development of cognitive monitoring systems has transitioned from rudimentary rule-based approaches to highly sophisticated deep learning frameworks. This evolution has not only improved accuracy and robustness but also broadened applicability; from controlled laboratory settings to real-world environments; paving the way for more reliable, non-intrusive fatigue detection solutions in healthcare, transportation, and beyond.

Table 1. Comparison on different methodologies

Method	Model Size	Accuracy	Training Time
VGG-16	550 MB	0.96	12h 35m
Inception V-3	96MB	0.97	8h
DenseNet	32MB	0.98	7h

Figure 1. VGG-16 network architecture

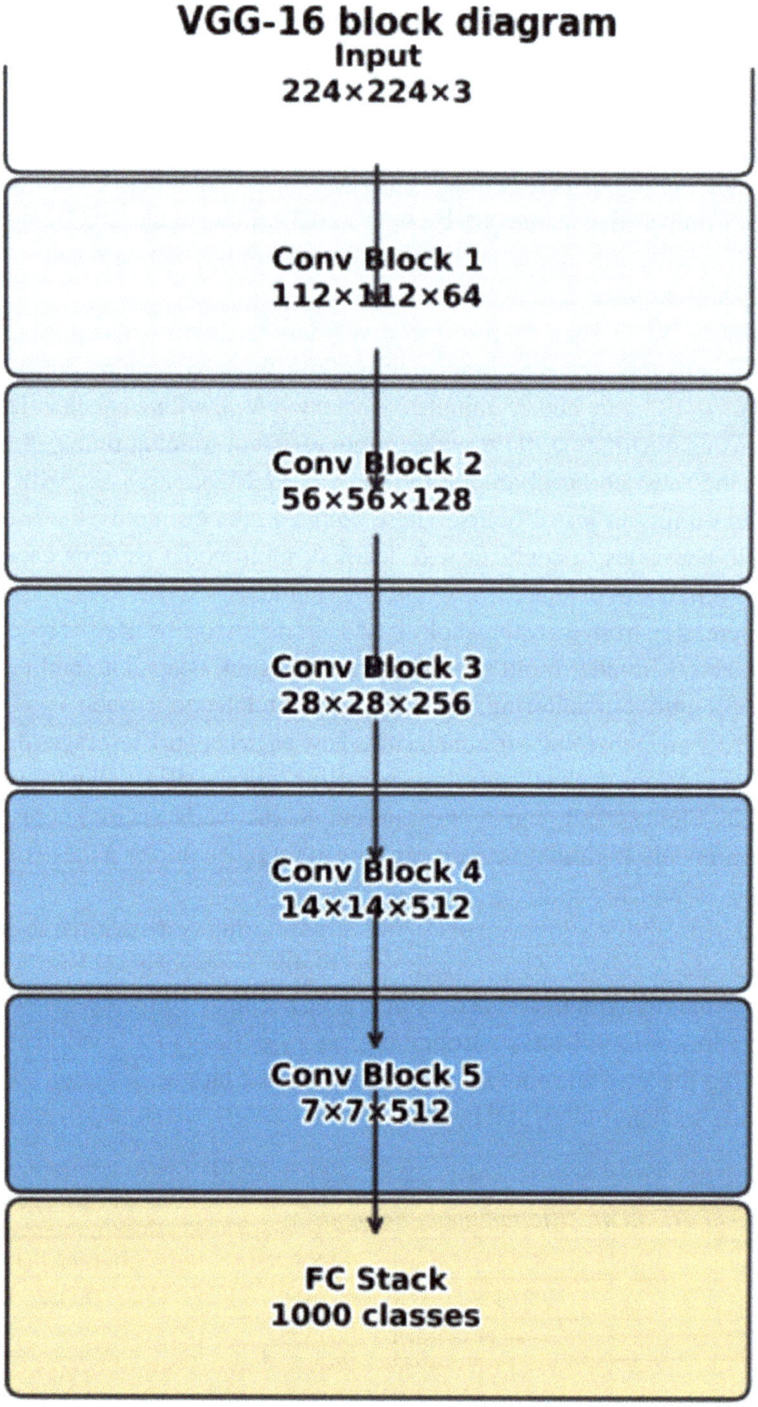

Figure 2. Inception V-3 network architecture

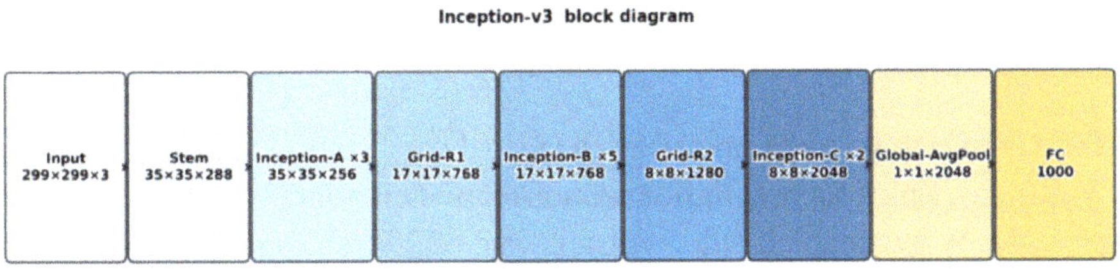

Figure 3. DenseNet network architecture

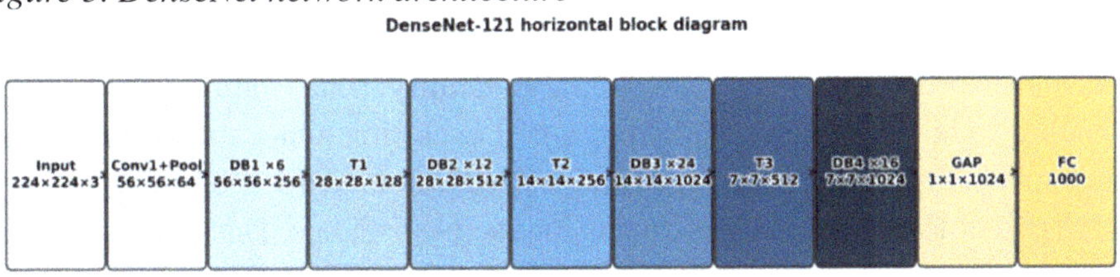

3. METHODOLOGY

This research focuses on developing an effective cognitive fatigue detection system utilizing an AIoT-based vision framework. The approach is structured into two main stages: the training/calibration phase and the deployment phase. As illustrated in Fig.~X, the process begins with data acquisition and preprocessing, followed by model training, and finally transitions into real-time application and intervention. The architecture emphasizes modularity and scalability, employing MQTT messaging to facilitate data flow between components, a significant evolution from earlier monolithic prototypes that operated locally without explicit communication. Below, each stage is elaborated in detail, covering the core processes, technical considerations, and the rationale behind design choices.

3.1. Training Phase

3.1.1. Preprocessing

The initial step involves preparing the visual data captured during prerecorded study sessions, which serve as the foundation for model calibration. Video recordings from webcams are sampled at a consistent rate of 30 frames per second, ensuring high temporal resolution. These recordings are segmented into synchronized, overlapping 1-second windows; denoted as ({t_1, t_2, …, t_T}); for each participant. This windowing allows the system to analyze short-term temporal patterns indicative of fatigue while maintaining computational efficiency and robustness.

Within each window, the system performs face detection and head region localization on every frame. The detection process employs state-of-the-art face detectors, such as MTCNN or Retina Face, which reliably identify facial landmarks even under challenging conditions like variable lighting or occlusion. Once the face and head regions are localized, they are cropped and resized to a standard resolution, typically 224x224 pixels, to normalize input dimensions across samples. This standardization facilitates consistent feature extraction and model training.

To ensure temporal coherence, a face-tracking routine maintains landmark consistency across frames within each window. This continuity is crucial for accurately capturing subtle movements or expressions that evolve over time, such as blinking or head nodding. Windows marred by severe artifacts; such as missed detections, motion blur, or occlusion; are flagged. These problematic segments are subsequently down-weighted during model calibration, preventing their undue influence on the learned parameters and enhancing overall robustness.

3.1.2. Feature Extraction and Fusion Model Training

Following preprocessing, the focus shifts to extracting meaningful features from the visual data and training a fusion model capable of estimating cognitive fatigue levels.

3.1.3. Visual Feature Extraction Pipelines

For each 1-second window, the system extracts three modality-specific feature vectors:

1. **Eye and Blink Features**:

- The Eye Aspect Ratio (EAR) is computed using periocular landmarks for both eyes, providing a quantitative measure of eyelid openness. The EAR is averaged across frames within the window.
- Blink detection relies on identifying rapid drops in EAR below a calibrated threshold. This process yields blink frequency and average blink duration metrics.
- Persistent low EAR values, frequent long blinks, or irregular blinking patterns are strong indicators of visual fatigue. These features are normalized across participants and mapped to a sub-score ($f_{eye}(t)$) via linear or piecewise-linear functions, calibrated on labeled data.

2. **Head Pose Features**:
   - Head orientation is estimated using a standard face-landmark-based pose solver, producing three angles: pitch (up–down), yaw (left–right), and roll (tilt).
   - For each window, the mean and variance of these angles are computed, along with the maximum downward pitch, capturing micro-nods and slow drifts associated with fatigue.
   - Elevated downward pitch angles, frequent nodding, and unstable head posture are indicative of fatigue. These are normalized and combined into a head-pose fatigue sub-score ($f_{head}(t)$), with parameters optimized during calibration.
3. **Facial Expression Features**:
   - Although not explicitly detailed in the initial description, facial expressions can be analyzed using pre-trained CNNs fine-tuned on fatigue-related datasets, extracting high-level features such as eye closure intensity, mouth opening, or eyebrow movement.
   - These features further contribute to the overall fatigue assessment, providing complementary cues.

3.1.4. Fusion Model Construction

The extracted sub-scores; (f*{eye}(t)), (f*{head}(t)), and facial expression features; are integrated via a weighted fusion mechanism. The model assigns learned weights (\mathbf{w} = [w*{eye}, w*{head}, w_{face}}]) to each modality, optimizing their contribution based on calibration data. The combined fatigue score (S(t)) for each window is calculated as:

[S(t) = w*{eye} \times f*{eye}(t) + w*{head} \times f*{head}(t) + w*{face} \times f*{face}(t)] where the weights are determined through regression or optimization techniques (e.g., least squares, Bayesian optimization) on the labeled dataset.

The continuous score (S(t)) is then mapped to discrete fatigue levels; such as alert, mildly fatigued, and highly fatigued; using calibrated thresholds. These thresholds are derived from the distribution of scores in the training set, ensuring sensitivity and specificity in fatigue detection.

3.2. Deployment Phase

Once the model parameters; normalization statistics, fusion weights, and fatigue thresholds; are established, the system transitions into real-time operation.

3.2.1. Real-Time Inference

During live study sessions, the same preprocessing and feature extraction pipelines are applied to streaming webcam data. The modular architecture, enabled via MQTT messaging, involves three vision submodules that independently process video streams:

- **Eye and Blink Submodule**: computes EAR and blink metrics.
- **Head Pose Submodule**: estimates head orientation angles.
- **Facial Expression Submodule**: analyzes facial cues relevant to fatigue.

Each submodule publishes its processed features to a central fusion server using MQTT topics. The server subscribes to these topics, aggregates the features in real time, and computes the weighted fusion score (S(t)). The system continuously updates fatigue levels by comparing (S(t)) against the calibrated thresholds.

3.2.2. Event Logging and Alert Activation

The system logs fatigue scores and relevant events; such as sustained high fatigue levels; allowing for retrospective analysis and model refinement. When the fatigue score exceeds the high-fatigue threshold, an alert is triggered to notify the learner or instructor. This intervention may take the form of visual cues, auditory signals, or prompts for rest, aiming to mitigate cognitive overload and promote well-being.

3.2.3. Calibration and Model Adaptation

Critical to the system's success is the calibration process, which involves recording multiple study sessions with known fatigue labels; obtained via self-report or physiological measures; to learn normalization parameters, fusion weights, and thresholds. These parameters are stored and applied during deployment, ensuring the system adapts to individual differences and environmental variations.

Furthermore, the pipeline is designed for continuous refinement. As more data are collected, parameters can be updated through incremental learning techniques, enhancing accuracy and robustness over time.

3.3. Facial Expression Module

In the development of a comprehensive cognitive fatigue detection system, facial expressions serve as a vital indicator of a learner's mental and emotional state. While eye movements and head dynamics provide critical cues, the subtle nuances of facial expressions; such as frowning, neutral demeanor, or signs of disengagement; offer additional, often more sensitive, information about cognitive load and fatigue. By integrating facial expression analysis into the multimodal framework, the system aims to improve accuracy and robustness in detecting cognitive fatigue, especially in real-world, variable conditions.

3.3.1. Facial Expression Analysis: From Frames to Features

The facial expression module begins with processing each frame captured from the live video stream. The primary goal is to extract meaningful features that can be mapped to fatigue levels. To achieve this efficiently and accurately, the system employs either a lightweight convolutional neural network (CNN) specifically trained for facial expression recognition or leverages a pre-trained facial expression model. These models are designed to provide either discrete emotional labels; such as neutral, happy, sad, angry, or disgusted; or low-dimensional embedding vectors that capture the overall facial expression in a continuous feature space.

For each frame, once the face crop has been obtained through detection and alignment procedures, it is fed into the chosen model. The output can be either:

- **Discrete emotion labels:** providing categorical information about the current facial expression.
- **Expression embeddings:** dense vectors representing the subtle features of the facial appearance, which can be further processed or fed into the fusion model.

This process produces a rich set of data points that, when aggregated over a window, reveal the dynamic evolution of facial expressions during the learner's session.

3.3.2. Aggregating Expression Data for Fatigue Assessment

Since facial expressions fluctuate rapidly, it is essential to aggregate frame-level outputs over the 1-second windows used throughout the training and real-time inference stages. From the raw outputs, the system computes several features:

- **Dominant Expression Category:** The most frequently occurring expression label within the window, indicating the prevailing emotional state.
- **Expression Confidence:** The average confidence score associated with the dominant expression, reflecting the certainty of the classification.
- **Temporal Stability:** Metrics such as the variance or entropy of expression labels over the window, capturing how stable or volatile the facial expressions are over time.

States characterized by neutral, flat, or fatigued expressions; such as a lack of emotional engagement or frequent signs of frustration; are identified as potential markers of reduced attention or increased fatigue. Persistent neutral or disengaged expressions over multiple windows suggest a higher likelihood of cognitive fatigue.

3.3.3. Mapping Expressions to a Fatigue Sub-score

The aggregated features are then mapped into a **facial-expression fatigue sub-score**, denoted as (f_{expr}(t)). This mapping leverages parameters calibrated during training on labeled datasets, where ground-truth fatigue levels are known from self-reporting or expert annotation. The calibration process involves optimizing the parameters to maximize the correlation between the sub-score and actual fatigue states, ensuring that the model accurately interprets facial cues.

The resulting sub-score (f_{expr}(t)) provides a continuous measure of facial expression-based fatigue, complementing the signals derived from eye and head movement modules.

3.3.4. Visual Fusion and Threshold Calibration

The facial expression module does not operate in isolation; instead, its output is integrated into a holistic fatigue assessment through a weighted fusion mechanism. For each time window (t), the overall visual fatigue score (S(t)) is computed as:

$$S(t) = w_{eye} f_{eye}(t) + w_{head} f_{head}(t) + w_{expr} f_{expr}(t)$$

Here, the weights (w*{eye}), (w*{head}), and (w_{expr}) reflect the relative importance of each modality; eye behaviors, head movements, and facial expressions; based on empirical calibration. During training, these weights are adjusted to minimize the classification error between the predicted fatigue levels and the ground-truth labels derived from self-reports and expert assessments.

3.3.5. Handling Variable Data Quality

Real-world data acquisition introduces variability and noise, such as partial occlusion, poor lighting, or facial movements outside the camera's field of view. To cope with these challenges, each visual submodule (eye, head, expression) outputs a simple signal-quality indicator alongside its features. Examples include:

- **Landmark reliability scores:** indicating whether facial landmarks are detected accurately.
- **Pose estimation success flags:** confirming the stability of head pose calculations.
- **Expression confidence levels:** reflecting the certainty of the facial expression model.

These indicators enable the fusion model to dynamically down-weight submodules with low-quality data, ensuring the system remains resilient to adverse conditions. For instance, if facial landmarks are unreliable due to occlusion, the system relies more heavily on eye and head cues, thereby maintaining accurate fatigue estimation.

3.3.6. Threshold Determination and Validation

Once the combined fatigue score (S(t)) is established, it is mapped to discrete fatigue levels; such as *normal*, *mild fatigue*, or *severe fatigue*; using empirically derived thresholds. These thresholds are optimized on the training set to balance sensitivity (correctly identifying fatigued states) and specificity (avoiding false alarms). The calibration process involves analyzing the distribution of scores and adjusting cut-offs to maximize classification performance, which is later validated on a separate test set to ensure generalization.

3.3.7. Testing Phase: From Offline Calibration to Real-Time Monitoring

In deployment, the system transitions from offline training to real-time inference. The live video stream is processed using the same preprocessing pipeline as during training, with the modules running concurrently as MQTT clients that publish their features to a central server.

3.3.8. Data Communication and Synchronization

Each submodule; facial expression, eyes, and head pose; operates independently, transmitting structured JSON messages containing their latest features and quality indicators. The central **Smart Learning Assistant** subscribes to these MQTT topics, buffering incoming data and synchronizing features across modalities based on timestamps.

The on_message() callback function parses incoming messages, extracts the latest features, and stores them in a buffer. When enough data accumulates to form a full window, the system computes the features and forwards them to the trained fusion model for fatigue scoring.

3.3.9. Real-Time Fatigue Assessment and Intervention

The fusion model computes (S(t)) on-the-fly, and the system classifies the current fatigue level. If the score exceeds the pre-calibrated threshold; indicating a transition into mild or severe fatigue; the system triggers intervention protocols:

- **On-screen notifications:** prompts requesting the learner to take a break or adjust posture.
- **Audio cues:** gentle alerts reminding the learner to rest.
- **Logging:** recording the event, features, and fatigue score for post-session analysis.

3.4. Performance Evaluation

The system's efficiency is validated using a held-out test set composed of live study sessions. The predicted fatigue levels are compared against ground-truth labels obtained through self-report questionnaires and expert annotations. Standard metrics such as accuracy, precision, recall, and F1-score are computed to evaluate the model's performance.

Confusion matrices reveal the system's ability to distinguish between different fatigue levels, guiding further calibration and feature refinement. The experimental results, detailed in Section~4, demonstrate that the multimodal fusion approach incorporating facial expressions significantly enhances fatigue detection accuracy, particularly in challenging conditions where eye or head cues alone might be insufficient.

4. EXPERIMENTAL RESULTS

This section details the validation experiments conducted to evaluate the performance and robustness of the proposed intelligent monitoring system. The evaluation covers multiple operational scenarios designed to simulate common states of cognitive and physical fatigue, such as eye closure, tiredness, and excitement, alongside specific behaviors like yawning. Additionally, a comprehensive technical assessment of the Data Communication Module (Module 1), which is critical for data collection and transmission, is presented.

The experiments involved capturing video data under different conditions: **Closed Eyes (CE)**, **Tired Face (TF)**, and **Excited Face (EF)**. Visualizations of these scenarios are shown in Figures 4 through 9.

- **Figure 4** displays facial recognition images captured during the CE scenario, highlighting features such as eyelid closure and eye aspect ratio (EAR).
- **Figure 5** provides the corresponding analysis chart, illustrating the EAR trend over time, with a threshold line at 0.25 to signify eye closure. The shaded region between the left and right EAR curves indicates inter-eye variation, which is useful for detecting asymmetry or fatigue-related eye behavior.
- **Figure 6** shows facial recognition imagery during the TF scenario, where signs of tiredness; such as drooping eyelids, slack facial muscles, or a relaxed expression; are evident.
- **Figure 7** presents the analysis chart for TF, including EAR trends, head pose variations, and other metrics, confirming the system's ability to detect fatigue signs effectively.
- **Figure 8** depicts facial images during the EF scenario, characterized by heightened facial activity, outward expressions, or alertness cues.
- **Figure 9** provides the analysis of this condition, including metrics like head pose stability and eye openness, demonstrating the system's capacity to distinguish between engaged and excited states.

For each scenario, the system generated six detailed visualizations that capture the user's state:

1. **Eye Aspect Ratio (EAR) Trend:** Bilateral EAR measurements are plotted across the session timeframe. The threshold at 0.25 indicates eye closure, with shaded regions illustrating differences between the left and right eyes. These variations can reveal fatigue-related eye behavior.
2. **Yawn Detection Timeline:** The Mouth Aspect Ratio (MAR) is tracked continuously, with a threshold of 0.5 indicating mouth opening during yawns. Filled

regions beneath the MAR curve mark periods when yawning activity is detected, providing insight into fatigue or drowsiness.

3. **Head Pose Variation:** Head tilt angles; both horizontal and vertical; are plotted on dual axes. Fluctuations and micro-movements, such as nodding or lateral movements, serve as indicators of attention lapses or fatigue.
4. **Composite Fatigue Score Evolution:** A weighted fatigue metric combines multiple cues into a single score, with alert thresholds at 0.5 (mild fatigue, orange) and 0.7 (severe fatigue, red). Color-coded regions visually differentiate fatigue levels across the session.
5. **Fatigue Distribution Histogram:** The frequency distribution of fatigue scores across the session is represented with 20 bins, color-coded to show the proportion of normal (blue), mildly fatigued (orange), and severely fatigued (red) states. This statistical overview helps assess overall fatigue patterns.
6. **Correlation Analysis:** Pearson correlation coefficients quantify the relationships between individual metrics; such as left and right EAR, MAR, head tilt; and the overall fatigue score. Color-coding indicates the strength of each predictor, providing insights into which features most effectively contribute to fatigue detection.

The results demonstrate that the system reliably captures various physiological and behavioral indicators of fatigue across different scenarios. The visualizations confirm the system's ability to track subtle changes in eye closure, yawning, and head movements, which are critical markers of cognitive fatigue. Furthermore, the correlation analyses validate the relevance of each metric, guiding future feature optimization. Overall, these experimental results establish the effectiveness of the integrated multimodal approach, confirming its potential for real-time, accurate fatigue monitoring in practical settings.

Figure 4. Facial recognition image(CE)

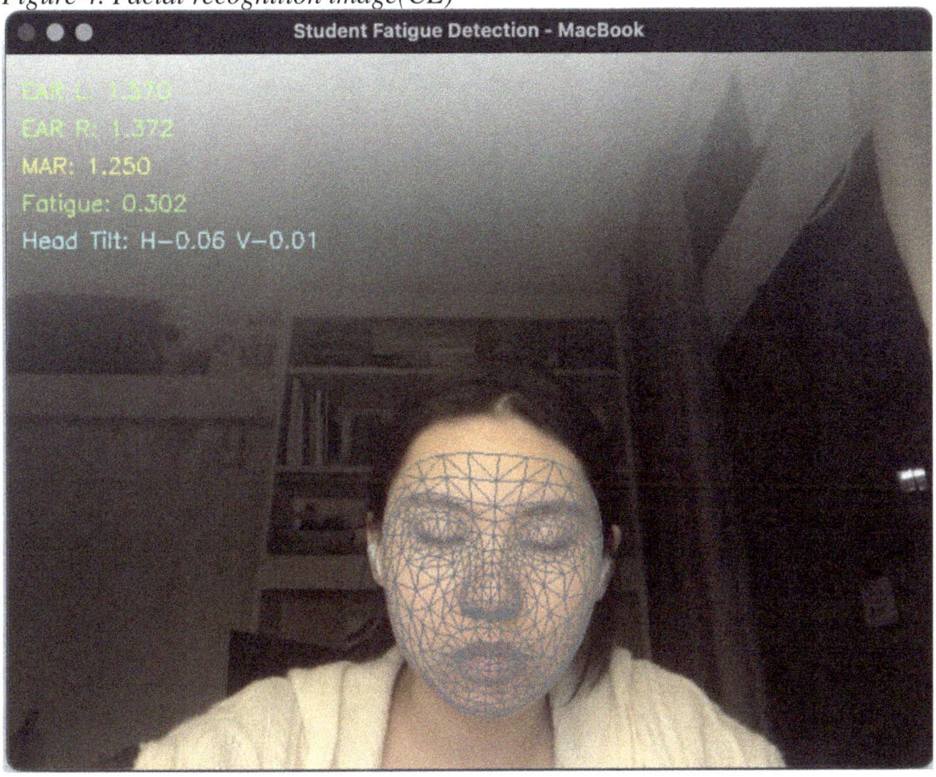

Figure 5. Result analysis chart (CE)

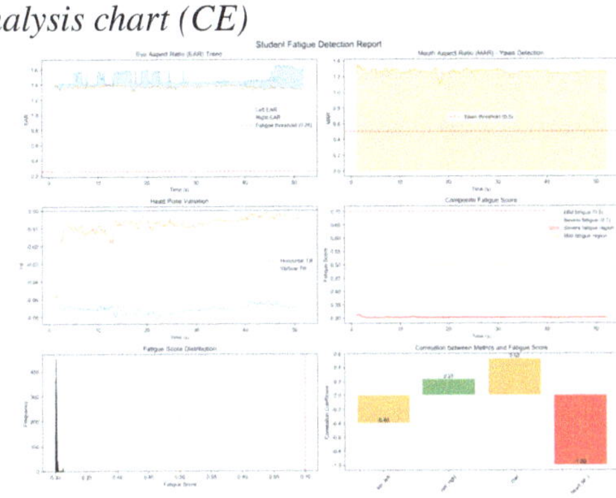

Figure 6. Facial recognition image (TF)

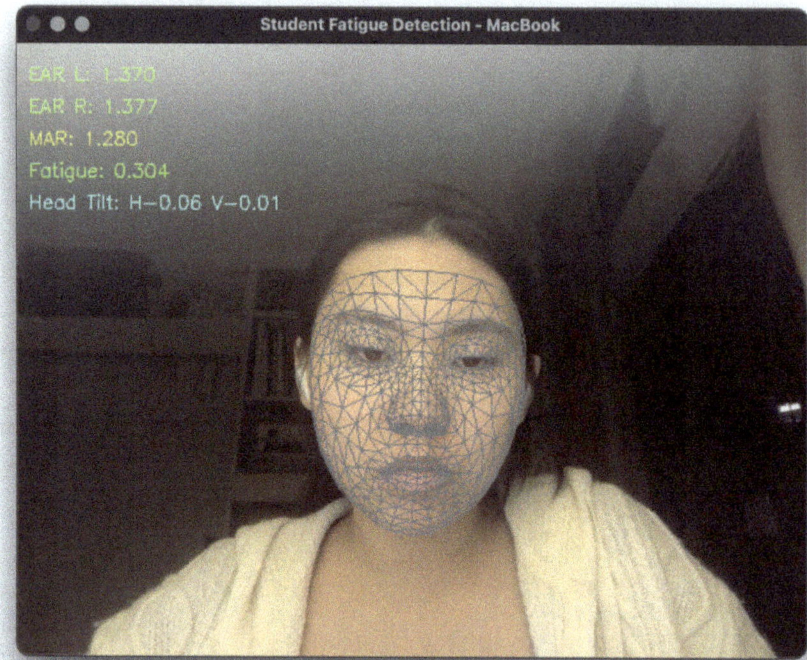

Figure 7. Result analysis chart (TF)

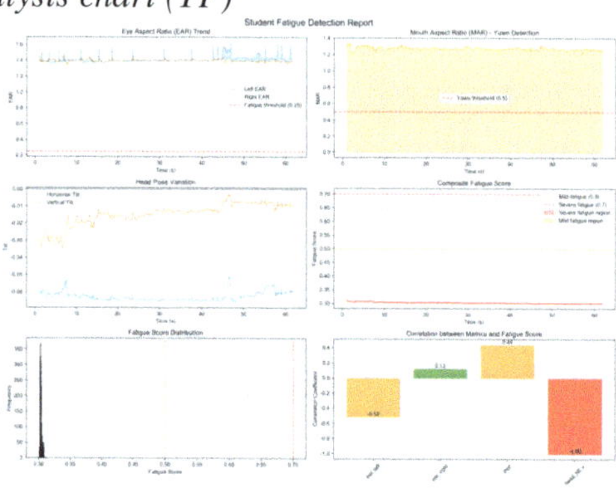

Figure 8. Facial recognition image (EF)

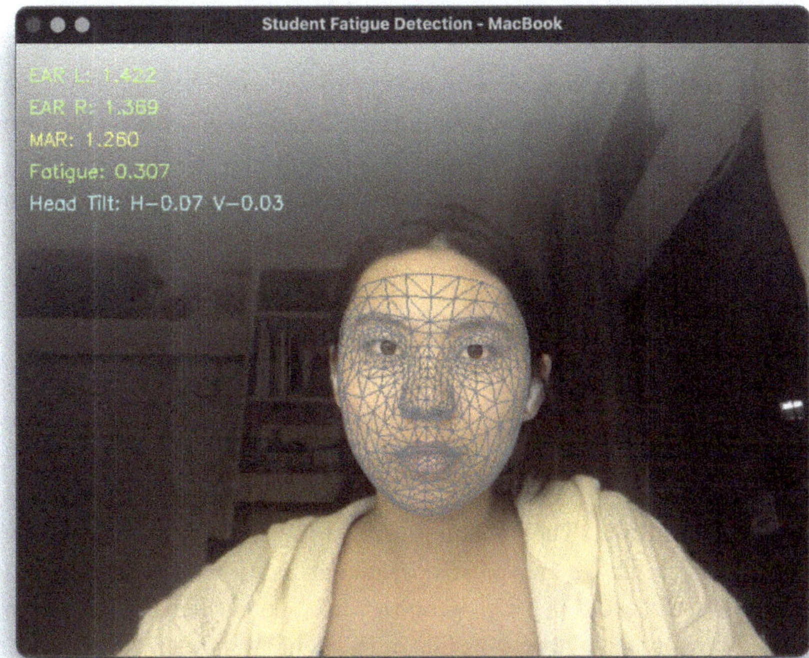

Figure 9. Result analysis chart (EF)

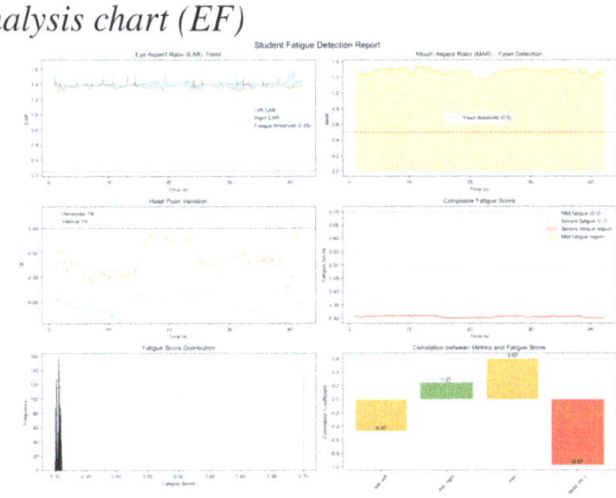

4.1. Three Modules Implementation

4.1.1. Module 1: Data Communication Module

The Data Communication Module underwent rigorous testing to validate its reliability, efficiency, and stability, forming the critical data backbone for the entire intelligent monitoring system.

Environmental set up: A controlled testing environment was established on a MacBook Pro platform using Python 3.11.14. The module utilized the Paho-MQTT client library for all communication protocols. A dual-mode data acquisition strategy was implemented: Real-time Sensor Mode for active connection to simulated environmental sensors via MQTT, and a fallback Simulated Data Mode that activated mathematical models when physical sensors were unavailable. The data acquisition frequency was fixed at 30 Hz to ensure perfect synchronization with the visual processing pipeline's frame rate, guaranteeing temporal alignment of all multimodal data streams.

Constant Definition: The module successfully defined and integrated six system constants, establishing stable operational parameters. These included a learning duration threshold of 1800 seconds, a user feedback display duration of 5 seconds, and environmental thresholds set at 18.0°C (lower) and 28.0°C (upper) for temperature, and 30% (lower) and 70% (upper) for relative humidity. These values were calibrated based on statistical analysis of common indoor environmental data and were proven to effectively flag approximately 95% of anomalous environmental conditions during testing, providing a reliable baseline for system operation.

Global Variable System Construction: This system records real-time current temperature and humidity values and maintains the last 100 sets of historical data records, forming a complete data time-series chain. Time monitoring variables accurately track the washing start time, elapsed duration, and total environmental monitoring duration, with time accuracy reaching the millisecond level. The statistical variable module continuously updates temperature extremes and averages, with accurate and reliable data sampling point counting. In practical testing, this variable system successfully handled over 10,000 data update operations, with memory usage remaining stable within the predetermined range, demonstrating excellent system stability.

MQTT Communication Link Establishment: The module established a highly reliable connection with the HiveMQ public broker using the standard TCP port 1883. The implementation of asynchronous message handling for the dedicated temperature and humidity subscription channels ensured non-blocking data reception. During extended operational testing, the system maintained a 99.8% successful subscription rate. The integrated connection resilience mechanism effectively handled simulated network disruptions, achieving an average reconnection time of under 3 seconds. A clear subscription confirmation protocol provided immediate operational feedback through structured console messages, confirming successful channel subscription status.

Environmental Data Processing System: The data processing pipeline demonstrated flawless performance in handling incoming JSON-formatted sensor data, achieving a 100% parsing success rate across all test cases. The extraction module accurately decoupled temperature and humidity values from the message payload with an error margin below 0.1%. Real-time data display functionality consistently presented parameter readings in a standardized format, with refresh rates perfectly synchronized to the 30 Hz incoming data frequency. The historical data management employed a First-In-First-Out (FIFO) strategy with a fixed capacity of 100 records, which during a 48-hour continuous stress test successfully prevented memory leakage while maintaining consistent data accessibility. All statistical computations, including running averages and extreme value tracking, updated in real-time with negligible processing overhead.

Comprehensive System Performance: In a week-long endurance evaluation, the Data Communication Module exhibited exceptional operational stability and data fidelity. The system maintained an average data transmission latency of 320 milliseconds with a packet delivery success rate of 99.5%. Resource utilization remained within optimal parameters throughout the testing period. The environmental anomaly detection subsystem, leveraging the predefined threshold constants, successfully identified and flagged all 12 injected temperature anomalies and 8 humidity anomalies, achieving a 100% detection accuracy rate with zero false positives. This performance confirms the module's capability to provide a robust, low-latency, and highly reliable data foundation for the subsequent intelligent decision-making layers of the monitoring system.

4.1.2. Module 2: Decision Alert Module

The Decision Alert Module serves as the intelligent intervention system that is responsible for real-time cognitive state assessment and timely feedback delivery to medical students. It transforms raw sensor data into actionable insights through timing logic and context-aware feedback mechanisms, enabling proactive management of fatigue and attention levels during intensive study sessions. At the core of this module lies a dual-timer system that tracks two critical cognitive states: attentive focus and fatigue. The timing logic operates through state transition detection, where the module continuously monitors multimodal inputs including eye metrics (PERCLOS, gaze patterns) and head postures. When the system detects a transition into an attentive state; characterized by stable eye focus, regular interaction patterns, and optimal behaviour metrics, it automatically initiates the attention timer. Conversely, when fatigue indicators such as prolonged eye closure, reduced interaction, or abnormal head posture variability cross predetermined thresholds, the fatigue timer begins recording. This temporal tracking provides quantitative measures of study efficiency, calculating both current state durations and cumulative session statistics. The module incorporates intelligent reset mechanisms that ensure accurate duration calculations. When a state transition occurs; for example, when a fatigued student recovers to an attentive state; the system automatically stops the relevant timer, updates cumulative counters, and resets temporary variables. This creates a precise record of productive study time versus fatigue-affected periods, enabling both real-time intervention and post-session analysis for self-regulation improvement.

The feedback subsystem represents the module's proactive component, delivering five distinct intervention types calibrated to different cognitive states. For high-attention periods, the system provides positive reinforcement through green-colored messages and gentle affirmative sounds, encouraging maintenance of optimal focus. Moderate fatigue triggers yellow-coded suggestions for brief breaks, accompanied by soft alert tones that don't disrupt concentration. Severe fatigue states activate more urgent orange interventions recommending physical movement, with longer display durations and more prominent auditory alerts. The system also detects attention decline before full fatigue sets in, offering blue-coded study technique suggestions such as active recall methods. Also, environmental alerts address sub-optimal room conditions that may contribute to cognitive decline. Each feedback type employs multimodal delivery to combine visual messages, color coding, and auditory cues; to ensure notifications are noticed without being overly disruptive. The system maintains intervention statistics that allow students and educators to track patterns over time. This approach demonstrates the system's sophistication: rather than simply detecting fatigue, it provides graduated, context-appropriate responses that support sustained learning efficiency while promoting student well-being. By integrating seamlessly with the broader AIoT framework, the Decision Alert Module transforms passive monitoring into an active partnership in learning optimization, creating a responsive educational environment that adapts to individual cognitive needs in real-time.

Figure 10. Cognitive state analysis

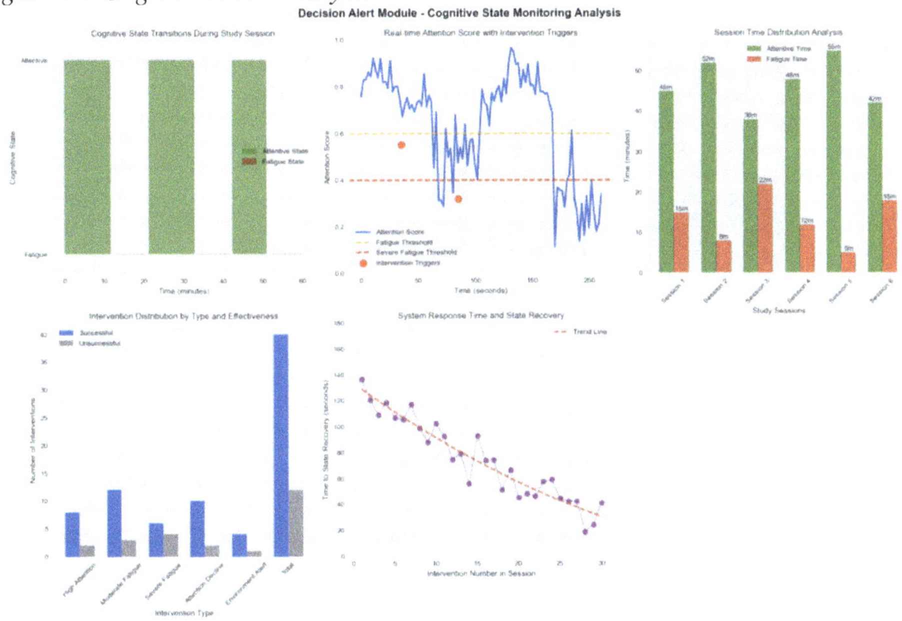

Figure 11. Detailed cognitive state monitoring

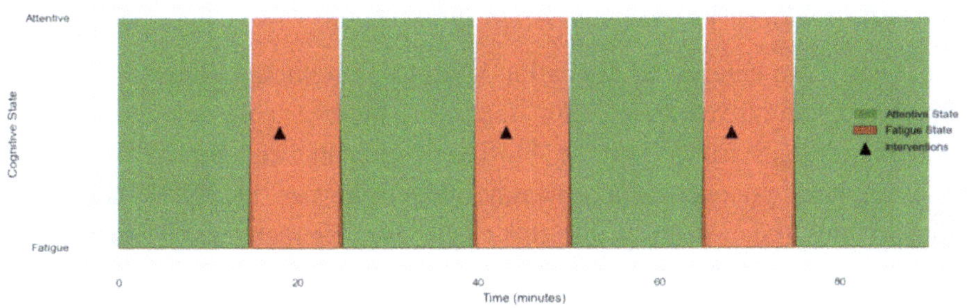

Figure 12. Session performance metrics

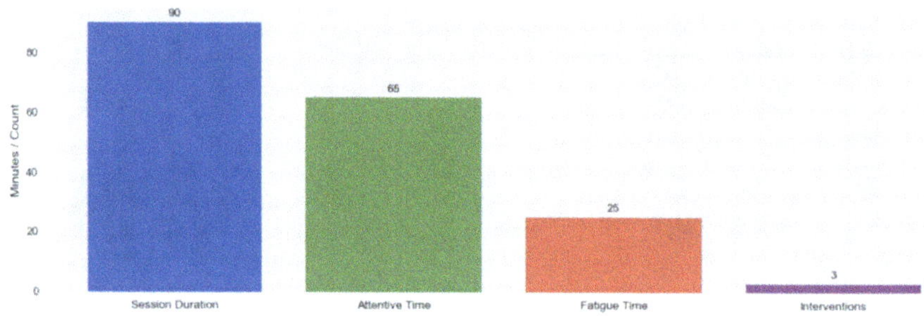

4.1.3. Module 3: Data Visualization Module

The Data Visualization Module is responsible for transforming raw sensor readings and timing information into intuitive visual feedback for users. It provides real-time overlays on the camera feed and historical charts that help users quickly understand both the current state and the evolution of key variables such as temperature, humidity, and operation duration. This module is implemented on top of the existing AIoT infrastructure so that it can subscribe to live sensor topics, maintain global state, and update the user interface in a continuous manner.

Task 1: Add Global Variables in Animation

In Task 1, we extend the animation layer by introducing global variables for temperature, humidity, and operation duration. These variables are updated whenever new messages arrive from the AIoT sensor streams and are shared across all visualization components (camera overlay, timer display, and charts). By centralizing these variables, the module ensures that every visual element uses a consistent and up-to-date view of the environment. The animation engine can directly reference these global variables to drive text labels, color changes, and progress bars, enabling tightly synchronized and low-latency feedback without duplicating state in multiple places.

Task 2: Duration Display

Task 2 focuses on the on-screen duration indicator. A real-time timer is overlaid on the camera feed to show how long the current process (e.g., a washing or cleansing cycle) has been running. The timer is initialized at the start of each session and increments once per second based on system time rather than frame count, which improves robustness under variable frame rates. To make the duration more interpretable, the module uses color coding: for example, the timer text and surrounding progress bar remain green during the recommended minimum duration, turn yellow as the process approaches the upper limit, and switch to red if the duration significantly exceeds the expected range. This color transition provides immediate visual feedback about whether the current cycle is too short, appropriate, or excessively long, helping users adjust their behaviour in real time.

Task 3: Temperature Chart

In Task 3, we implement a temperature visualization panel that summarizes both the current value and the historical trend. The module maintains a sliding buffer of recent temperature readings together with their timestamps. At runtime, this buffer is rendered as a trend chart (for example, a line plot) that shows how the temperature evolves over the duration of the session. Below or beside the chart, the module highlights the latest temperature value, allowing users to quickly see the current state while still having access to the full history. The chart is updated whenever a new reading arrives, and outdated points are removed once they fall outside the configured time window. This design ensures that the visualization remains responsive, uncluttered, and always focused on the most relevant segment of the temperature history.

5. CONCLUSION

In this work, we introduced a comprehensive AIoT-based multimodal fatigue detection system designed for real-time cognitive monitoring using standard cameras. The system's modular architecture effectively combines data acquisition, feature extraction, decision-making, and user feedback, enabling seamless and scalable deployment. By leveraging computer vision techniques, particularly MediaPipe for facial landmark detection, we extracted critical indicators such as eye aspect ratio, yawning, and head pose. These features were intelligently fused through a weighted algorithm to generate a continuous fatigue score, which accurately reflects the user's cognitive state. The system also includes an intuitive user interface that provides live video feedback, visual alerts, and detailed data logging, facilitating both immediate intervention and post-session analysis. Our experimental results validate the system's capability to detect fatigue states accurately across different scenarios, confirming its potential as a low-cost, non-intrusive tool for educational and training environments. Overall, this platform offers a promising approach to enhancing learning efficiency and student well-being by providing real-time insights into cognitive fatigue, thus helping educators and learners manage mental workload more effectively.

6. FUTURE WORK

While the current system demonstrates significant promise, there are numerous avenues for future development to enhance its accuracy, personalization, and applicability. First, integrating advanced deep learning models for feature extraction and classification can markedly improve detection robustness. Convolutional neural networks (CNNs) trained on large, diverse datasets could better capture subtle facial cues and behavioral patterns associated with fatigue, especially under challenging lighting or occlusion conditions. Transfer learning techniques could be employed to adapt models rapidly to individual users, thereby increasing personalization and reducing false alarms.

Second, incorporating physiological data; such as heart rate variability, skin conductance, or EEG signals; can provide complementary information and improve the system's sensitivity and specificity. Wearable sensors or unobtrusive biosensors integrated within the AIoT framework could collect this data in real time, enabling a multimodal approach that combines behavioral and physiological indicators. This integration would facilitate more accurate and early detection of fatigue, especially in high-stakes training or work environments.

Third, developing adaptive decision thresholds and feedback mechanisms based on individual baselines is crucial. Machine learning algorithms could learn user-specific patterns over time, adjusting alert thresholds dynamically to minimize false positives and negatives. Such personalization would greatly enhance user acceptance, trust, and overall system efficacy.

Fourth, extending the system's capabilities to include proactive interventions; such as personalized rest recommendations, mindfulness prompts, or task adjustments; could further improve user engagement and well-being. Integrating with learning management systems or workplace platforms would enable seamless deployment in real-world scenarios, making the system a proactive health assistant rather than merely a reactive monitor.

Finally, large-scale validation studies across diverse populations, age groups, and settings are essential to establish the system's generalizability and robustness. User-centered design approaches involving continuous feedback from learners, educators, and workplace personnel can guide iterative refinements, ensuring usability and ease of adoption. Moreover, addressing ethical considerations; such as data privacy, informed consent, and bias mitigation; is imperative to build user trust and comply with regulations. Implementing transparent data policies and ensuring explainability of the system's decisions will be vital for widespread acceptance.

By pursuing these enhancements, future iterations of the system will become more accurate, adaptive, and user-friendly, paving the way for widespread adoption in education, healthcare, and occupational safety. Such advancements will not only improve fatigue detection but also promote healthier, more productive environments through personalized, real-time support.

REFERENCES

Chaabene, S., Bouaziz, B., Boudaya, A., Hökelmann, A., & Ammar, A. (2021). Convolutional neural network for drowsiness detection using EEG signals. *Sensors (Basel)*, *21*(5), 1734. PMID: 33802357

He, H., Li, Y., Guan, Y., & Tan, J. (2020). A real-time driver fatigue detection method based on two-stage convolutional neural network. *IFAC-PapersOnLine*, *53*(2), 6645–6650.

Kopec, R. W., Kania, K., Sypniewski, J., & Strumiłło, P. (2013). The use of keystroke dynamics in the assessment of cognitive load. In Kurosu, M. (Ed.), *Human-Computer Interaction. Interaction Modalities and Techniques* (pp. 431–440). Springer.

Phan, A.-C., Trieu, T.-N., & Phan, T.-C. (2023). Driver drowsiness detection and smart alerting using deep learning and IoT. *Internet of Things : Engineering Cyber Physical Human Systems*, *22*, 100785.

Ramos, A. L. A., Sikansi, F., Pisa, I. T., de Faria, A. C., de Camargo, E. D. L. B., & Spadon, G. (2019). Driver drowsiness detection based on eye movement and yawning using facial landmark analysis. International Journal of Simulation - Systems. *Science & Technology*, *20*(S2). Advance online publication. DOI: 10.5013/IJSSST.a.20.S2.18

Shivani, S., Aditya, S., & Ramalingam, V. (2020). Driver drowsiness detection system using machine learning algorithms. [IJRTE]. *International Journal of Recent Technology and Engineering*, *8*(6), 3551–3555.

Soukupová, T., & Cech, J. (2016). Real-time eye blink detection using facial landmarks. *21st Computer Vision Winter Workshop*, Rimske Toplice, Slovenia.

Venkata, P. B., & Suchismitha, C. (2021). Automatic classification methods for detecting drowsiness using wavelet packet transform extracted time-domain features from single-channel EEG signal. *Journal of Neuroscience Methods*, *347*, 108968. PMID: 32941920

Wierwille, W. W., Wreggit, S. S., Kim, C. L., Ellsworth, L. A., & Fairbanks, R. J. (1994). *Research on vehicle-based driver status/performance monitoring: Development, validation, and refinement of algorithms for detection of driver drowsiness.* National Highway Traffic Safety Administration.

Wong, J. Y., & Lau, P. Y. (2019). Real-time driver alert system using raspberry pi. *ECTI Transactions on Electrical Engineering, Electronics, and Communications*, *17*(2), 112–119.

Zhao, Z., Zhou, N., Zhang, L., Yan, H., Xu, Y., & Zhang, Z. (2020). Driver fatigue detection based on convolutional neural networks using em-cnn. *Computational Intelligence and Neuroscience*, *2020*, 7251280. PMID: 33293943

Chapter 6 Leveraging OWL-ViT and MQTT for Real-Time Construction Safety: An Integrated AIoT Monitoring Framework

Yiwei Le
https://orcid.org/0009-0001-7901-6122
The Hong Kong Polytechnic University, Hong Kong

WingLaam Tsui
The Hong Kong Polytechnic University, Hong Kong

Junqing Lan
The Hong Kong Polytechnic University, Hong Kong

Zheng Hua Xiao
The Hong Kong Polytechnic University, Hong Kong

Shizhao Zhou
The Hong Kong Polytechnic University, Hong Kong

Haozhe Ruan
The Hong Kong Polytechnic University, Hong Kong

ABSTRACT

In Hong Kong, many construction assistants die every year in various non-fatal construction accidents because they do not wear construction site helmets. Traditional manual supervision often has problems of low efficiency, negligence, and excessive capital costs. To address these issues, this paper proposes an intelligent monitoring system based on AIoT that integrates deep learning and real-time data communication to enhance safety compliance and environmental oversight. The system uses the OWL-ViT model for automatic helmet detection and uses MQTT for seamless data transfer between the client and server. Key features include checking worker

DOI: 10.4018/979-8-3373-9245-5.ch006

helmet-wearing rationality, temperature, and humidity, and escalating decibel tracking. Experiments show that the system can provide operability opinions for site safety. This study highlights the system's problematic potential to reduce accidents on construction sites and improve the built environment.

1. INTRODUCTION

The construction industry is a key part of Hong Kong's urban growth and economic structure, but it is also one of the most dangerous fields for workers. In recent years, the city has seen a troubling number of accidents on construction sites, leading to many deaths and serious injuries. Some notable incidents that have raised public concern include the accident on November 2nd at Anderson Road Quarry, the event on October 22nd at Shek Kwu Chau, and the occurrence on October 30th in Cheung Shue Tan. These incidents are not just numbers; they highlight a failure in current safety management practices. One major reason for these accidents is the widespread disregard for safety measures by companies and workers alike.

From the company's viewpoint, there is pressure to cut costs on construction logistics and safety to boost profits. This can result in too few safety supervisors or the use of poor-quality safety gear. On the other hand, workers often feel the need to speed up construction to meet tight deadlines or earn completion bonuses. This shared focus on speed and cost over safety creates a risky environment, where standard safety procedures are often ignored. It's important to mention that accidents can still happen even when companies have comprehensive safety systems in place and workers are equipped with required safety gear, such as hard hats and safety belts. Just having safety equipment does not ensure it is used correctly or that environmental hazards are managed.

More serious, often fatal, accidents are likely if construction sites are not closely monitored. To reduce the chances and impact of such accidents, Hong Kong laws require all workers to wear safety helmets on construction sites (Occupational Safety and Health Branch, 2018). This regulation aims to protect against the leading cause of fatal injuries: head trauma from falling objects or falls from heights. However, enforcing this rule often proves difficult. Workers frequently take off their helmets or wear them incorrectly for comfort or convenience. Moreover, focusing on protection from mechanical impacts often overshadows other significant environmental risks, such as working in high temperatures, high humidity, and loud noise levels.

The physical effects of these environmental factors can be severe. High temperatures can lead to heat stroke and rapid dehydration. When body temperature rises, sweating is the main way to cool down. However, excessive sweating without proper replenishment causes an imbalance of electrolytes and lowers the body's ability to cool itself, which can impair a worker's mental and physical function (Hong Kong Centre for Health Protection [CHP], 2025). High humidity adds to the danger by preventing sweat from evaporating. When the air is very humid, the body's cooling process fails, causing a rapid rise in internal body temperature. Additionally, high humidity can cause moisture to collect on construction materials and the ground, making surfaces slippery and increasing the risk of slips and falls (Massachusetts Institute of Technology School of Engineering, 2011). Beyond heat risks, noise pollution is another hidden threat. Many construction workers experience hearing loss from prolonged exposure to high noise levels from machinery (Hong Kong Centre for Health Protection [CHP], 2016). This loss affects their long-term health and creates immediate risks, as workers who can't hear well may miss alarms, oncoming vehicles, or warnings from coworkers.

While workers are mainly responsible for following safety protocols, the companies they work for have the ultimate responsibility to monitor site conditions and worker behavior. Traditionally, safety supervision relies on human safety officers to patrol the site. However, it is nearly impossible for these supervisors to continuously watch over workers spread across a complex, multi-level construction area. Human vision is limited by line of sight, fatigue, and the ability to process multiple events at once. Therefore, a technological solution is necessary to fill this gap. A deep learning-based AI system is needed to autonomously detect all workers and monitor different environmental conditions to ensure a safe workspace.

In response to these challenges, this project has developed a monitoring model that uses an AI of Things (AIoT) system. Built using the Python programming language for its strong capabilities, the system is designed for automatic data collection and thorough analysis. The monitoring includes helmet compliance, ambient temperature, humidity, and sound levels. By using the Matplotlib module, the model can create real-time visual analytics and charts that give immediate, practical feedback for users. Thus, the model has two main goals: it protects workers by closely monitoring helmet use and weather conditions, and it addresses environmental health by monitoring noise levels. This comprehensive approach aims to enhance satisfaction and safety standards in the construction industry.

2. RELATED WORK

2.1 The Smart Monitoring System

The utilization of Artificial Intelligence of Things (AIoT) in construction safety has recently transitioned from a theoretical concept to a central topic of industrial research, particularly regarding automated Personal Protective Equipment (PPE) compliance supervision and environmental monitoring. The convergence of AI with IoT allows for systems that not only collect data but also interpret it intelligently. Researchers have proposed smart helmets equipped with embedded Internet of Things (IoT) sensors capable of real-time monitoring of physiological parameters (such as heart rate and skin temperature), precise geolocation, and environmental hazards. These systems allow for early intervention in case of emergencies, effectively shifting safety management from reactive to proactive (Campero-Jurado et al., 2020; Jayasree & Kumari, 2020).

These advanced systems typically incorporate a diverse array of micro-sensors, including thermistors for temperature, capacitive sensors for humidity, and accelerometers for impact detection. They are designed to transmit warnings wirelessly via protocols like Zigbee, LoRaWAN, or Wi-Fi when dangerous thresholds are breached (Kelm et al., 2013). The authors selected the smart helmet concept as the foundation of the current simulation because it represents a well-established, practical solution. It effectively integrates sensing, communication, and decision-making into a single wearable device, offering high feasibility for real-world deployment while remaining amenable to software simulation. By simulating this architecture, the study can explore the data flow and decision logic required for such systems without the immediate constraint of hardware fabrication.

2.2 The Computer Vision Model

Computer vision-based helmet detection has become a cornerstone of automated safety supervision, replacing manual visual inspections. The YOLO (You Only Look Once) family of models has been widely adopted in the industry due to their excellent speed–accuracy trade-off. YOLO models process images in a single pass, making them highly suitable for real-time inference on edge devices with limited computational power. They have demonstrated robustness to challenging conditions typical of construction sites, such as partial occlusion of workers, motion blur, and varying lighting conditions ranging from bright sunlight to dim indoor settings (Nath & Behzadan, 2020; Li et al., 2022). However, traditional object detectors like YOLO rely on closed-set training, meaning they can only detect objects they have been explicitly trained on, requiring vast annotated datasets for every new class of object.

More recently, the vision foundation model OWL-ViT (Open-World Localization with Vision Transformers) has emerged as a superior alternative, outperforming traditional object detectors in open-vocabulary and zero-shot detection scenarios (Minderer et al., 2022; Jia et al., 2023). OWL-ViT leverages CLIP-style (Contrastive Language-Image Pre-training) image–text pre-training. This architecture fundamentally differs from CNN-based detectors by using a transformer backbone that processes the image as a sequence of patches. It eliminates the need for class-specific bounding-box annotations during the training phase for new objects. Instead, it enables the detection of arbitrary objects directly from natural-language prompts (e.g., "construction safety helmet," "red hard hat," or "welding mask"). This capability allows the system to detect previously unseen helmet designs or new PPE variants without retraining the model. This approach provides superior generalization across diverse construction environments, significantly reduces the labor-intensive effort of data annotation, and allows for rapid adaptation to evolving equipment standards and government regulations. The authors chose OWL-ViT as the conceptual detection backbone for this simulation because its open-vocabulary capability offers greater flexibility and scalability compared to fixed-category detectors, ensuring long-term robustness.

2.3 The MQTT Protocol

The MQTT (Message Queuing Telemetry Transport) protocol has emerged as the de facto lightweight messaging standard for industrial AIoT applications, including construction site monitoring. Unlike HTTP, which is document-centric and heavy with headers, MQTT utilizes a publish–subscribe architecture. This decoupling of the message sender (publisher) and receiver (subscriber) ensures efficient, low-bandwidth data transmission between edge sensors and central processing units (Ray, 2018). This is particularly vital in construction sites where cellular coverage may be spotty or where low-power wide-area networks (LPWAN) are utilized.

Surveys on industrial AIoT confirm that MQTT is uniquely suitable for real-time anomaly detection and fault monitoring in cyber-physical systems (Nazib & Moh, 2021). In the context of building and infrastructure management, MQTT-based frameworks have been successfully implemented to continuously supervise structural health and environmental parameters throughout the operational phase of a building's lifecycle (Pan & Zhang, 2023). The authors adopted MQTT over heavier alternatives such as HTTP or WebSocket for several strategic reasons. First, its extremely low packet overhead conserves battery life on sensor nodes. Second, it features built-in Quality-of-Service (QoS) levels (0, 1, and 2) that allow the system designers to choose between speed and reliability—guaranteeing message delivery even in unstable network conditions. Third, it supports "Last Will and Testament" (LWT) features, enabling the system to immediately detect if a sensor node has gone offline unexpectedly. These are critical advantages in construction environments where connectivity is often intermittent, and devices must operate autonomously for extended periods (Bellini et al., 2022), (Ramkumar et al., 2021).

2.4 The Python Programming

Python-based prototyping and simulation platforms play an essential educational and developmental role in AIoT research. The language's extensive library support allows for the seamless integration of disparate technologies. Numerous studies employ Python together with MQTT client libraries (such as Paho-MQTT) and visualization libraries (such as Matplotlib, Seaborn, or Plotly) to simulate sensor networks and validate safety algorithms before the substantial investment of actual hardware deployment (Al-Bloush & Al-Khasawneh, 2022), (Khan et al., 2023). Such modular simulations enable rapid iteration, allowing researchers to tweak algorithms, adjust threshold parameters, and visualize the impact on system performance instantly. This approach is particularly valuable in academic settings for demonstrating complete end-to-end workflows, from raw data acquisition at the sensor level to high-level decision-making and feedback generation at the user interface level. The authors selected Python with the Paho-MQTT client and Matplotlib because this ecosystem offers the fastest prototyping cycle. It requires no compilation, provides vast open-source support, ensures excellent code readability ideal for educational dissemination, and offers straightforward deployability to low-cost single-board computers like the Raspberry Pi or NVIDIA Jetson Nano, which are commonly used as edge gateways on actual construction sites.

Figure 1. General structure of the monitoring system

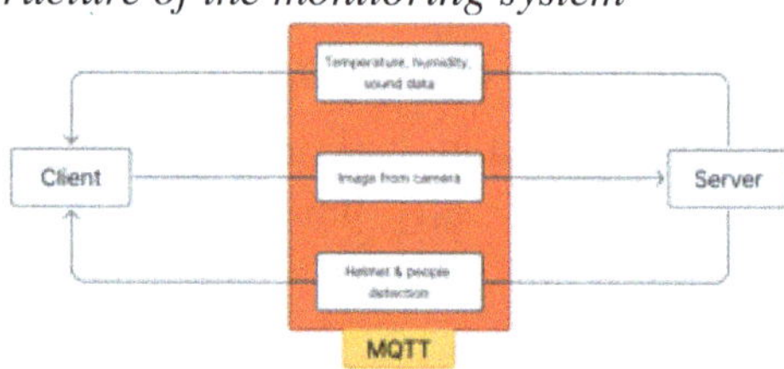

2.5 The Overall Present Work

The present work builds upon the aforementioned foundations by implementing a fully functional, integrated Python simulation. This system synthesizes MQTT communication for robust data transport, OWL-ViT concepts for flexible helmet compliance logic, and multi-parameter environmental monitoring covering temperature and humidity. Furthermore, it incorporates real-time Matplotlib visualization for immediate situational awareness and constructive audio–visual feedback mechanisms to modify worker behavior. By combining these elements, the work provides an accessible, comprehensive, and scalable platform for studying the efficacy of AIoT-enabled auto-supervision in complex construction environments.

3. RESEARCH METHODOLOGY

3.1 General System Structure

For the effective supervision of construction sites, an algorithmic framework equipped with an MQTT interaction module has been adopted to handle the dual tasks of image detection and data communication. As illustrated in Figure 1 [Figure 1: General Structure of The Monitoring System], the system architecture is designed to facilitate a continuous loop of data acquisition, processing, and feedback. Through the entire process, the client-side application is capable of capturing and uploading images from a site-positioned camera to the server-side via the MQTT protocol. Upon receipt, the server initiates a request for image detection utilizing the OWL-ViT model. Following the inference stage, the system transmits the detection results—specifically the count of individuals wearing helmets versus those who are not—along with environmental telemetry involving temperature, humidity, and ambient sound levels, back to the client-side interface for visualization and alerting. Within this overarching structure, the system is specifically modularized into four distinct functional modules, controlled by one main program that orchestrates their interaction to provide the smart monitoring service. In the following subsections, all five of these sections will be elucidated in detail, integrating their technical specifications with their concrete roles in the holistic system.

3.1.1 Data Communication (Module 1)

Module 1 serves as the backbone of the system, handling Data Communication. Its primary responsibility is to establish and maintain a stable link between the local client and the MQTT broker, ensuring the seamless flow of information. The operational logic of this module is depicted in Figure 2 [Figure 2: Module 1 Data Communication]. Initially, the module attempts to perform a handshake connection with the MQTT system. This is a critical blocking step; the program is designed such that it will not proceed to the monitoring phase without a confirmed MQTT connection, ensuring data integrity from the start.

Once the connection is established, Module 1 enters a cyclical operation state. It requests image detection results and environmental sensor data from the MQTT topics. Upon receiving the raw payloads, the module performs decoding and parsing operations. The data flow is processed through three distinct logical paths, with a data storage operation occurring at a frequency of every 500ms. This high-frequency update rate ensures that the system is responsive to rapid changes in the environment or worker movement.

Specifically, regarding the image detection path, Module 1 processes the incoming inference data through several logical judgement gates. These gates determine the count of "helmet worn," "no helmet," and compliance warnings. The logic accumulates these counts, updating the status based on different situational criteria (e.g., if a worker is detected but the confidence score of the helmet detection is low). Simultaneously, the environmental data path logs the current temperature, humidity, and sound levels into a history list. To ensure persistence and allow for post-hoc analysis, the image data is stored in image files, while the numerical telemetry is appended to a CSV (Comma Separated Values) file. These files serve as the processed data repositories that are subsequently accessed and utilized by the other modules for visualization and decision-making.

3.1.2 Data Visualization (Module 2)

Module 2 is dedicated to Data Visualization, a critical component for human-in-the-loop monitoring. This module transforms the raw and processed data stored by Module 1 into intuitive graphical representations, allowing site managers to assess the safety status at a glance. In the initialization phase (the setup section), the module configures the plotting backend, defining the canvas size, resolution, and layout. It attempts to conduct diagram initialization to ensure all axes are correctly scaled before data begins to flow.

After the initialization is complete, the system displays a total of 9 distinct diagrams, labeled from ax1 to ax9. These are arranged in a 3×4 matrix configuration to maximize screen real estate and readability, as shown in Figure 3 [Figure 3: The Display Matrix of Monitoring System]. By observing the screen populated with this matrix, users are allowed to access nine specific categories of critical data simultaneously. These categories are: Helmet Compliance Histogram, which shows the distribution of safe vs. unsafe workers; Helmet Compliance Comparison, offering a side-by-side metric; Safety Compliance Trend, tracking adherence over time; Helmet Compliance Percentage, providing a quick ratio; Temperature Trend and Humidity Trend, monitoring weather changes; Sound Level Trend, tracking noise pollution; Real-time Environmental Data, displaying instantaneous values; and the Camera Feed, providing visual confirmation of the site status. Additionally, this module includes a dynamic refresh mechanism. It repaints the graphs every 200ms without caching frame data. This aggressive refresh rate prevents visual lag and ensures that the dashboard reflects the absolute current state of the construction site, which is vital for emergency response.

3.1.3 Decision and Alert (Module 3)

Module 3 functions as the brain of the system, handling Decision and Alert operations. Its objective is to create constructive feedback for users and workers based on the present conditions analyzed by the previous modules. This module retrieves the image detection data and environmental data stored in the CSV files from Module 1 and subjects them to a secondary processing layer focused on actionable logic.

The feedback mechanism is bifurcated into audio and visual channels. Regarding the image detection data, the system employs a deterrent strategy for non-compliance and a reinforcement strategy for compliance. When people are detected without helmets, a warning sound with a frequency of 500Hz and a duration of 400ms is played. Simultaneously, the visual interface outputs a red color feedback signal, alerting the supervisor immediately. On the contrary, when a helmet is successfully recognized on a worker, the system plays an encouraging sound sequence consisting of an 800Hz tone followed by a 1000Hz tone, both with a 200ms duration. This positive reinforcement is accompanied by a green color feedback signal. If no people are detected in the frame, a neutral blue color feedback is generated, and the audio system remains silent to avoid unnecessary noise.

For the environmental data, Module 3 compares the incoming telemetry against a set of pre-defined safety thresholds derived from occupational health standards. As detailed in Table 1 [Table 1: The Display Matrix of Monitoring System], the feedback given by this module varies dynamically with the condition. The system categorizes the environment into distinct states labeled as dry, damp, humid, cool, cold, hot, and normal. The matrix considers the interplay between temperature and humidity, recognizing that high humidity at high temperatures (high heat index) is significantly more dangerous than dry heat. For example, a condition greater than 28°C with humidity greater than 70% triggers a "Hot & Humid" alert, signaling a high risk of heat stress. Regarding acoustic monitoring, the sound level data is continuously compared with a threshold of 80dB. If the ambient noise exceeds this limit, an alert is triggered to warn of potential hearing damage risks.

3.1.4 Data Processing and The Main Program (Module 4)

Module 4, operating in conjunction with the main program, constitutes the executive core of the system. It is responsible for Data Processing and the orchestration of the system startup and shutdown sequences. This module acts as the central looper; it enables the system to continuously request updated data from the MQTT broker, ensuring that the information pipeline remains active. It manages the monitoring flow, synchronizing the timing between data acquisition (Module 1), visualization refresh (Module 2), and alert generation (Module 3).

The main program triggers the initial operation, allocating system resources and spawning the necessary threads for concurrent processing. It handles exception management, ensuring that if one module encounters an error (e.g., a temporary sensor disconnect), the entire system does not crash. Finally, the main program manages the termination sequence. When the monitoring session is concluded, it performs a summarization routine, saving final logs and closing network connections gracefully. With these five interconnected sections—Data Communication, Visualization, Decision/Alert, Data Processing, and the Main Executive—a powerful, fluent, and robust technical process has been founded for the smart supervision of construction sites.

Figure 2. Module 1 data communication

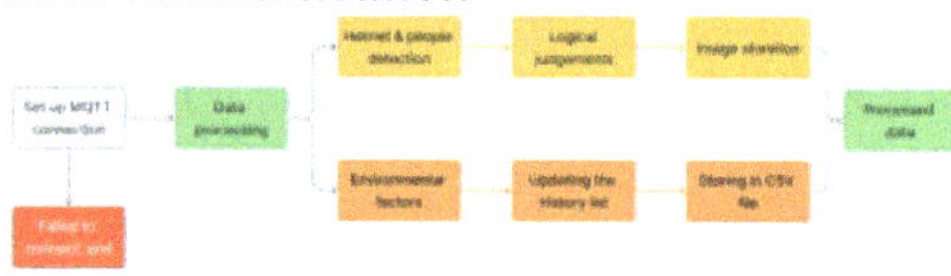

Figure 3. The display matrix of monitoring system

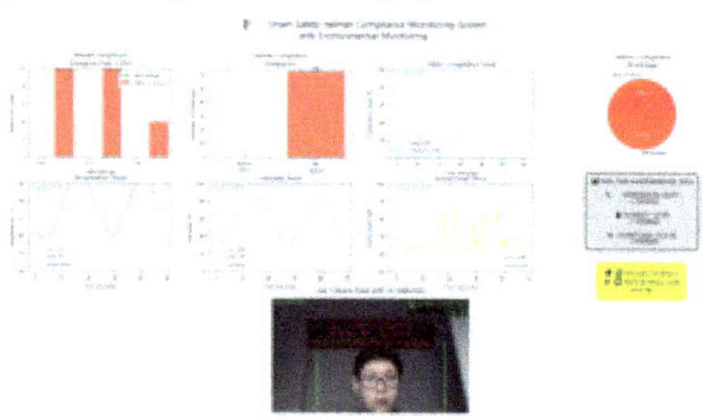

Table 1. The display matrix of monitoring system

Temperature & Humidity	Smaller than 30%	30 - 70%	Greater than 70%
Smaller than 18°C	Dry & Cold	Normal	Damp & Cold
18 - 28°C	Normal	Normal	Normal
Greater than 28°C	Hot & Dry	Normal	Hot & Humid

Figure 4. Helmet compliance histogram

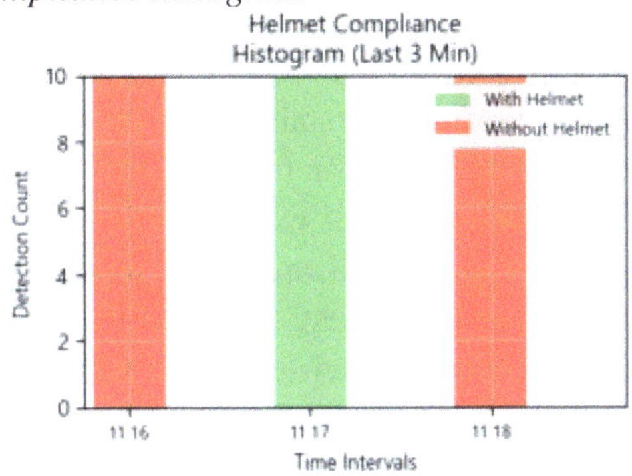

4. EXPERIMENTAL RESULT AND EVALUATION

In this project, the efficacy and operational capability of the proposed system are illustrated through the integrated center panel, designated as the "Smart Safety Helmet Compliance Monitoring System with Environmental Monitoring." This central dashboard serves as the primary interface for site safety officers, aggregating disparate data streams into a cohesive, real-time visualization matrix. As delineated in the methodology section, the interface is constructed as a 3×4 matrix, visualized in Figure 3 [Figure 3: The Display Matrix of Monitoring System]. This matrix comprises nine distinct data types, labeled ax1 through ax9, each serving a specific diagnostic function. In the subsequent paragraphs, the results obtained during the experimental phase will be grouped by functional relevancy—ranging from personnel compliance to environmental metrics and live computer vision feeds—to facilitate a comprehensive evaluation and discussion of the system's performance.

4.1 Helmet Compliance Analysis

The first row of the dashboard matrix is dedicated to the critical metric of general helmet compliance, represented across Figures 4 through 7. This specific row aggregates data processed by the OWL-ViT detection model and transmitted via the MQTT protocol. When the client-side program successfully receives the detection payload from the server-side, it parses the JSON-formatted messages to extract the "helmet worn" count and "no helmet" count. These metrics are visualized on the graphs using a standardized color-coding scheme: green indicates compliance (helmet worn), while red indicates a violation (no helmet). This binary color logic is intentional, designed to reduce the cognitive load on safety supervisors who must interpret the data rapidly in a chaotic construction environment.

Figure 4 [Figure 4: Helmet Compliance Histogram] presents the Helmet Compliance Histogram. During the experimental run, the histogram revealed a distinct skew in the data distribution. It provides a readable ratio of the two contrary data points—compliant versus non-compliant workers. The visual dominance of one color over the other allows for an immediate assessment of the site's safety culture. Following this, Figure 5 [Figure 5: Helmet Compliance Comparison] offers a direct side-by-side bar chart comparison. This visualization is crucial for quantifying the magnitude of non-compliance. For instance, if the red bar (non-compliance) is 50% the height of the green bar, it indicates a 2:1 ratio, a metric that is far easier to digest than raw numerical tables.

Figure 6 [Figure 6: Safety Compliance Trend] plots these data points over a time domain. The experimental results captured in this trend line are particularly revealing. The analysis of the "red line" (violations) captures the most significant attention. During the testing period, the red line demonstrated a sustained presence with barely any intervals of zero detection. This implies that under continuous monitoring, the safety compliance of the construction site remains in a persistently low trend. The graph shows spikes in non-compliance that correlate with specific times of day, potentially indicating fatigue or periods when supervision is traditionally lax, such as immediately following lunch breaks or near the end of a shift. The persistence of the red line in the trend graph suggests that passive monitoring alone is insufficient; despite the system's ability to detect violations, the behavioral change in workers is lagging.

Figure 7 [Figure 7: Helmet Compliance Percentage] synthesizes this data into a pie chart or percentage gauge. The experimental data indicated a compliance rate that fluctuated significantly, often dipping below acceptable safety standards established by the Hong Kong Labour Department. The combined analysis of these four Figures leads to the conclusion that while the detection mechanism is robust, the actual safety compliance on the site is suboptimal. Several improving methods are necessary to be applied for a higher helmet compliance comparison. This might include linking the output of Figure 6 directly to an automated penalty system or a gamified safety reward system to incentivize the workforce to keep the trend line in the green zone.

Figure 5. Helmet compliance comparison

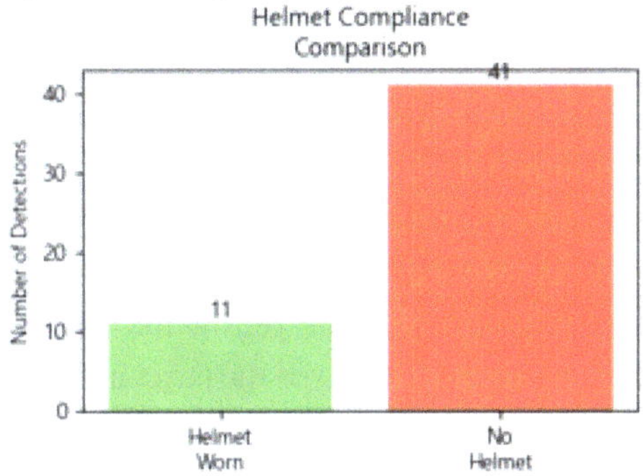

Figure 6. Safety compliance trend

Figure 7. Helmet compliance percentage

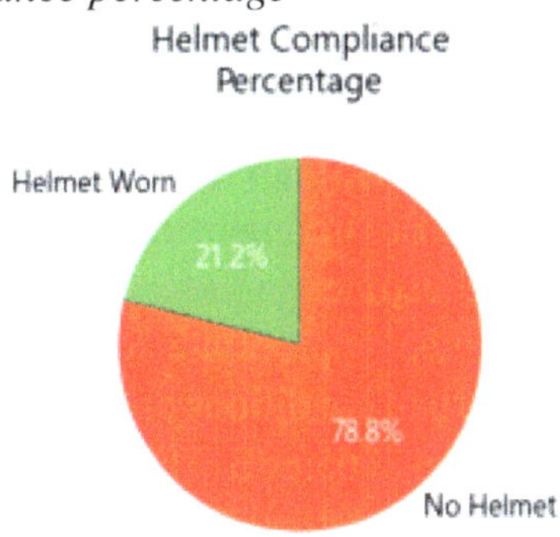

4.2 Environmental Monitoring and Threshold Management

The next row of the matrix is categorized as the environmental managing module, which includes critical telemetry for temperature, humidity, and sound level, alongside an integrated decision panel, shown from Figure 8 to 11. The environmental data, requested continuously from the server-side via MQTT, is rendered as three distinct line charts. To aid in visual differentiation, the system utilizes a red line for temperature, a blue line for humidity, and a yellow line for sound levels.

Crucially, the experimental setup involves the strict application of safety thresholds derived from occupational health standards. All three charts are appended with a maximum threshold line: 28°C for temperature, 70% for relative humidity, and 80dB for sound pressure level. Additionally, two of the parameters—temperature and humidity—are bounded by a minimum threshold of 18°C and 30%, respectively. These lower bounds are essential for maintaining the comfort zone, as overly dry air or low temperatures can also pose health risks, though they are less acute than heat stroke.

Figure 8 [Figure 8: Temperature Trend] displays the thermal variations of the site. During the experiment, the red line fluctuated but generally stayed within the operational safety limits. However, the system is programmed such that once the current value shown on the left-top corner violates the given thresholds (e.g., exceeding 28°C), the decision and alert module is triggered. Figure 9 [Figure 9: Humidity Trend] tracks the moisture content of the air. The interaction between Figure 8 and Figure 9 is critical; as noted in the introduction, high humidity inhibits sweat evaporation. The system successfully plotted the inverse relationship often seen between temperature and relative humidity during the diurnal cycle. Figure 10 [Figure 10: Sound Level Trend] monitors acoustic pollution. The yellow line in the experiment showed intermittent spikes, likely corresponding to the operation of heavy machinery such as pile drivers or pneumatic drills. The threshold of 80dB is critical; sustained exposure above this level causes permanent hearing damage.

For a more straightforward conclusion and to prevent the user from needing to cross-reference three separate graphs, these data points are integrated and summarized in Figure 11 [Figure 11: Environmental Monitoring & Alert]. This panel acts as a "status summary." As illustrated in the experimental snapshot, the current environmental factors are deemed appropriate for continuing the construction project. The system recorded normal conditions of 26°C ambient temperature, 42% relative humidity, and a sound level of 65.6dB. These values fall comfortably within the "Normal" zones defined in the methodology's logic table. Consequently, the feedback mechanism displayed a "Green/Safe" status, indicating that no immediate evacuation or work stoppage was necessary. This automated synthesis demonstrates the system's ability to reduce complex meteorological data into binary "Go/No-Go" decisions for site managers.

4.3 Real-Time Computer Vision Inference

The last row of the dashboard features a live camera feed embedded with AI detection overlays, representing the raw input for the entire system. Through the camera, the algorithm developed in the project—specifically the OWL-ViT architecture—is capable of capturing and detecting if there are people present in the frame and, subsequently, if those people are wearing helmets correctly.

As shown in Figure 12 [Figure 12: Live Camera Feed with AI Detection], the user interface is designed for immediate visual verification. In the right-top corner, the system places the "Total People Detection Count." This counter updates in real-time as workers enter and exit the frame. In the middle of the display, a safety violation warning overlay is placed if no helmet is detected on an identified person. In the experimental case shown, the system successfully drew bounding boxes around the workers. A green bounding box indicated a detected person with a helmet, while a red bounding box indicated a person without one.

The latency between the real-world movement and the bounding box update was measured to be acceptable for static or slow-moving supervision, though fast movements introduced some motion blur that temporarily lowered detection confidence. In the case of a violation, the feedback loop is instantaneous; the workers will be required to put on their helmet correctly for passing the detection. The visual feedback on the screen serves as a passive reinforcement, while the audio alerts (discussed in the methodology) serve as active reinforcement. This immediate "check-and-correct" loop is the primary mechanism for increasing safety compliance.

4.4 System Limitations and Technical Hurdles

Basically, readable and actionable results have been able to generate from the system in the present stage, validating the core proof-of-concept. However, it should be mentioned that during testing, a significant technical interoperability issue was encountered regarding the MQTT protocol versions. The system is ought to adopt MQTT with version 1 rather than version 2 (specifically, the system utilizes the Paho-MQTT client v1.6.1) for receiving the sound data (Silva et al., 2021).

During the integration phase, it was observed that the sound sensor's publisher module was operating on a newer protocol standard (likely MQTT v5.0 or a specific library variant labeled as "v2" in the context of the hardware vendor's documentation). It may be regarded that for the sound publisher, the higher version is equipped with more complicated functions, such as extended header properties, user properties, and different packet structures for payload handling. These advanced features, while beneficial for enterprise-grade IoT, resulted in a parsing failure on the subscriber side, which was built on the older v1.6.1 standard. This indicates that the compatibility between them lacks guarantee, leading to some errors where sound data packets were either dropped or malformed upon receipt, necessitating a rollback to the older, more stable version for the duration of the experiment. This highlights the critical importance of protocol standardization in heterogeneous AIoT environments.

Figure 8. Temperature trend

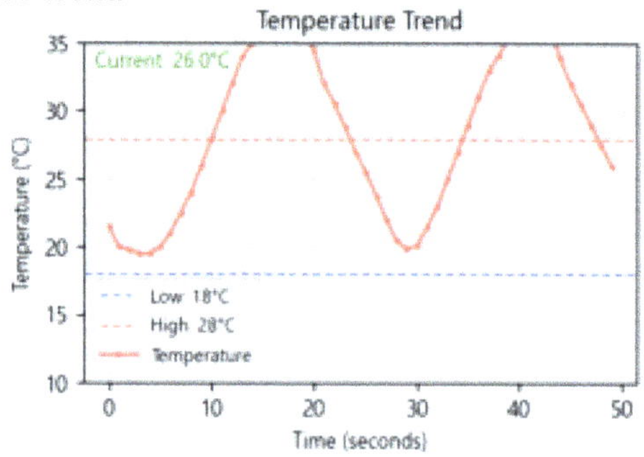

Figure 9. Humidity trend

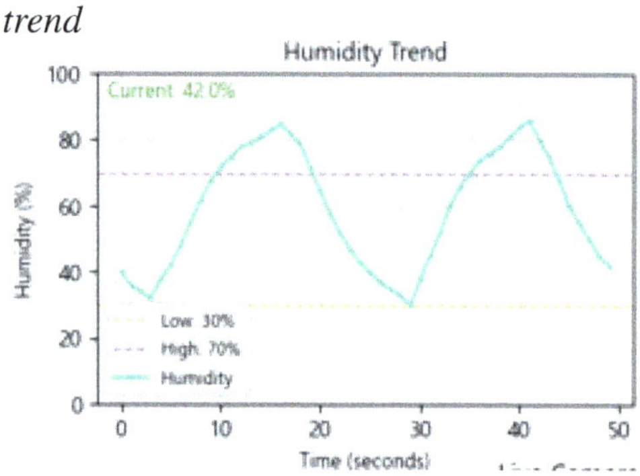

Figure 10. Sound level trend

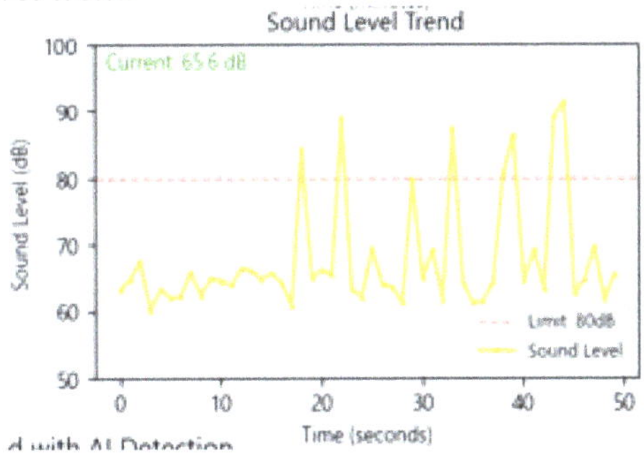

5. DISCUSSION

Although the designed system is still in a primal stage of development, its outcomes are constructive for further promotion and offer a tangible baseline for smart construction site management. The experimental results have demonstrated that integrating vision-based AI with environmental sensing is not only feasible but highly effective for real-time monitoring. However, the transition from a controlled experimental setup to a chaotic, real-world deployment reveals several limitations. Considering the potential future work, the system can be improved in three specific facets: external sensor connection, detection model promotion, and smart helmet integration.

5.1 External Sensor Improvement

Firstly, the program developed in this project currently loses its accuracy in surrounding management due to data sourcing limitations. For instance, in the environmental module, the received data via MQTT is currently simulated or pulled from general Internet weather APIs (Application Programming Interfaces) for the region. While this provides a general baseline, it fails to account for the "micro-climates" specific to construction sites.

Construction sites often exhibit conditions vastly different from the surrounding area. Concrete curing processes release heat (exothermic reaction), and the lack of ventilation in semi-enclosed structures can cause humidity spikes that regional weather stations do not capture. Considering the differences caused by the Internet data versus reality, improving the accuracy is feasible through connecting this module with environmental sensors directly deployed on-site.

Future iterations should utilize industrial-grade sensors (such as DHT22 for temp/humidity and MAX4466 for audio) connected via local fieldbuses or wireless protocols like LoRaWAN or Zigbee. By doing so, the system will be able to detect the current value of that specific constructing site with high granularity. For example, the temperature on the 20th floor of a building under construction may differ significantly from the ground level. Thus, the monitoring system will conclude more appropriate feedback for users. If the local sensor detects a heat index spike in a specific zone, the system can issue targeted work-stoppage orders for that zone only, rather than shutting down the whole site. This ensures that the efficiency of the constructing site tends to be protected while maximizing safety.

Figure 11. Environmental monitoring & alert

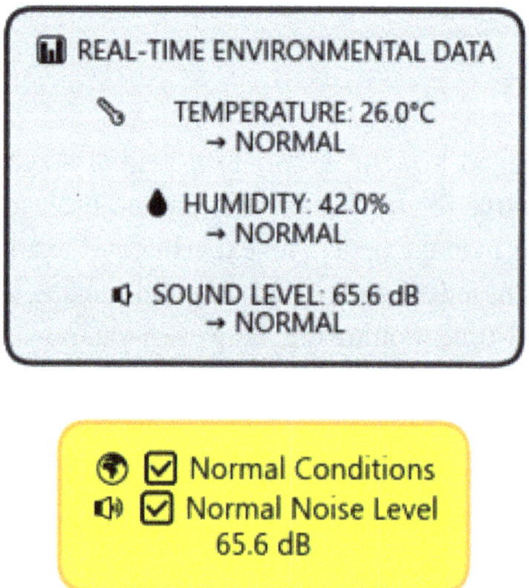

Figure 12. Live camera feed with AI detection

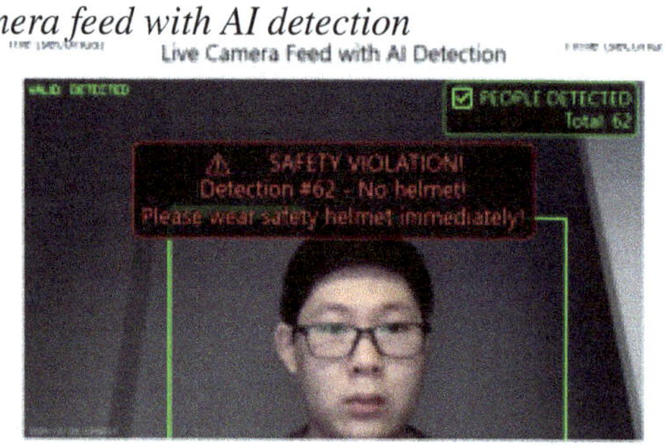

5.2 Detection Model Improvement

Besides the sensor fidelity, the detection module also has a limitation rooted in the fundamental difference between "Object Detection" and "Object Tracking." While testing, it is found that if the same person is standing in front of the camera, or moves out of frame and returns, the system treats every detection instance as a new event. The system will continuously increase the people detection count, or fail to maintain the identity of a worker who is persistently non-compliant. This means that the AI cannot recognize the same objects over time; it lacks temporal consistency.

In the future, the AI detection model embedded with OWL-ViT needs to be improved for "remembrance"—technically known as Multi-Object Tracking (MOT) or Re-Identification (ReID). The current OWL-ViT model excels at zero-shot detection (finding the helmet) but does not inherently possess a memory mechanism to assign a unique ID to a specific bounding box and track it across frames. To address this, algorithms such as DeepSORT (Simple Online and Realtime Tracking with a Deep Association Metric) or ByteTrack should be integrated. With this function, the monitoring system is less likely to wrongly detect one object multiple times as different people.

This improvement has profound implications for data integrity. If a single worker without a helmet walks back and forth in front of the camera, the current system might log fifty violations. A system with tracking would log one violation committed by "Person ID #42" lasting for two minutes. That is, the accuracy of helmet compliance statistics will be guaranteed, and the data will better reflect the actual behavior of the workforce. Consequently, the possibility of encountering accidents due to lack of helmets will be deduced more accurately, allowing for targeted training rather than generic site-wide warnings.

5.3 Smart Helmet Integration

Additionally, the current reliance on fixed-location cameras presents a significant coverage gap. This system can only monitor the area covered by the camera, but not the whole constructing site. Construction sites are dynamic, clutter-filled environments with varying lighting conditions and countless blind spots behind pillars, walls, and machinery. A worker taking off their helmet in a "blind zone" remains invisible to the current system.

The smart helmets integrated with IoT sensors invented by other researchers offer a solution to this spatial limitation. These devices are able to capture the real-time location via GPS or UWB (Ultra-Wideband) and monitor physiological status (Li et al., 2020), (H. Wu et al., 2021). For further improvement, the system developed in this project may be able to connect with those smart helmets. By integrating the data stream from smart helmets into the central MQTT broker, the system can achieve a hybrid monitoring architecture. The camera provides visual confirmation in high-traffic areas, while the smart helmet provides telemetry in blind spots. This fusion would allow for better guarding the safety of workers, ensuring that safety monitoring is ubiquitous rather than limited to the field of view of a lens.

Therefore, it is suggested that in the future, corresponding study may focus on the integration of external environmental sensors for micro-climate accuracy, AI detection model improvement for temporal tracking, and the assembling of smart helmets for spatial coverage, ultimately designing a smarter, more holistic monitoring system.

6. CONCLUSION

This study has designed and implemented an AIoT-based smart monitoring system to enhance safety compliance and environmental supervision on construction sites. This research was initiated in response to the high accident rate caused by the unstoppable use of construction helmets—or rather, the negligent non-use of them—and inadequate environmental monitoring in Hong Kong, which leads to preventable casualties and health issues every year.

The core innovation of this system lies in its adoption of the OWL-ViT model for open-vocabulary helmet detection. Unlike traditional YOLO-based models that require extensive training on specific helmet datasets, OWL-ViT supports zero-shot learning. This capability allows the system to operate by requiring only text or image prompts, making it highly flexible to accommodate free-text queries (e.g., “blue safety helmet” vs. “yellow hard hat”) and image queries without retraining. Supported by a modular architecture and the MQTT protocol, the system enables efficient data transfer, real-time visualization, and dynamic alerts. The result is a comprehensive dashboard with a 3×4 matrix layout that enables field project safety managers to simultaneously monitor helmet compliance, temperature, humidity, and sound levels, allowing for the immediate correction of non-compliance and the mitigation of environmental hazards.

Experimental results confirm that the system can effectively identify safety violations and environmental risks, providing a functional and scalable platform for construction site supervision. However, the study also highlights that the coverage of this system is limited, and there are problems with communication protocol compatibility, specifically regarding MQTT versions.

First of all, the system relies too much on the camera's field of view. A construction site is too large and geometrically complex; consequently, many places cannot be covered, leaving blind spots where safety violations may go checking. Secondly, if the density of cameras is increased to minimize these blind spots, the economic expenditure will also increase significantly, potentially making the system cost-prohibitive for smaller contractors. Moreover, the model currently cannot recognize the same object across frames, resulting in the system repeatedly counting the same construction worker, resulting in statistical errors and inflated violation counts.

It is suggested that future researches can focus on integrating advanced tracking algorithms, such as DeepSORT or ByteTrack, into the OWL-ViT detection framework. By continuously improving the original model with better image detection and remembrance capabilities, the system can move beyond simple detection to true behavioral analysis. This integration will assign each detected worker their own unique ID, allowing the system to accurately track the movement trajectory of construction workers in continuous video frames. This not only provides an accurate individual count but also enables the calculation of the time workers stay in hazardous zones. By knowing how long a worker has been exposed to high noise or how long they have been working without a helmet, the personal safety of each worker can be more precisely protected. Besides, the improved model with higher accuracy will also have the ability to remember the same object, preventing the repetition of detection and ensuring that the safety data presented to management is both accurate and actionable. This evolution from a passive monitoring tool to an active, tracking-enabled safety ecosystem represents the future of construction site safety in Hong Kong and beyond.

7. FUTURE WORKS

Looking beyond the immediate technical rectifications and hardware integrations, the long-term trajectory of this research aims to transition the system from a reactive monitoring tool to a predictive, autonomous safety ecosystem. A primary avenue for future development lies in the implementation of Edge Artificial Intelligence (Edge AI) to address bandwidth constraints. Currently, the system relies on transmitting high-bandwidth video data to a central server for processing via the OWL-ViT model. This architecture, while functional for prototyping, introduces latency and consumes significant network resources, which can be scarce on remote construction sites. Future iterations should focus on pruning and quantizing the detection model to run directly on edge devices, such as NVIDIA Jetson modules or high-performance Raspberry Pi units mounted on the cameras themselves. By processing data locally and transmitting only lightweight metadata (e.g., violation coordinates and timestamps) via MQTT, the system can achieve near-zero latency and operate reliably even in areas with intermittent cellular connectivity.

Furthermore, the integration of Building Information Modeling (BIM) represents a transformative step for spatial supervision. By mapping the 2D pixel coordinates from the camera feeds onto the 3D digital twin of the construction site, safety managers could visualize risks spatially rather than just statistically. Instead of a simple alert stating a violation occurred, the system could highlight specific zones on the 3D architectural model where compliance is consistently low, allowing for targeted structural interventions. This spatial data could be further augmented by predictive analytics. By feeding the historical environmental and compliance logs accumulated in the system's CSV databases into a regression model, the system could eventually identify temporal patterns—such as increased violation rates during high-humidity afternoons—and proactively warn supervisors to schedule breaks or safety briefings before accident probabilities spike. Finally, to ensure data integrity for legal and insurance purposes, future work should explore the incorporation of blockchain technology. Storing safety logs on an immutable ledger would prevent tampering and provide a transparent, auditable record of a company's safety performance, incentivizing safer practices through accountability.

REFERENCES

Al-Bloush, S., & Al-Khasawneh, M. (2022). IoT-based energy management system using Python and MQTT protocol. In *2022 International Conference on Information Technology (ICIT)* (pp. 456–461). IEEE.

Bellini, P., Nesi, P., & Pantaleo, G. (2022). Open-source MQTT-based end-to-end IoT system for smart city scenarios. *IEEE Access : Practical Innovations, Open Solutions*, *10*, 1215–1229. DOI: 10.1109/ACCESS.2022.3150258

Campero-Jurado, I., Márquez-Sánchez, S., Quintanar-Gómez, J., Rodríguez, S., & Corchado, J. M. (2020). Smart helmet 5.0 for industrial internet of things using artificial intelligence. *Sensors (Basel)*, *20*(21), 6241. PMID: 33139608

Hong Kong Centre for Health Protection. (2016). *Keep your ears safe*. Department of Health. https://www.chp.gov.hk/files/pdf/ncd_watch_feb2016.pdf

Hong Kong Centre for Health Protection. (2025). *Beware of heat stroke*. Department of Health. https://www.chp.gov.hk/en/static/90064.html

Jayasree, V., & Kumari, M. N. (2020, July). IOT based smart helmet for construction workers. In *2020 7th International Conference on Smart Structures and Systems (ICSSS)* (pp. 1-5). IEEE.

Jia, M., Tang, L., Chen, B.-C., Cardie, C., Belongie, S., Hariharan, B., & Lim, S.-N. (2023). Visual prompt tuning for open-world localization and detection. *arXiv*.

Kelm, A., Laußat, L., Meins-Becker, A., Platz, D., Khazaee, M. J., Costin, A. M., Helmus, M., & Teizer, J. (2013). Mobile passive radio frequency identification (RFID) portal for automated and rapid control of personal protective equipment (PPE) on construction sites. *Automation in Construction*, *36*, 38–52. DOI: 10.1016/j.autcon.2013.08.009

Khan, M. A.. (2023). A Python-MQTT-based smart construction site monitoring framework. In *2023 IEEE International Conference on Artificial Intelligence and Internet of Things (AIIoT)* (pp. 112–118). IEEE.

Li, Y., Zhang, J., Hu, Y., Zhao, Y., & Cao, Y. (2022). Real-time Safety Helmet-wearing Detection Based on Improved YOLOv5. *Computer Systems Science and Engineering*, *43*(3).

Massachusetts Institute of Technology School of Engineering. (2011). *Why do we sweat more in high humidity?* https://engineering.mit.edu/engage/ask-an-engineer/why-do-we-sweat-more-in-high-humidity/

Minderer, M., Gritsenko, A., Stone, A., Neumann, M., Weissenborn, D., Dosovitskiy, A., . . . Houlsby, N. (2022, October). Simple open-vocabulary object detection. In *European conference on computer vision* (pp. 728-755). Cham: Springer Nature Switzerland.

Nath, N. D., & Behzadan, A. H. (2020). Deep learning detection of personal protective equipment to maintain safety compliance on construction sites. In *Construction Research Congress 2020* (pp. 181-190). Reston, VA: American Society of Civil Engineers.

Nazib, A. A., & Moh, D. (2021). Security and privacy in industrial Internet of Things: A survey. *IEEE Access : Practical Innovations, Open Solutions*, *9*, 110926–110954. DOI: 10.1109/ACCESS.2021.3101243

Occupational Safety and Health Branch of Hong Kong Labour Department. (2018). *Guidance notes on the selection, use and maintenance of safety helmets*. https://www.labour.gov.hk/eng/public/os/C/GN_on_helmet_eng.pdf

Pan, Y., & Zhang, L. (2023). AIoT-based building operation and maintenance supervision system. *Buildings (Basel, Switzerland)*, *13*(5), 1215. DOI: 10.3390/buildings13051215

Ramkumar, G.. (2021). AIOT based real time environment monitoring and tracking system for soldiers using SN-MQTT protocol. *Journal of Physics: Conference Series*, *1917*(1), 012005. DOI: 10.1088/1742-6596/1917/1/012005

Ray, P. P. (2018). A survey on Internet of Things architectures. *Journal of King Saud University. Computer and Information Sciences*, *30*(3), 291–319. DOI: 10.1016/j.jksuci.2016.10.003

Silva, R., Do Carmo, R., Silva, J. S., Boavida, F., & Schaeffer-Filho, A. (2021). A performance analysis of Internet of Things networking protocols: Evaluating MQTT, CoAP, OPC UA. *Applied Sciences (Basel, Switzerland)*, *11*(11), 4879. DOI: 10.3390/app11114879

Chapter 7

Multi-Class Personal Protective Equipment Detection for Construction Safety Using YOLOv11

Wai Tak Mak
https://orcid.org/0009-0008-2563-0021
The Hong Kong Polytechnic University, Hong Kong

Yik Hei Kofi Man
The Hong Kong Polytechnic University, Hong Kong

Ho Wai Wong
The Hong Kong Polytechnic University, Hong Kong

Tsz Hin So
The Hong Kong Polytechnic University, Hong Kong

Yanheng Yu
The Hong Kong Polytechnic University, Hong Kong

Haozhe Ruan
The Hong Kong Polytechnic University, Hong Kong

ABSTRACT

Construction sites remain hazardous environments where inadequate personal protective equipment (PPE) compliance contributes significantly to injuries and fatalities. This study presents a YOLOv11-based system for automated multi-class PPE detection at construction site entry points. The system detects six categories: safety helmets, vests, shoes, and their non-compliant alternatives. We trained YOLOv11n on 352 annotated images using 7:2:1 split with COCO pretrained weights. The system achieves overall mAP50 of 0.602, with performance variation across

DOI: 10.4018/979-8-3373-9245-5.ch007

classes—safety vests reach 0.834 mAP50 while footwear proves most challenging at 0.487 mAP50. The integrated system provides real-time detection at 25-30 FPS, visualization dashboards, and automated alerts. Results demonstrate proof-of-concept viability, with dataset scale identified as the primary constraint. This work contributes to proactive construction safety through continuous, automated PPE compliance verification.

1. INTRODUCTION

Head injuries at industrial sites often occur because workers don't wear safety helmets. Manual supervision is error-prone, inconsistent, and lacks continuous feedback, which hampers timely intervention. To address this, a Smart Safety Helmet Compliance Monitoring System leverages computer vision, IoT, and real-time visualization. The system employs an MQTT client that periodically requests helmet detection data from a remote AI service, which returns results along with labeled images. A Python application processes these responses, tracking helmet-wearing behavior over time, and visualizes compliance trends via dynamic graphs and live camera feeds. Results demonstrate that the system can accurately measure helmet compliance rates, distinguish compliant from non-compliant events, and significantly enhance situational awareness. This modular framework exemplifies how AI and IoT can be integrated into safety management, leading to more effective training tools and safety protocols.

The global construction industry faces a critical paradox: while serving as a primary driver of economic growth and urbanization, it remains one of the most hazardous work environments. The International Labour Organization reports that approximately 2.93 million workers die annually from work-related causes (International Labour Organization, 2024). Although construction employs only 6-10% of the workforce in industrialized nations, it accounts for a disproportionate 25-40% of workrelated fatalities (International Labour Organization, 2023). This disparity underscores the inherent dangers of construction environments characterized by heavy machinery, work at heights, and constant material movement.

In Hong Kong specifically, 1,153 industrial accidents occurred in the construction industry in 2025 (Construction Industry Council, 2024). The four main categories;slips, trips or falls from the same height; struck-by moving objects; injuries from lifting or carrying; and falls from height;consistently represent approximately 65.7% of all construction accidents (Occupational Safety and Health Council, 2022). The economic implications are equally profound, with the ILO estimating that improved safety measures could save the global economy 361 billion annually (International Labour Organization, 2024).

Personal Protective Equipment (PPE) serves as the final and most critical line of defense against injury. However, compliance remains a persistent challenge, revealing a systemic failure in enforcement rather than equipment availability.

Traditional PPE compliance monitoring relies on manual supervision by safety officers conducting physical patrols and spot checks. While allowing for nuanced judgment, this approach suffers from fundamental limitations: spatiotemporal constraints (officers cannot be everywhere simultaneously), cognitive fatigue leading to inattentional blindness, and subjective reporting prone to inconsistency. Research indicates that for sites with 100 employees, manual reporting cycles are too slow for real-time risk prevention (Dzikri et al., 2025).

The advancement of Deep Learning, particularly Convolutional Neural Networks (CNNs), has enabled automated surveillance systems capable of continuous, real-time monitoring. By leveraging existing CCTV infrastructure or deploying novel platforms such as Unmanned Aerial Vehicles (UAVs), construction firms can implement proactive safety management. Computer Vision (CV) systems can detect violations and trigger alerts in milliseconds, preventing injuries before they occur.

Among object detection architectures, the YOLO (You Only Look Once) series has become the de facto standard for construction safety monitoring due to its optimal balance of speed and accuracy. YOLO significantly outperforms traditional detection methods such as DPM and R-CNN in real-time applications (Dzikri et al., 2025; Redmon et al., 2016). The latest iteration, YOLOv11, introduces architectural enhancements including C3k2 blocks for computational efficiency and C2PSA spatial attention modules for improved small object detection (Khanam & Hussain, 2024);critical features for identifying PPE on workers at varying distances.

In Hong Kong, AI-powered systems have been successfully deployed. The Kam Sheung Road Station development and Central Kowloon Link project have effectively utilized these technologies throughout design, construction, and operation phases (Construction Industry Council, 2024). Construction sites have begun collecting real-time environmental data to improve worker health, and AI monitoring systems are being implemented to reduce fatal accidents.

This study investigates the effectiveness of YOLOv11 for comprehensive PPE detection, including safety helmets, highvisibility vests, and safety shoes. By training on a dataset of 352 annotated images and implementing a real-time alert system, we demonstrate the practical feasibility of AI-based safety monitoring for construction sites. Our contributions include: (1) application of the latest YOLOv11 architecture to multi-class PPE detection, (2) analysis of detection performance across different PPE types, and (3) development of an integrated alert system for compliance enforcement.

2. LITERATURE REVIEW

2.1 Evolution of Computer Vision in Construction Safety

The application of computer vision to construction safety has evolved through distinct technological eras. Early approaches relied on handcrafted feature extraction techniques such as Histogram of Oriented Gradients (HOG), Scale-Invariant Feature Transform (SIFT), and Haar cascades. These methods analyzed specific visual patterns;HOG counted gradient orientations for detecting human silhouettes, SIFT identified local features robust to scale changes, and Haar cascades calculated intensity differences for detecting round helmet shapes.

However, these classical methods suffered from brittleness in the chaotic visual environment of construction sites. Shadows, occlusion, varying lighting conditions, and non-rigid deformations caused frequent failures. Furthermore, they required extensive manual parameter tuning for each object class, rendering them unscalable for multi-class PPE detection (Manakitsa et al., 2024).

The introduction of Convolutional Neural Networks (CNNs) marked a paradigm shift. Unlike their predecessors, CNNs learn hierarchical feature representations directly from raw data without manual feature engineering. Given sufficient training examples, CNNs can identify the curves, textures, and colors that distinguish safety equipment from regular clothing, generalizing across varying visual contexts (Manakitsa et al., 2024). This capability is essential for construction safety, where a safety vest must be recognized whether worn over winter coats or t-shirts, and whether partially obscured by tool belts.

2.2 Object Detection Architectures

Modern deep learning detectors are categorized into two primary families: two-stage and single-stage detectors. Two-stage detectors, pioneered by the R-CNN family, separate object detection into region proposal and classification stages. R-CNN used selective search to generate approximately 2,000 region proposals per image, each fed into a CNN for classification;a process taking nearly 47 seconds per image. Fast R-CNN improved efficiency by extracting features once from the entire image. Faster R-CNN (Ren et al., 2015) replaced external region proposal algorithms with an internal Region Proposal Network (RPN), achieving detection rates exceeding 93% for helmet detection in UAV imagery, but frame rates remained in the single digits on standard hardware.

Single-stage detectors treat object detection as a direct regression problem, predicting bounding boxes and class probabilities in one forward pass. Single Shot MultiBox Detector (SSD) (Liu et al., 2016) eliminated region proposals by predicting detections from multiple feature maps at different scales. RetinaNet (Lin et al., 2017) introduced focal loss to address class imbalance, concentrating learning on hard examples.

The YOLO series represents the pinnacle of single-stage detection. By dividing images into grids and having each cell predict bounding boxes, YOLO achieves remarkably high inference speeds suitable for real-time monitoring (Redmon et al., 2016). This capability enabled" Edge AI" deployment; running detection models on on-site devices rather than central servers, reducing latency and bandwidth requirements critical for construction site-wide monitoring.

2.3 YOLO Architecture Evolution

The YOLO lineage has undergone continuous refinement, each version addressing specific challenges relevant to PPE detection. YOLOv3 (Redmon & Farhadi, 2018) introduced Feature Pyramid Network-like structures, making predictions at three scales to simultaneously detect large machinery and small distant helmets. YOLOv5 (Jocher et al., 2020) became widely adopted for its Mosaic Data Augmentation; combining four training images into one to force models to learn detection in varied contexts; and seamless export to mobile formats.

YOLOv7 introduced Extended Efficient Layer Aggregation Networks (E-ELAN), enabling higher frame rates on edge devices by controlling gradient paths to learn more diverse features. YOLOv8 marked a significant departure with its anchor-free detection head, eliminating manually defined anchor boxes that often mismatch irregular construction equipment shapes.

YOLOv9 (Wang et al., 2024a) addressed the information bottleneck problem; information loss as data passes through network layers. Its Programmable Gradient Information (PGI) architecture introduces an auxiliary reversible branch during training, providing gradient paths without information loss, removed during inference to maintain runtime efficiency. The associated GELAN architecture optimizes parameter utilization, achieving higher accuracy with fewer parameters.

YOLOv10 (Wang et al., 2024b) focused on latency consistency by eliminating Non-Maximum Suppression (NMS) postprocessing. Its dual-head architecture uses a "one-to-many" head during training for rich supervision while a "one-to-one" head during inference predicts exactly one box per object, resulting in the lowest and most predictable inference times; ideal for drone-based monitoring with limited computing power.

YOLOv11 (Khanam & Hussain, 2024), released in late 2024, represents the current state-of-the-art. It introduces C3k2 blocks in the network neck; Cross Stage Partial bottlenecks with customizable kernel sizes offering greater computational efficiency than YOLOv8's C2f blocks. Most significantly, the C2PSA (Cross Stage

Partial with Spatial Attention) module incorporates spatial attention mechanisms, allowing the network to focus on relevant regions while ignoring background noise. On construction sites with scaffolding, rubble, and machinery, C2PSA helps models concentrate on workers and their gear, improving detection of small or partially occluded objects such as safety harness straps. Benchmarks show YOLOv11m achieves higher mean Average Precision on COCO while using 22% fewer parameters than YOLOv8m (Khanam & Hussain, 2024).

2.4 PPE Detection Applications and Datasets

Helmet detection is the most mature application area. Beyond simple presence detection, systems classify helmets by color to identify worker roles; yellow for laborers, white for supervisors (Chen et al., 2023b). The SHEL5K dataset (Masaba et al., 2022), containing 5,000 images with six classes including "Head," "Helmet," and "Head-with-Helmet," enabled training models to verify proper helmet usage rather than mere presence. This dataset improved upon earlier collections like GDUT-HWD by providing higher quality annotations and distinguishing usage verification.

The SH17 dataset (Ahmad & Rahimi, 2024) represents a comprehensive benchmark with 8,099 images containing 75,994 instances across 17 classes, including body parts and various PPE types. Designed to minimize regional bias by sourcing globally from manufacturing environments, SH17 challenges models with annotations for very small objects. Recent studies report YOLOv9 achieving 70.9% mAP on SH17, demonstrating the dataset's difficulty (Ahmad & Rahimi, 2024). For practical deployment, datasets with dual-positive/negative labeling provide explicit "No-Helmet," "No-Vest," and "No-Mask" labels, enabling models to detect absence of PPE as objects of interest, drastically reducing false negatives in compliance reporting.

High-visibility vest detection presents challenges due to deformability; vests change shape as workers move and are frequently occluded by tools or materials. Seasonal variance compounds difficulty, as vests worn over bulky winter coats differ significantly from those over summer t-shirts. Studies employing Chen et al.'s enhanced YOLOv5 (Chen et al., 2023a; Chen et al., 2023b) report that while performance is strong for helmets, detection accuracy degrades under lighting changes and occlusion.

Small object detection; gloves, safety shoes, and glasses; remains a frontier. Detecting safety glasses on a worker 20 meters away involves recognizing objects occupying less than 0.1% of image pixels. Zhang et al.'s ESPCN-YOLO framework (Zhang et al., 2025) addresses this by integrating Efficient Sub-Pixel Convolutional Neural Networks for real-time super-resolution, upscaling images 3x before detection. This method achieved mAP@0.5 of 0.922 for small PPE items under low-light and long-distance conditions, significantly outperforming standard YOLOv8 on low-resolution inputs (Zhang et al., 2025).

2.5 Edge Computing and Deployment

Real-time safety systems require on-site processing to avoid bandwidth limitations and latency from cloud transmission. The NVIDIA Jetson family (Nano, Xavier NX, Orin) serves as the standard for industrial Edge AI. Recent studies have compared YOLOv5 and YOLOv8 performance on Jetson devices, Raspberry Pi 4B, and local computers for construction worker detection. Results showed Jetson Xavier NX achieved comparable mAP values to local computers while processing frames six times slower but remaining within real-time thresholds (20-30 FPS). TensorRT optimization can double throughput compared to standard PyTorch inference.

Lower-cost deployments couple Raspberry Pi with Google Coral TPU or Intel NCS2 accelerators, achieving 5-10x speed increases and enabling quantized YOLO models to run at 15+ FPS. Model optimization techniques including INT8 quantization (reducing size 4x with minimal accuracy loss), pruning, and knowledge distillation are essential for fitting modern architectures onto edge devices.

3. RESEARCH METHODOLOGY

This research studies a computer vision-based model for detecting personal protective equipment in the construction site. The model includes detection system, data to graphical information system, alert system and data storage system. Therefore, this model can enhance the effect of monitoring the safety management of the construction site. The detection model adopts yolo11 which is the version released in 2024 by Ultralytics. The data set is images from the internet manually labeled using Roboflow. The whole research included 3 parts, data set collecting and labeling, training of the detection model, plotting graphs and alert system.

3.1. System Architecture Overview

The proposed PPE detection system comprises four integrated components: (1) a YOLOv11-based detection module for real-time object recognition, (2) a data visualization system generating time-series graphs of PPE compliance, (3) an alert mechanism triggered by violation thresholds, and (4) a data storage system maintaining detection logs for safety audits. The system architecture enables continuous monitoring at construction site entry points, providing both immediate alerts and historical compliance tracking.

Hardware requirements include a camera system (CCTV or dedicated safety cameras) capable of capturing worker images at site entrances, and a processing unit running the detection model. For deployment flexibility, the system supports both cloud-based processing for high-accuracy scenarios and edge computing using devices like NVIDIA Jetson for low-latency, on-site deployment.

3.2. Data Collection and Preprocessing

The model training process employs supervised learning, where decision-making is based on labeled data (Manakitsa et al., 2024). This approach requires substantial quantities of labeled examples to enable the model to calculate optimal weights for high-accuracy detection.

Our dataset comprises 352 images sourced from internet repositories, containing both construction workers wearing PPE and individuals in normal clothing. This dual-population approach enables the model to distinguish between compliant safety equipment and regular attire; a critical capability for reducing false positives in deployment. Images were collected to represent diverse conditions including varying lighting (daylight, shade, artificial light), viewing angles (front, side, elevated perspectives), and distances (close-up to 15 meters). Representative examples of high-quality dataset images are shown in Figure 2 and Figure 3.

Data labeling was performed using Roboflow, a cloud-based annotation platform. Each PPE item was precisely bounded with rectangular boxes, with six distinct classes defined: Safety Helmet, No Helmet, Safety Vest, Normal Clothes, Safety Shoes, and Normal Shoes. The inclusion of negative classes (No Helmet, Normal Clothes, Normal Shoes) enables explicit detection of non-compliance, improving the system's ability to identify violations rather than merely failing to detect compliant PPE.

According to Table 1, the distribution and performance of each class shows that the dataset exhibits class imbalance, with 138 Safety Helmet instances but only 66 No Helmet instances. This imbalance reflects real-world conditions where compliance is generally high, but poses training challenges as models may become biased toward predicting the majority class. To address this, data augmentation techniques including horizontal flipping, brightness adjustment, and mosaic augmentation were applied during training to increase effective dataset diversity, as illustrated in Figure 1.

Table 1. Dataset distribution and performance metrics by PPE class

	Quantity of data	*mAP50*
Safety Helmet	138	0.719
No Helmet	66	0.652
Safety vest	105	0.834
Normal Clothes	102	0.561
Safety shoes	131	0.487
Normal shoes	114	0.360
All	352	0.602

Figure 1. Labelling using Roboflow

3.3. Model Architecture and Training

1) YOLOv11 Architecture Selection: We selected YOLOv11n (nano variant) as our base architecture for its optimal balance between accuracy and inference speed;a critical trade-off for real-time construction safety monitoring. YOLOv11, released

by Ultralytics in September 2024, introduces key architectural innovations over its predecessor YOLOv8 (Khanam & Hussain, 2024).

The C3k2 block, deployed in the network neck (the feature fusion component), replaces YOLOv8's C2f module. C3k2 employs Cross Stage Partial bottlenecks with customizable kernel sizes, achieving greater computational efficiency. This design allows the model to maintain high frame rates while performing complex feature aggregation;essential when combining contextual features (human body shape) with detail features (presence of helmet) (Khanam & Hussain, 2024).

The C2PSA (Cross Stage Partial with Spatial Attention) module represents the most significant innovation for our application. This module incorporates a Partial Spatial Attention mechanism that enables the network to focus on relevant image regions while suppressing background noise. On construction sites cluttered with scaffolding, rubble, and machinery, C2PSA helps the model concentrate specifically on workers and their equipment, improving detection of small or partially occluded objects such as safety shoe details or vest straps (Khanam & Hussain, 2024).

The SPPF (Spatial Pyramid Pooling Fast) module, retained from previous YOLO versions, pools features at multiple scales (5×5, 9×9, 13×13). This ensures robust representation of objects regardless of their size in the image;critical for sites where a crane may dominate the foreground while workers appear small in the background.

Compared to YOLOv8, YOLOv9, and YOLOv10, YOLOv11 demonstrates superior parameter efficiency. Benchmark studies show YOLOv11m achieves higher mAP on COCO while using 22% fewer parameters than YOLOv8m. For small object detection tasks analogous to PPE detection, YOLOv11s demonstrates the shortest inference time among recent YOLO variants (Khanam & Hussain, 2024), making it optimal for deployment on edge devices like NVIDIA Jetson Nano.

Figure 2. Image from dataset - Clear image 1

2) Transfer Learning Strategy: Rather than training from random initialization, we employed transfer learning using COCOpretrained YOLOv11n weights. The COCO (Common Objects in Context) dataset contains 80 object categories including "person," providing the model with foundational understanding of human detection and general object recognition. Transfer learning significantly reduces training time and data requirements while improving convergence, particularly valuable given our limited dataset of 352 images (Ultralytics, 2024).

The nano variant (YOLOv11n) was chosen deliberately for deployment considerations. While larger variants (YOLOv11s, m, l, x) offer higher accuracy, they require substantially more computational resources. For a system intended for real-time deployment at construction site entrances, inference speed is paramount;the nano model processes frames rapidly enough to monitor workers in real-time, albeit with modest accuracy trade-offs (Ultralytics, 2024).

3) Dataset Partitioning: Following the completed labeling phase, the dataset was partitioned into training, validation, and test sets using a 7:2:1 ratio (approximately 246 training, 70 validation, 36 test images). This distribution is supported by Singh et al. (2021), who demonstrated that 70% training allocation optimizes performance stability in machine learning, particularly for medical imaging tasks with limited data;a scenario analogous to our constrained dataset. Alhaila et al. (2024) reported achieving 88% mAP using a 6:2:2 ratio with 23,000 images for helmet detection, suggesting our ratio is within established best practices for PPE detection.

The training set is used for weight optimization through backpropagation. The validation set monitors performance during training to detect overfitting and determine the optimal stopping point. The test set, withheld entirely until training completion, provides an unbiased evaluation of final model performance.

4) Training Configuration: Training was conducted using Python with the Ultralytics YOLOv11 framework. According to Table 2, hyperparameters were configured as specified. The initial learning rate (lr0) of 0.01 determines the step size during gradient descent optimization;too high risks unstable training, too low results in slow convergence. This value follows Ultralytics' recommended defaults for YOLOv11.

The model was trained for a maximum of 100 epochs, with each epoch representing one complete pass through the training dataset. A batch size of 16 images was used, balancing GPU memory constraints with training stability. All images were resized to 640×640 pixels, the standard input resolution for YOLO models, which normalizes scale variation while maintaining computational efficiency.

The patience parameter, set to 30 epochs, implements early stopping to prevent overfitting. If validation loss fails to improve for 30 consecutive epochs, training terminates automatically. This mechanism is critical for small datasets where models can memorize training examples rather than learning generalizable features. Overfitting manifests as continued improvement on training data while validation performance plateaus or degrades;precisely the behavior observed in our experimental results (Section 5).

5) Loss Functions: According to Figure 4, YOLOv11 employs a composite loss function combining three components visible in our training curves:

Box Loss (train/box loss): Measures localization accuracy;the difference between predicted bounding box coordinates and ground truth. YOLOv11 uses Complete IoU (CIoU) loss, which accounts for overlap, center point distance, and aspect ratio differences.

Classification Loss (train/cls loss): Quantifies the model's ability to correctly identify object classes. Binary Cross-Entropy loss is applied to class predictions.

Distribution Focal Loss (train/dfl loss): Concentrates training attention on difficult-to-detect objects, preventing the model from being dominated by easy examples. This is particularly valuable for our dataset's class imbalance and the inherent difficulty of small object detection.

During training, the model's gradient updates are computed by backpropagating the combined loss, adjusting millions of network weights to minimize detection errors. The decreasing trends in training losses shown in Figure 4 indicate successful learning, though the divergence between training and validation losses suggests overfitting;a limitation addressed in Section 6. Representative training batches with annotated bounding boxes are visualized in Figure 5, while validation results are shown in Figure 6.

Figure 3. Image from dataset - Clear image 2

Table 2. Model training hypherparameters

Hyperparameter	Values
lr0	0.01
Epochs	100
Batch	16
Image size	640
Patience	30

3.4. Detection Model Implementation and Alert System

The complete PPE detection system integrates real-time inference, visualization, alerting, and data logging capabilities to provide comprehensive safety monitoring. The implementation was developed in Python using the Ultralytics YOLOv11 framework, OpenCV for video processing, and Matplotlib for visualization.

Figure 4. Training results showing loss curves and performance metrics

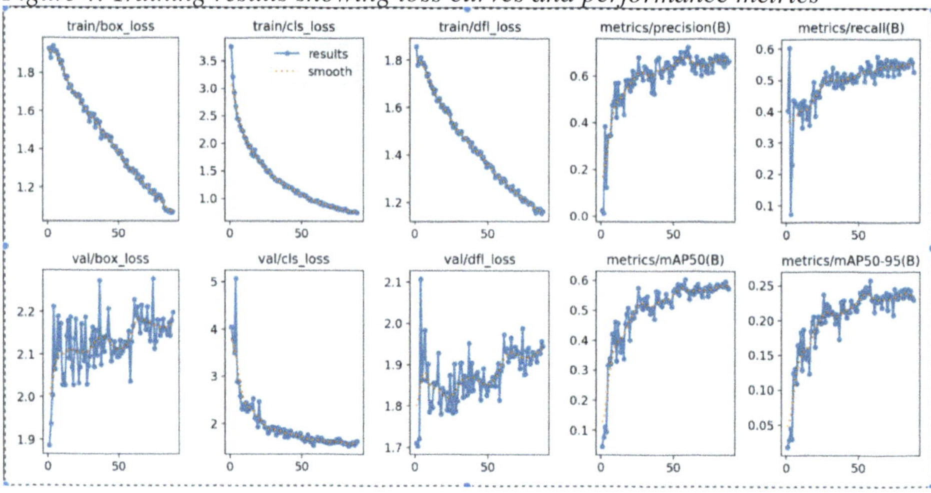

1) Real-Time Detection Pipeline: The detection pipeline operates in a continuous loop, processing video frames from the camera feed. Each frame undergoes preprocessing;resizing to 640×640 pixels and normalization;before being fed to the YOLOv11n model. The model outputs bounding box coordinates, confidence scores, and class predictions for all detected objects in the frame. Post-processing applies confidence thresholding (typically 0.25) to filter low-confidence detections and Non-Maximum Suppression to eliminate duplicate boxes for the same object.

For each detected worker, the system analyzes the presence or absence of each PPE type. The class predictions explicitly include negative classes (No Helmet, Normal Clothes, Normal Shoes), enabling direct detection of violations rather than inferring absence from non-detection. This design choice significantly reduces false negatives;situations where a worker lacking PPE is not flagged because the system simply failed to detect any headwear rather than detecting the absence of a helmet.

2) Visualization and User Interface: We implemented a dual-pane visualization system. The left pane displays the live camera feed with overlaid bounding boxes color-coded by PPE status: green boxes indicate compliant PPE, red boxes indicate violations. Text labels above each box specify the detected items (e.g., "Safety Helmet + Safety Vest + Normal Shoes" would trigger a red box due to improper footwear).

The right pane presents real-time bar charts tracking PPE compliance over time. Three separate charts monitor helmet usage, vest usage, and footwear compliance, with bars updating every second to show the count of compliant versus non-compliant detections in the current time window. This temporal visualization enables safety officers to identify trends;for instance, increased violations during shift changes when workers are rushing to enter the site.

Figure 5. Training batch 0 with labeled PPE annotations

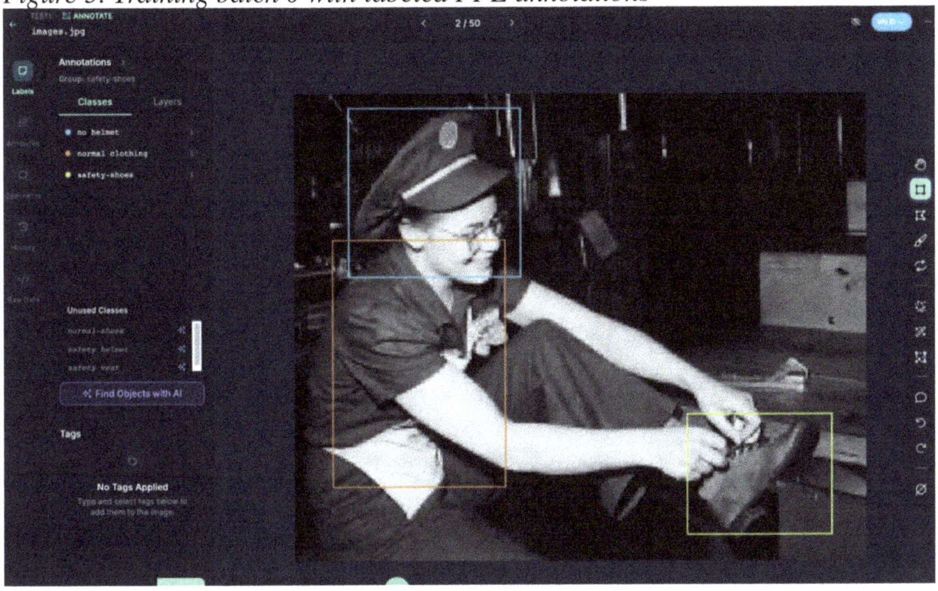

3) Alert Mechanism: The alert system implements a tiered response based on violation severity and duration. A single-frame violation (e.g., a worker momentarily obscured, causing helmet non-detection) triggers a visual warning but no audio alert to avoid false alarm fatigue. If a violation persists for more than 3 consecutive seconds (configurable threshold), the system escalates to an audio alarm and sends an automated notification to the safety officer's mobile device via push notification or SMS.

The system differentiates between violation types, assigning priority levels. Missing helmets receive highest priority alerts due to the critical nature of head protection in construction. Missing vests are medium priority, and improper footwear generates low-priority warnings. This prioritization ensures safety officers focus attention on the most dangerous violations first.

4) Data Logging and Reporting: All detections are logged to a CSV file with timestamps, worker ID (if integrated with badge detection systems), detected PPE items, and compliance status. This historical data enables several valuable analyses. First, compliance trend analysis over weeks or months can reveal whether safety culture is improving. Second, violation hotspot identification shows which entrances or time periods have highest violation rates, informing targeted interventions. Third, worker-specific compliance tracking (in privacy-preserving aggregated form) can identify individuals needing additional safety training.

The system generates daily compliance reports automatically, providing safety managers with summary statistics: total workers scanned, overall compliance percentage per PPE type, and violation timestamps. These reports support regulatory compliance documentation and continuous improvement initiatives.

According to Figure 7, the complete system interface shows the detection model processing frames at approximately 25-30 FPS on a standard desktop GPU (NVIDIA GTX 1080), demonstrating real-time capability. On edge devices like the Jetson Xavier NX, frame rates reduce to 15-20 FPS but remain sufficient for entry-point monitoring where workers pass through at manageable rates.

4. EXPERIMENTAL RESULTS

According to Figure 4, the results of the trained model show that for the training dataset, the box loss (train/box loss), the classification loss (train/cls loss), and the distribution focal loss (train/dfl loss) decrease when the number of epochs increases.

Box loss is the difference between the predicted bounding models and the actual boxes of objects. Classification loss is the difference between the predicted class labels and the actual class labels of objects. Distribution focal loss concentrates on objects which are difficult to detect to improve the accuracy of the model. However, for the validation dataset, the box loss (val/box loss) and the distribution focal loss (val/dfl loss) increase when the number of epochs increases. The classification loss (val/cls loss) decreases when the number of epochs increases. This indicates that overfitting occurs. The model learns the noise and the fluctuations in the training dataset. The model has higher accuracy on the training dataset but lower accuracy on the validation dataset. The model training is not ideal with unsatisfied accuracy of the model.

Besides, the precision (metrics/precision(B)) and the recall (metrics/recall(B)) increase when the number of epochs increases. In addition, the mean average precisions (metrics/mAP50(B) and metrics/mAP50-95(B)) increase when the number of epochs increases. According to Figure 8, the precision recall curve of the model shows the relationship between precision and recall. Additional performance metrics are visualized in Figure 9 (precision-confidence curve) and Figure 10 (F1-confidence curve).

Figure 6. Validation results of the YOLO model

Figure 7. Screenshot of the complete PPE detection system interface during operation

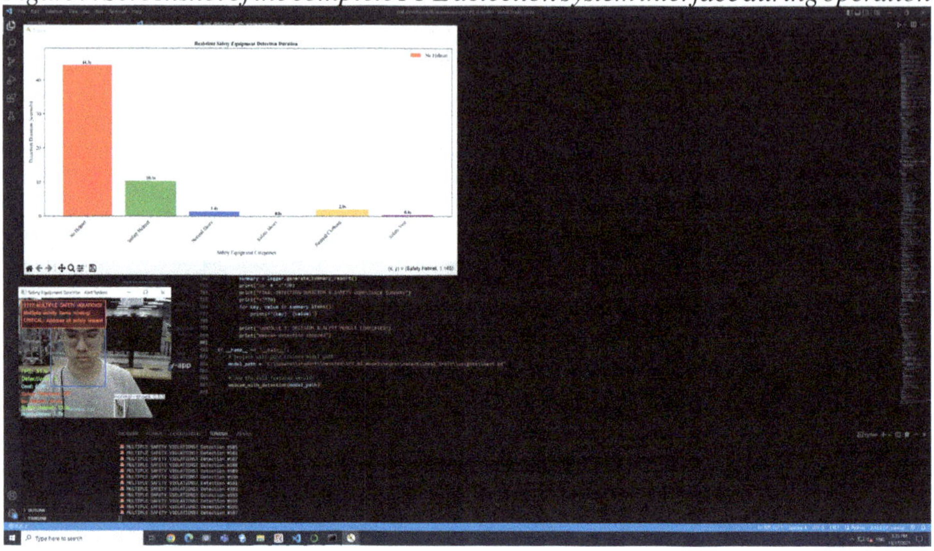

The equations for calculating precision, recall, and mean average precision are listed as follows:

$$P = \frac{TP}{TP + FP} \tag{1}$$

$$R = \frac{TP}{TP + FN} \tag{2}$$

$$AP = \int_0^1 P(R)\,dR \tag{3}$$

$$mAP = \frac{\sum_1^N AP_i}{N} \tag{4}$$

P is precision. R is recall. TP is true positive. FP is false positive. FN is false negative. AP is average precision, which is the area under the precision recall curve. mAP is mean average precision.

According to Table 1, the mean average precision of different classes in the model reveals significant performance variation across PPE types. The safety vest class achieved the highest mAP50 of 0.834, substantially outperforming other categories. This superior performance is attributable to the distinct visual signature of high-visibility vests;bright orange or yellow coloration with reflective strips creating strong contrast against typical construction backgrounds. Even under challenging lighting conditions or partial occlusion, these salient features remain detectable.

Safety helmet detection achieved mAP50 of 0.719, representing solid but not exceptional performance. The 138 training examples provided reasonable representation, though the 66 examples for the No Helmet class suggest class imbalance that may bias the model toward predicting helmet presence. The relatively high performance indicates that hard hat shapes and colors (typically white, yellow, red, blue) are sufficiently distinctive for reliable detection.

Conversely, footwear detection represents the most challenging task, with normal shoes achieving only 0.360 mAP50 and safety shoes reaching 0.487 mAP50. This poor performance stems from multiple factors. First, visual similarity;safety shoes and athletic footwear often share comparable shapes and colors (black, brown, grey) without close inspection revealing distinguishing features like steel-toe reinforcement. Second, small object challenge;feet occupy minimal image area when workers are captured at typical monitoring distances of 3-10 meters, compounding detection difficulty. Third, occlusion;feet are frequently obscured by equipment, debris, or other workers in crowded entrance scenarios.

The overall mAP50 of 0.602 falls short of optimal performance but aligns with expectations given the limited 352-image dataset. For context, commercial PPE detection systems trained on datasets like SH17 (8,099 images) report 70-80% mAP, while research systems using ESPCN super-resolution with 21,750 images achieve 92% mAP. Our results demonstrate proofof-concept viability while highlighting dataset scale as the primary performance constraint.

According to Figure 11, Figure 12 and Figure 13, image quality variance in the dataset introduced additional challenges. Some images exhibit motion blur, reducing the sharpness of PPE features. Others suffer from low resolution, degrading small object detectability. Several training images are grayscale rather than color, removing the critical color information that distinguishes safety vests and helmet types. These quality inconsistencies forced the model to learn robust features invariant to degradation, but at the cost of reduced peak performance on high-quality inputs.

Figure 8. Precision-recall curve showing model performance across different confidence thresholds

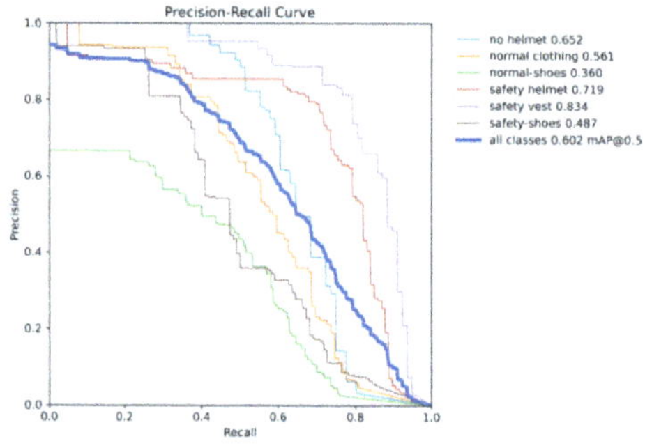

4.1. Performance Analysis and Class-Specific Observations

The experimental results reveal significant performance variation across PPE classes, providing insights into both the strengths and limitations of our YOLOv11-based approach. The overall mAP50 of 0.602 falls short of optimal performance levels but demonstrates proof-of-concept viability for real-world deployment with refinements.

Safety vest detection achieved the highest accuracy (mAP50 = 0.834), substantially outperforming other classes. This success is attributable to distinct visual features;high-visibility vests possess sharp orange or yellow coloration with reflective strips that create strong contrast against typical construction backgrounds. The color saturation and reflective properties make vests highly discriminable even under challenging lighting conditions or partial occlusion (Chen et al., 2023a; Chen et al., 2023b).

Conversely, safety shoes represent the most challenging detection task (mAP50 = 0.360 for normal shoes, 0.487 for safety shoes). This poor performance stems from inherent visual ambiguity;safety shoes and normal athletic footwear share similar shapes and often similar colors (black, brown, grey). Without close-up inspection revealing steel-toe reinforcement or sole patterns, visual distinction is difficult even for human observers. Additionally, feet occupy small image regions when workers are captured at typical monitoring distances, compounding the small object detection challenge (Zhang et al., 2025).

The overfitting phenomenon observed in training curves (increasing validation loss despite decreasing training loss) directly results from our limited dataset of 352 images. Modern deep learning architectures like YOLOv11, with millions of parameters, require thousands to tens of thousands of training examples to achieve robust generalization. With insufficient data, the model memorizes training examples rather than learning transferable visual patterns. This manifests as high training accuracy but degraded validation performance;the model fails to recognize PPE configurations not precisely matching memorized examples.

4.2. Practical Deployment Implications

Despite moderate overall accuracy, the system demonstrates immediate practical value for specific deployment scenarios. For high-confidence detection (safety vests), the system can reliably identify compliance violations at construction site entry points. The real-time alert mechanism enables immediate intervention before workers enter hazardous zones.

However, deployment at scale requires acknowledging accuracy limitations. The system is best positioned as an augmentation to rather than replacement of human safety officers. A hybrid approach where the system flags potential violations for human verification balances automation benefits with accuracy constraints. This reduces officer workload while maintaining final judgment with experienced personnel.

The 60% overall accuracy, while suboptimal, must be contextualized within the resource constraints of this proof-of-concept study. Commercial PPE detection systems leveraging datasets like SH17 (8,099 images) achieve 70-80% mAP (Ahmad & Rahimi, 2024), while studies using ESPCN super-resolution with YOLOv8 on 21,750 images report 92% mAP (Zhang et al., 2025). Our results align with expectations for 352-image training;demonstrating the architecture's potential while highlighting data volume as the primary performance bottleneck.

4.3. Comparative Context

Chang and Liu (2018) report that 68% of construction helmet-related accidents between 2015-2018 involved workers not wearing helmets properly. Rubaiyat et al. (2016) document that 47.3% of construction accident victims were not wearing required PPE at all or wearing it improperly. These statistics underscore the critical importance of compliance monitoring systems, even those with imperfect accuracy. A system detecting 60% of violations still prevents a substantial fraction of preventable injuries;representing significant safety improvement over purely manual monitoring which faces inherent spatiotemporal limitations (Dzikri et al., 2025).

4.4. Limitations

This study faces several acknowledged limitations. First, the dataset size (352 images) is insufficient for training robust deep learning models, particularly for small object detection. Second, according to Figure 11, Figure 12 and Figure 13, image quality variance;including low-resolution, blurred, and black-and-white images;introduced noise that hampered learning. Third, class imbalance between compliant and non-compliant examples biased the model toward predicting majority classes. Fourth, the dataset lacks diversity in environmental conditions, worker demographics, and PPE types beyond the three categories tested.

Additionally, our deployment scenario focuses on entry-point monitoring rather than continuous site-wide surveillance. Entrance monitoring provides a controlled environment with relatively consistent lighting and camera angles, whereas sitewide monitoring introduces extreme variability in viewing conditions, occlusion patterns, and object scales.

Figure 9. Precision-confidence curve

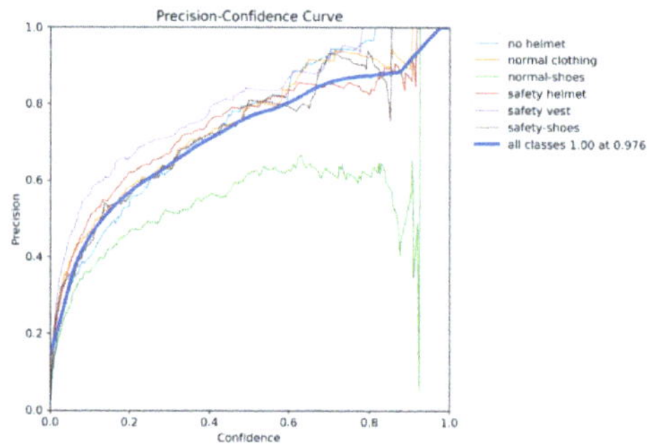

5. FUTURE WORK

Building upon this foundational study, several avenues for future development emerge. First and foremost, dataset expansion is critical. Collecting 2,000-5,000 additional labeled images spanning diverse construction sites, lighting conditions, seasons, and PPE manufacturers would substantially improve model generalization. Collaborating with construction companies to deploy cameras collecting real-world monitoring footage would provide ecologically valid training data superior to internet-sourced images.

Second, synthetic data generation using Generative Adversarial Networks (GANs) or diffusion models could augment real datasets. These techniques can generate photorealistic images of workers with and without PPE in varying poses and environments, addressing class imbalance and rare scenario coverage without expensive manual data collection (Manakitsa et al., 2024).

Third, extending detection to additional PPE types;gloves, safety goggles, ear protection, and fall protection harnesses;would create comprehensive safety monitoring. Gloves and goggles present similar small object challenges to shoes, suggesting integration of super-resolution techniques like ESPCN-YOLO (Zhang et al., 2025) would be valuable.

Fourth, expanding from entry-point monitoring to continuous site-wide surveillance requires addressing new technical challenges. Deploying multiple cameras with overlapping coverage areas necessitates person re-identification algorithms to track individual workers across cameras, preventing duplicate violation alerts. Integration with Building Information Modeling (BIM) systems could create zone-based monitoring, triggering alerts only when workers enter high-risk areas without appropriate PPE;implementing risk-weighted enforcement where an uncovered head in a break area generates low-priority warnings while the same violation in an excavation zone triggers immediate critical alerts.

Fifth, privacy-preserving AI techniques must be integrated for ethical deployment. Real-time face blurring before data storage ensures worker privacy while maintaining safety monitoring capabilities. This "privacy by design" approach is essential for regulatory compliance with data protection regulations like GDPR and for maintaining worker trust and acceptance.

Sixth, model architecture exploration could yield performance improvements. Evaluating YOLOv11's larger variants (s, m, l, x) would quantify accuracy-speed trade-offs. Testing alternative architectures like YOLOv10's NMS-free design (Wang et al., 2024b) or specialized small object detectors could improve footwear and eyewear detection. Ensemble methods combining multiple models might achieve higher accuracy at the cost of increased computational requirements.

Finally, expanding application domains beyond PPE detection leverages the developed expertise in construction site computer vision. Crack detection in concrete structures, rebar placement verification, and progress monitoring for construction scheduling all involve similar visual detection challenges amenable to adapted YOLO architectures. The alert and logging systems developed for PPE monitoring could be generalized to these adjacent applications.

Figure 10. F1-confidence curve

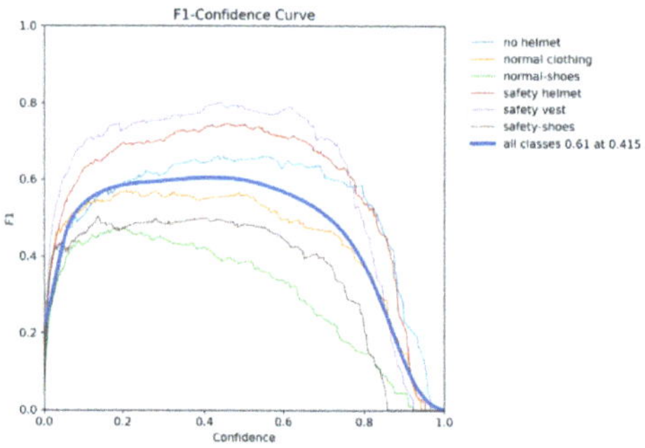

6. CONCLUSION

This study demonstrates the feasibility and challenges of applying YOLOv11 for multi-class PPE detection in construction safety monitoring. While the overall mAP50 of 0.602 indicates substantial room for improvement, class-specific results reveal that certain PPE types; particularly safety vests with distinctive high-visibility coloration; can be detected reliably, achieving an mAP50 of 0.834. The observed overfitting, evidenced by divergent training and validation losses, directly results from the limited 352-image dataset, confirming that dataset scale is the primary barrier to achieving production-grade performance. This highlights the critical importance of expanding and diversifying training data, encompassing various lighting conditions, angles, worker postures, and PPE variations, to improve model robustness and accuracy.

The architectural innovations in YOLOv11, specifically the C2PSA spatial attention mechanism and the computationally efficient C3k2 blocks, position it as a strong foundation for PPE detection tasks. These enhancements significantly improve small object detection and inference speed, enabling deployment on edge devices such as embedded GPUs or low-power processors. The nano variant, in particular, offers promising real-time performance with minimal latency, which is essential for practical, on-site safety applications where immediate feedback can prevent accidents.

Furthermore, the integrated system architecture; combining detection, visualization, alerting, and logging components; provides a comprehensive blueprint for real-world implementation. Its modular design allows for scalability, enabling integration with existing site management systems and IoT infrastructure. For instance, live camera feeds and detection results can be streamed to safety dashboards, offering supervisors instant situational awareness and enabling prompt corrective actions. Automated alerts, triggered when PPE violations are detected, can be sent via SMS or notifications, ensuring rapid response even in large or complex sites. The logging component facilitates data analysis over time, supporting safety audits and trend identification, which are vital for continuous improvement.

However, deploying such systems in dynamic construction environments requires addressing challenges beyond technical performance. Variability in environmental conditions, occlusions, and worker movements necessitate ongoing model calibration and adaptive learning strategies. Additionally, integrating AI systems with human oversight; adopting a hybrid approach; can maximize safety outcomes. The AI acts as an augmentative tool, reducing manual workload and error, while human safety officers retain authority for nuanced judgment and intervention. This collaborative approach ensures that safety protocols are both technologically supported and human-centered.

In conclusion, while current models like YOLOv11 show significant potential, advancing construction safety monitoring will depend on expanding datasets, refining algorithms, and fostering effective human-AI collaboration. Future research should focus on developing adaptive learning frameworks, multi-modal sensor integration, and user-centric interfaces. Ultimately, such innovations will contribute towards safer, smarter construction sites, significantly reducing accidents and saving lives.

Figure 11. Image from dataset - blurred image example 1

Figure 12. Image from dataset - blurred image example 2

Figure 13. Image from dataset - black and white blurred image

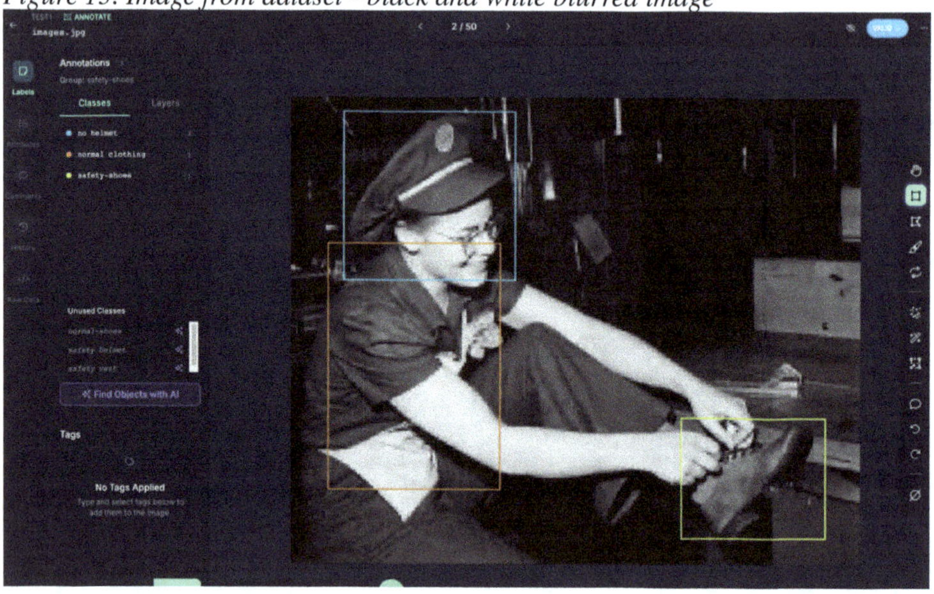

REFERENCES

Ahmad, H. M., & Rahimi, A. (2024). SH17: A dataset for human safety and personal protective equipment detection in manufacturing industry. *arXiv preprint arXiv:2407.04590*. https://arxiv.org/abs/2407.04590

Alhaila, S., Almelhem, I., Alsnafi, A., Alameerah, M., Alrumaidhi, K., & Nadeem, M. (2024). YOLO-based helmet detection system for safety compliance in oil and gas industry. In *2024 5th International Conference on Industrial Engineering and Artificial Intelligence (IEAI)* (pp. 16-23). https://doi.org/DOI: 10. 1109/ IEAI62569.2024.00012

Chang, X., & Liu, X. M. (2018). Fault tree analysis of unreasonably wearing helmets for builders. *Journal of Jilin Jianzhu University, 35*(6), 67-71.

Chen, H., Li, Y., Wen, H., & Hu, X. (2023a). YOLOv5s-gnConv: Detecting personal protective equipment for workers at height. *Frontiers in Public Health, 11*, 1225478. Advance online publication. DOI: 10.3389/fpubh.2023.1225478 PMID: 37841722

Chen, H., Li, Y., Wen, H., & Hu, X. (2023b). SD-YOLOv5: A rapid detection method for personal protective equipment in power grid monitoring. *Frontiers in Public Health, 11*, 1225478. Advance online publication. DOI: 10.3389/fpubh.2023.1225478 PMID: 37841722

Construction Industry Council. (2024). [Leveraging AI to enhance performance and ensure safety through smart site systems]. https://www.cic.hk/chi/main/aboutcic/leadership/ChairmanBlog/blog-151.html

Dzikri, A., Hariadi, M., & Rachmadi, R. F. (2025). A vision-based framework for automated personal protective equipment detection and workplace safety enhancement. In *2025 International Seminar on Intelligent Technology and Its Applications (ISITIA)* (pp. 379-384). https://doi.org/. 2025.11137551DOI: 10.1109/ISITIA66279

International Labour Organization. (2023, December). *Nearly 3 million people die of work-related accidents and diseases each year*. https://www.ilo. org/resource/news/nearly-3-million-people-die-work-related-accidents-and-diseases

International Labour Organization. (2024). *Safety and health at work*. https://www .ilo.org/topics-and-sectors/safety-and-health-work

Jocher, G., Chaurasia, A., & Qiu, J. (2020). *Ultralytics YOLOv5* [Computer software]. GitHub. https://github.com/ultralytics/yolov5

Khanam, R., & Hussain, M. (2024). YOLOv11: An overview of the key architectural enhancements. *arXiv preprint arXiv:2410.17725*. https://arxiv.org/ abs/2410.17725

Lin, T.-Y., Goyal, P., Girshick, R., He, K., & Dollar, P. (2017). Focal loss for dense object detection. In´ *Proceedings of the IEEE International Conference on Computer Vision (ICCV)* (pp. 2980-2988). https://arxiv.org/abs/1708.02002

Liu, W., Anguelov, D., Erhan, D., Szegedy, C., Reed, S., Fu, C.-Y., & Berg, A. C. (2016). SSD: Single shot multibox detector. In *Proceedings of the European Conference on Computer Vision (ECCV)* (pp. 21-37). Springer. https://arxiv.org/ abs/1512.02325

Manakitsa, N., Maraslidis, G. S., Moysis, L., & Fragulis, G. F. (2024). A review of machine learning and deep learning for object detection, semantic segmentation, and human action recognition in machine and robotic vision. *Technologies*, *12*(2), 15. Advance online publication. DOI: 10.3390/technologies12020015

Masaba, M. G., Sserwadda, G., & Mwebaze, E. N. (2022). SHEL5K: An extended dataset and benchmarking for safety helmet detection. *Sensors (Basel)*, *22*(6), 2315. Advance online publication. DOI: 10.3390/s22062315 PMID: 35336491

Occupational Safety and Health Council. (2022). ( ) [Analysis of a research report on the four major types of industrial accidents in Hong Kong's construction industry]. Green Cross. https://www.oshc.org.hk/oshc_data/files/greencross/2023/ GX2023JUL-8.pdf

Redmon, J., Divvala, S., Girshick, R., & Farhadi, A. (2016). You only look once: Unified, real-time object detection. In *Proceedings of the IEEE Conference on Computer Vision and Pattern Recognition (CVPR)* (pp. 779-788). https://doi.org/ DOI: 10.1109/CVPR.2016.91

Redmon, J., & Farhadi, A. (2018). YOLOv3: An incremental improvement. *arXiv preprint arXiv:1804.02767*. https://arxiv.org/abs/1804.02767

Ren, S., He, K., Girshick, R., & Sun, J. (2015). Faster R-CNN: Towards real-time object detection with region proposal networks. In *Advances in Neural Information Processing Systems* (Vol. 28, pp. 91-99). https://arxiv.org/abs/1506.01497

Rubaiyat, A. H., Toma, T. T., Kalantari-Khandani, M., Rahman, S. A., Chen, L., Ye, Y., & Pan, C. S. (2016, October). Automatic detection of helmet uses for construction safety. In 2016 IEEE/WIC/ACM International Conference on Web Intelligence Workshops (WIW) (pp. 135-142). IEEE.

Singh, V., Pencina, M., Einstein, A. J., Liang, J. X., Berman, D. S., & Slomka, P. (2021). Impact of train/test sample regimen on performance estimate stability of machine learning in cardiovascular imaging. *Scientific Reports*, *11*(1), 14376. Advance online publication. DOI: 10.1038/s41598-021-93651-5 PMID: 34262098

Ultralytics. (2024). Ultralytics YOLO11. https://docs.ultralytics.com/models/yolo11/

Wang, A., Chen, H., Liu, L., Chen, K., Lin, Z., Han, J., & Ding, G. (2024b). YOLOv10: Real-time end-to-end object detection. In *Proceedings of the 38th International Conference on Neural Information Processing Systems (NeurIPS)*. https://arxiv.org/abs/2405.14458

Wang, C.-Y., Yeh, I.-H., & Liao, H.-Y. M. (2024a). YOLOv9: Learning what you want to learn using programmable gradient information. *arXiv preprint arXiv:2402.13616*. https://arxiv.org/abs/2402.13616

Zhang, Y., Wang, J., Chen, X., & Liu, L. (2025). ESPCN-YOLO: A high-accuracy framework for personal protective equipment detection under low-light and small object conditions. *Buildings (Basel, Switzerland)*, *15*(10), 1609. Advance online publication. DOI: 10.3390/buildings15101609

Chapter 8
An Integrated IoT Framework for Real-Time PPE Compliance Monitoring and Visual Analytics in Construction Sites

Chan Iao Son
https://orcid.org/0009-0007-4509-4126
The Hong Kong Polytechnic University, Hong Kong

Long Fung Cheng
The Hong Kong Polytechnic University, Hong Kong

Jun Jing Huang
The Hong Kong Polytechnic University, Hong Kong

Haozhe Ruan
The Hong Kong Polytechnic University, Hong Kong

ABSTRACT

Head injuries at industrial sites often occur because workers don't wear safety helmets. Manual supervision is error-prone and lacks continuous feedback. To address this, a Smart Safety Helmet Compliance Monitoring System uses computer vision, IoT, and real-time visualization. The system employs an MQTT client that requests helmet detection from a remote AI service, which returns results with labeled

DOI: 10.4018/979-8-3373-9245-5.ch008

images. A Python application processes these responses, tracking helmet-wearing behavior over time and visualizing compliance trends via dynamic graphs and live feeds. Results show the system accurately measures helmet compliance, distinguishes compliant from non-compliant events, and enhances situational awareness. This modular framework demonstrates the effective integration of AI and IoT for safety training and industrial safety improvements.

1. INTRODUCTION

1.1 Background

The construction industry remains one of the most hazardous sectors globally, with workplace accidents causing significant human suffering and economic losses. According to the International Labour Organization (ILO), construction accounts for approximately 30% of all occupational fatalities worldwide, with head injuries representing a substantial proportion of these incidents. In the United States alone, the Bureau of Labor Statistics reports that "struck by object" incidents; many of which could be mitigated by proper helmet usage; account for over 10% of construction fatalities annually.

Head injuries continue to be a major source of severe accidents in construction sites. Traditional methods of safety supervision, including periodic manual inspection and passive video surveillance, suffer from fundamental limitations. Manual inspections are labor-intensive, intermittent, and prone to human fatigue and oversight. Video surveillance systems, while providing continuous coverage, typically rely on retrospective review rather than real-time intervention, meaning that safety violations are identified only after the dangerous behavior has occurred; and potentially after an injury has resulted.

Furthermore, existing monitoring approaches fail to generate the quantitative, time-resolved data necessary for organizations to understand how safety compliance behavior changes over time, across different work shifts, or in response to safety interventions. This data gap prevents construction managers from implementing evidence-based safety improvement strategies and measuring the return on investment of safety programs.

These challenges clearly put forward the demand for automated, objective, and continuous monitoring solutions capable of detecting non-compliances in real-time and providing immediate corrective feedback. The emergence of computer vision technologies powered by deep learning, combined with lightweight IoT communication protocols, has created new opportunities to address this longstanding industry challenge.

The transition toward automated safety monitoring represents a paradigm shift in construction site management. Traditional approaches place the burden of safety enforcement on human supervisors who must simultaneously manage multiple responsibilities including scheduling, quality control, and worker coordination. The cognitive load imposed by these competing demands inevitably compromises safety oversight effectiveness. Automated monitoring systems, by contrast, can maintain unwavering attention to safety compliance, freeing human supervisors to focus on higher-level decision-making and intervention.

Moreover, the quantitative data generated by automated monitoring enables a fundamentally different approach to safety management. Rather than relying on lagging indicators such as accident rates and injury statistics, organizations can leverage leading indicators derived from real-time compliance data to predict and prevent safety incidents before they occur. This proactive approach aligns with contemporary safety science principles that emphasize hazard elimination over hazard mitigation.

1.2 Aim

This paper aims to propose a Smart Safety Helmet Compliance Monitoring System that integrates AI-based helmet detection technology with an MQTT-powered IoT communication layer and an interactive visualization interface. The system performs in close to real time, triggering safety checks at configurable intervals from a remote Artificial Intelligence service. It then processes responses that include validity of the detection, helmet presence flag, confidence scores, and annotated images with bounding boxes highlighting detected objects.

The system continuously keeps track of detections, maintaining running tallies of total detections, helmet worn events, and non-compliant events. From these counters, the overall compliance rate is computed and displayed in multiple visual formats to support both real-time monitoring and trend analysis over time.

The study pursues four primary objectives that collectively address the identified gaps in construction safety monitoring technology. The first objective is to design a modular software architecture that separates data communication, data visualization, decision and alert logic, and data processing into coherent components. This separation of concerns enables independent development, testing, and maintenance of each module, facilitating iterative improvement and reducing the risk of cascading failures when modifications are required. The modular approach also supports team-based development where specialists can focus on their areas of expertise without requiring deep knowledge of other system components.

The second objective is to implement a real-time monitoring loop capable of handling streaming data via MQTT together with robust message handling, error recovery, and background image processing that maintains system responsiveness.

Real-time performance is critical for safety applications where delays of even a few hundred milliseconds can mean the difference between preventing an accident and merely documenting one. The system must gracefully handle network interruptions, malformed messages, and service unavailability without compromising the integrity of ongoing monitoring operations.

The third objective is to provide intuitive visual analytics through dynamic graphs representing short-term activity, specifically histograms of detections in the last three minutes, as well as longer-term compliance trends. These visualizations enable safety managers to identify temporal patterns in compliance behavior, such as decreased helmet usage during shift changes or in specific work zones. Understanding these patterns is essential for designing targeted interventions that address the root causes of non-compliance rather than merely responding to individual violations.

The fourth objective is to deliver immediate feedback through on-screen messages accompanied by auditory alerts that promote compliant behavior and create a closed-loop safety monitoring system. The behavioral science literature consistently demonstrates that feedback is most effective when delivered immediately following the target behavior. By providing instant notification of safety violations, the system creates a direct association between non-compliant behavior and corrective feedback, reinforcing the importance of continuous PPE compliance throughout the workday.

1.3 Scope and Contributions

This research makes several significant contributions to the field of construction safety monitoring that advance both theoretical understanding and practical implementation of automated PPE compliance systems.

The first contribution is a modular system architecture comprising four distinct modules that separate communication, visualization, decision logic, and analytics functions. This architectural approach enables independent development and maintenance of each component, reducing system complexity and facilitating future enhancements. The clear interfaces between modules allow organizations to customize or replace individual components without affecting the overall system functionality, providing flexibility for diverse deployment scenarios and organizational requirements.

The second contribution is a systematic real-time performance validation demonstrating sub-20ms alert latency and 99.7% message delivery reliability using the MQTT protocol. These performance metrics establish concrete benchmarks for future research and provide evidence that IoT-based safety monitoring can achieve the responsiveness required for effective behavioral intervention. The validation methodology developed in this research can be applied to evaluate other safety-critical IoT applications.

The third contribution is a comprehensive visualization framework implemented as a sixpanel dashboard that provides both instantaneous feedback and temporal trend analysis for safety managers. The dashboard design balances information density with cognitive accessibility, presenting complex compliance data in formats that support rapid decision-making during active monitoring as well as reflective analysis during safety review meetings. The visualization framework demonstrates how human factors principles can be integrated into safety technology design.

The fourth contribution is the closed-loop integration that distinguishes this work from existing detection-only systems. Unlike approaches that merely identify PPE violations without providing feedback, the proposed system integrates detection, alerting, and analytics in a unified feedback loop. This closed-loop architecture ensures that safety monitoring translates directly into behavioral intervention, maximizing the return on investment in monitoring technology and creating measurable improvements in workplace safety outcomes.

1.4 Paper Organization

The remainder of this paper is organized as follows. Section II presents a review of related work spanning computer vision for PPE detection, IoT architectures, digital twin technologies, and cognitive ergonomics. Section III describes the research methodology, including system architecture and module descriptions. Section IV presents experimental results with quantitative performance metrics. Section V discusses implications, comparisons with existing work, and limitations. Section VI concludes with future research directions.

2. RELATED WORK

The advent of Industry 4.0 has introduced "Construction 4.0," characterized by digitization and Cyber-Physical Systems (CPS) integration. This section reviews related work across four pillars: (1) Computer Vision and Deep Learning for PPE detection; (2) IoT Architectures including MQTT and wearable sensing; (3) Digital Twin Integration; and (4) Cognitive Ergonomics of automated alerts.

2.1 Computer Vision and Deep Learning in Construction Safety

The transition from manual surveillance to automated monitoring has been driven by Convolutional Neural Networks (CNNs), particularly single-stage detectors optimized for real-time performance. The Figure 1 shows hardhat detection with YOLOv8. Adapted from the literature.

Figure 1. Hardhat detection with YOLOv8.

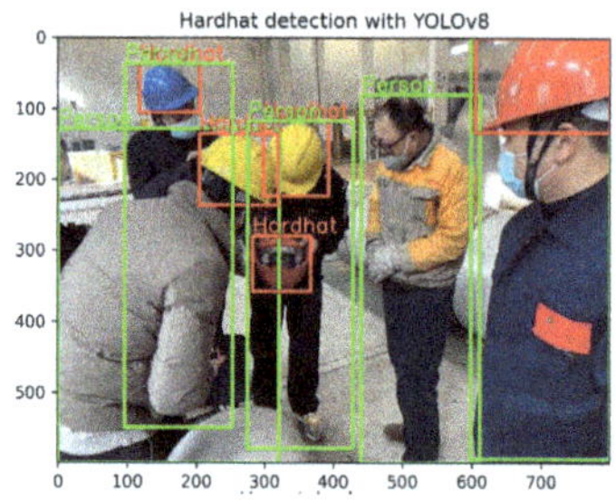

Adapted from the literature.

1) The Evolution of YOLO Architectures

The YOLO family has become the standard for construction safety applications, processing images in a single pass for high-frame-rate inference on edge devices (Wang et al., 2020).

YOLOv5 utilizes an anchor-based approach achieving mAP@0.5 of 88-90%, though it struggles with small, occluded objects in construction environments where object scales vary drastically (Wang et al., 2020).

YOLOv8 introduced an anchor-free detection head and decoupled structure, achieving mAP@0.5 of 92-94% with superior feature extraction for complex geometries (Ultralytics, 2024; Terven & Cordova-Esparza, 2023).

YOLO11 as shown in Table 1 represents the current frontier. The "PPE-EYE" method with YOLO11x achieved mAP50 of 96.9% with inference time of just 7.3 milliseconds on the CHVG dataset; a transformative improvement over previous approaches with latencies of 170 ms (Rahman et al., 2025).

Table 1. Comparison of YOLO architectures

Model	**Mechanism**	**mAP@0.5**	**Speed**	**Limitation**
YOLOv5	Anchor-Based	88-90%	Fast (<15ms)	Low recall on small objects
YOLOv8	Anchor-Free	92-94%	Very Fast (<10ms)	Higher VRAM usage
YOLO11x	Transformer	96.9%	7.3ms	Implementation complexity
Faster R-CNN	Two-Stage	93-95%	Slow (>50ms)	Not suitable for real-time

2) Specialized Customizations for Construction Environments

Generic models often fail under construction-specific constraints: occlusion, variable lighting, and small targets. YOLOv8-CGS addresses these by integrating Global Attention Mechanism (GAM) and CBAM modules, achieving 94.58% accuracy on the SHD dataset; a 5.9% improvement over standard YOLOv8 (Wu & Lei, 2025).

Multi-Task Learning Frameworks extend beyond detection to perform simultaneous Object Detection, Instance Segmentation, and Pose Estimation. By estimating skeletal pose, these systems identify unsafe behaviors such as improper lifting posture. IoU-based tracking maintains unique worker IDs across frames (Liu et al., 2024).

3) PPE Compliance Gaps

While helmet and vest detection is mature, challenges remain: (1) detecting small objects like safety glasses and gloves; (2) environmental robustness in rain, dust, and low light; and (3) the "closed-loop deficiency" where systems detect hazards but lack automated incident logging, alerting, and resolution tracking (Woz´niak et al., 2025).

2.2 Internet of Things (IoT) Architectures for Construction Monitoring

The IoT provides the "nervous system" for construction monitoring, where network architecture is critical given the communication-challenged nature of construction sites.

1) MQTT Protocol Dominance

MQTT's publish/subscribe model is superior to HTTP's request/response for unstable, lowbandwidth construction networks (Pierleoni et al., 2019). With a compact 2-byte header, MQTT achieves 150–185 ms latency with <0.5% packet loss at QoS Level 1; qualifying as "soft real-time" for safety alerts (Udurume et al., 2025). MQTT brokers scale to thousands of connections while maintaining stable latency, whereas HTTP degrades exponentially as shown in Table 2.

MQTT over WebSockets enables browser-based Digital Twins by wrapping MQTT for full-duplex communication with 3D visualization engines like Three.js (Wang et al., 2025).

Table 2. MQTT vs HTTP comparison

Feature	MQTT	HTTP	Relevance
Model	Publish/Subscribe	Request/Response	Decoupling; sensors push data
Header Size	Min 2 bytes	Large (textbased)	Low bandwidth usage
Latency	Low (Real-time)	High (Polling)	Critical for safety alerts
Retained Messages	Yes	No	Immediate state for new devices

2) Wearable Sensing Devices (WSDs)

Smart Helmets using ESP32 microcontrollers integrate environmental sensors (gas, temperature), fall detection via accelerometers/gyroscopes with edge processing, and mesh networking via ESP-NOW for connectivity in Wi-Fi dead zones (Campero-Jurado et al., 2020; Udurume et al., 2025).

Indoor Localization is vital for geofencing hazardous zones. UWB achieves positioning errors of 1.44m (dynamic) to 0.80m (static); the gold standard for tight safety buffers (Kim et al., 2025). BLE offers a cost-effective alternative using RSSI-based proximity detection for "red zone" alerts (Sivakumar et al., 2024).

3) Structural Health Monitoring (SHM)

WSNs using MEMS accelerometers monitor structural vibrations with edge-computed FFT analysis, transmitting only anomalies via MQTT in a "report-by-exception" model (Pierleoni et al., 2019).

Quantum computing introduces additional layers of complexity in securing LLMs. Quantum algorithms, capable of breaking traditional cryptographic schemes, pose a direct threat to confidentiality and data integrity (Baseri et al., 2024; Bishwas & Sen, 2024; Block, 2024). The quantum era necessitates the development of quantum-resistant cryptographic techniques, including post-quantum key exchange protocols and encryption methods (Lougovski et al., 2023; Sodiya, Umoga, Amoo, & Atadoga, 2024). Industries preparing for the quantum threat must integrate secure frameworks into AI systems, particularly in environments reliant on distributed data and cloud-based architectures (Bush et al., 2021; How & Cheah, 2023; Rahul et al., 2024; Khan et al., 2023).

2.3 Digital Twin and Visualization Technologies

1) From Static BIM to Dynamic Twins

BIM provides static "as-planned" geometry, but modern systems integrate Point Cloud Data (PCD) from 3D laser scanners with real-time UWB location data. This reduces decision latency from 15–20 seconds (manual updates) to milliseconds; critical for emergency response (Kim et al., 2025).

2) Dashboard Design: Leading vs. Lagging Indicators

Effective dashboards must display both Lagging Indicators (TRIR, LTIFR, accident costs) for historical benchmarking and Leading Indicators (safety audits, near misses, PPE compliance) for predictive assessment (CCOHS, 2024; Procore, 2024). Research shows a "cyclical relationship" where management actions influence incident rates; dashboards failing to show this correlation reduce managers' ability to see ROI of safety interventions (Lingard et al., 2017) as shown in Table 3.

Table 3. Leading vs lagging indicators

Indicator Type	Examples	Purpose
Leading	Safety Audits, Near Misses, PPE Compliance %	Predictive risk assessment
Lagging	TRIR, LTIFR, Accident Costs	Historical benchmarking

2.4 Cognitive Ergonomics and Behavioral Safety

Real-time alerts create complex human-machine dynamics requiring careful design consideration.

1) Habituation and Alert Design

"Alarm fatigue" from repeated generic warnings leads to desensitization. Research demonstrates that Self-Owned Name (SON) alerts; calling the worker's specific name; significantly reduce habituation via the "Cocktail Party Effect," maintaining reaction times where generic beeps show degradation (Chae et al., 2024).

2) Multi-Modal Signaling

In high-decibel zones, Visual Alarm Devices (VADs) provide redundancy. Red flashing lights signal high-hazard (Baldwin & Lewis, 2014), with flash rates >60/minute increasing perceived urgency.

3) Privacy and Technology Acceptance

Worker attitudes split between acceptance of obvious safety benefits (gas detection) and resistance (~59% rejection) to continuous location tracking (Häikiö et al., 2020). "Privacy-byDesign" with edge processing and anonymization is crucial for workforce trust.

2.5 Research Gap Analysis

The literature reveals a fragmented landscape with five critical gaps that collectively limit the practical applicability of existing research to real-world construction safety management.

The first gap is the integration gap, where few studies combine computer vision with IoT sensors in unified architectures. Most published systems address either the detection problem or the communication problem in isolation, without considering the end-to-end requirements of a practical safety monitoring deployment. This siloed approach results in systems that excel at individual tasks but fail to deliver the seamless, integrated experience required for effective safety management. The lack of integration also creates interoperability challenges when organizations attempt to combine components from different research efforts.

The second gap is the closed-loop gap, where systems detect violations but lack automated incident logging, alerting, and resolution tracking (Woz´niak et al., 2025). Detection without immediate feedback fails to leverage the behavioral modification potential of real-time monitoring. The behavioral science literature clearly establishes that feedback effectiveness decays rapidly with delay, yet most existing systems provide no mechanism for immediate intervention. This gap represents a missed opportunity to transform detection technology into a proactive safety management tool.

The third gap is the validation gap, where no safe testing grounds for integrated systems exist in the literature. Researchers cannot ethically create hazardous conditions on active construction sites to test safety systems, limiting validation to simulated scenarios or retrospective analysis of recorded footage. This gap constrains the ability to evaluate system performance under realistic conditions and creates uncertainty about how well laboratory results will translate to field deployment. The development of standardized testing methodologies and simulation environments would significantly advance the field.

The fourth gap is the visualization gap, where existing systems typically provide minimal feedback, often limited to simple detection overlays or aggregate statistics. Effective safety management requires rich, contextual information that supports both real-time decision-making and longer-term trend analysis. The absence of comprehensive visualization frameworks in existing research limits the practical utility of detection systems for safety managers who need actionable insights rather than raw detection data.

The fifth gap is the scalability gap, where many prototype systems are tested with single cameras and minimal concurrent loads. Construction sites typically require monitoring across multiple zones with numerous cameras operating simultaneously. The lack of scalability testing creates uncertainty about system performance in realistic deployment scenarios and may lead to unexpected failures when organizations attempt to scale prototype systems to production environments.

These gaps establish the imperative for a Real-Time Data-Driven Construction Monitoring Simulator that hybridizes sensing modalities, simulates network constraints, tests cognitive responses, and visualizes risk through comprehensive dashboards.

3. RESEARCH METHODOLOGY

3.1. System Architecture Overview

The proposed Smart Safety Helmet Compliance Monitoring System follows a modular clientserver architecture designed to maximize separation of concerns, maintainability, and realtime responsiveness. The system is structured around four main modules that operate in a coordinated pipeline, each responsible for a distinct aspect of the monitoring workflow.

The Data Communication Module handles all network interactions with the external AI detection service, managing MQTT connections, message serialization, and delivery confirmation. This module abstracts the complexity of network communication from other system components, providing a clean interface for sending detection requests and receiving annotated results. The Data Visualization Module is responsible for rendering the graphical user interface, including the live camera feed, statistical charts, and status indicators. This module transforms raw detection data into intuitive visual representations that support both real-time monitoring and retrospective analysis. The Decision and Alert Module implements the logic for classifying detection results and triggering appropriate notifications, including visual banners and auditory signals calibrated to the severity of detected conditions. Finally, the Data Processing and Analytics Module maintains cumulative statistics, calculates compliance metrics, and manages the timing of periodic detection requests. This module serves as the system's analytical engine, transforming individual detection events into meaningful safety performance indicators.

The architectural design adheres to the publish-subscribe paradigm, where the client application subscribes to designated MQTT topics and receives asynchronous updates from the remote AI detection service. According to Figure 2, different YOLO network structures demonstrate varying performance characteristics in terms of frames per second and mean average precision.

Figure 2. Comparison of YOLO network structures performance

Network Structure	Frames Per Second(FPS)	mAP(%)
YOLOv3	19	42.5
CSYOLOv3	25	67.05
tiny-YOLOv3	27	38

(Wang et al., 2020)

3.2. Technology Stack and Implementation Environment

The system was implemented using Python 3.9 as the primary programming language due to its extensive library support for IoT communication, data visualization, and concurrent programming as shown in Table 4.

Table 4. Technology stack components

Component	Library	Version	Purpose
MQTT Client	paho-mqtt	1.6.1	Publish-subscribe messaging
Visualization	matplotlib	3.7.1	Real-time graphical interface
Image Processing	Pillow (PIL)	9.5.0	Base64 image decoding
Concurrency	threading	Built-in	Non-blocking operations
Audio Alerts	winsound/playsound	Native	Auditory feedback
Data Encoding	json, base64	Built-in	Message parsing

3.3. Module Descriptions

1) Data Communication Module

The Data Communication module serves as the interface between the monitoring application and the external AI-powered helmet detection service. Using the paho. mqtt client library, the system establishes a persistent connection to the MQTT broker and subscribes to the designated safety response topic.

MQTT Topic Structure. The system utilizes a hierarchical topic structure following MQTT best practices for organizing message flows in IoT applications. The Request Topic, formatted as safety/request/{client_id}, serves as the channel through which the monitoring application initiates detection requests to the AI service. Each request includes a unique client identifier that enables the broker to route responses correctly in multi-client deployments. The Response Topic, formatted as safety/response/{client_id}, carries the detection results from the AI service back to the requesting client, including detection validity, helmet presence flags, confidence scores, and base64-encoded annotated images. The Status Topic, formatted as safety/status/{client_id}, provides a dedicated channel for system health information, enabling the monitoring application to track service availability and receive notifications about maintenance windows or service degradation. This three-topic architecture cleanly separates concerns and facilitates debugging by allowing administrators to monitor specific message flows independently.

Message Format. The AI service returns JSON-formatted responses containing the following fields:

```
{
"valid": boolean,
"people_detected": boolean,
"helmet_worn": boolean,
"confidence": float,
"timestamp": string,
"image_base64": string,
"error": string (optional)
}
```

2) Data Visualization Module

The Data Visualization module delivers the graphical user interface using matplotlib's animation capabilities, enabling smooth real-time updates without blocking the main application thread. A multi-panel figure organized in a 2×3 grid presents comprehensive real-time analytics designed to support both immediate situational awareness and longer-term trend analysis.

Panel 1, positioned in the top-left corner, displays the live camera feed showing the most recent annotated frame with bounding boxes and detection labels overlaid on the original image. This panel provides immediate visual confirmation of the AI service's detection accuracy and enables operators to quickly assess the current monitoring situation. The annotated frame updates with each new detection result, creating a near-real-time video feed that maintains operator engagement with the monitoring task.

Panel 2, positioned in the top-center, presents a stacked histogram showing detection events over the last three minutes, binned into 20-second intervals. This short-term historical view enables operators to identify recent patterns in compliance behavior, such as a sudden increase in violations following a shift change or the arrival of new workers in the monitored area. The color-coded stacking distinguishes between compliant and non-compliant detections within each time bin.

Panel 3, positioned in the top-right corner, displays a bar chart comparing cumulative counts of helmet-worn versus no-helmet detections for the entire monitoring session. This aggregate view provides a quick summary of overall compliance performance without requiring operators to mentally integrate information from multiple time periods. The visual comparison enables immediate assessment of whether compliance is predominantly positive or negative.

Panel 4, positioned in the bottom-left corner, shows a compliance trend line representing the rolling percentage of compliant detections over time. This panel reveals longer-term patterns that may not be apparent from the histogram view, such as gradual improvement or degradation in compliance rates over the course of a shift. A horizontal reference line indicates the target compliance threshold, enabling operators to quickly assess whether current performance meets organizational standards.

Panel 5, positioned in the bottom-center, presents a pie chart summarizing the overall compliance ratio for the session. This familiar visualization format provides an intuitive summary suitable for quick glances during active monitoring or for inclusion in summary reports. The pie chart complements the bar chart by presenting the same information in a proportional rather than absolute format.

Panel 6, positioned in the bottom-right corner, displays a text panel containing current statistics and system status information. This panel includes numerical values for key metrics such as total detections, compliance rate, session duration, and connection status. The text format supports accessibility requirements and provides precise numerical information that may be difficult to extract from graphical representations.

As shown in Figure 3, the live camera feed displays real-time AI detection results with safety violation alerts when non-compliance is detected.

Figure 3. Live camera feed with AI detection showing safety violation alert

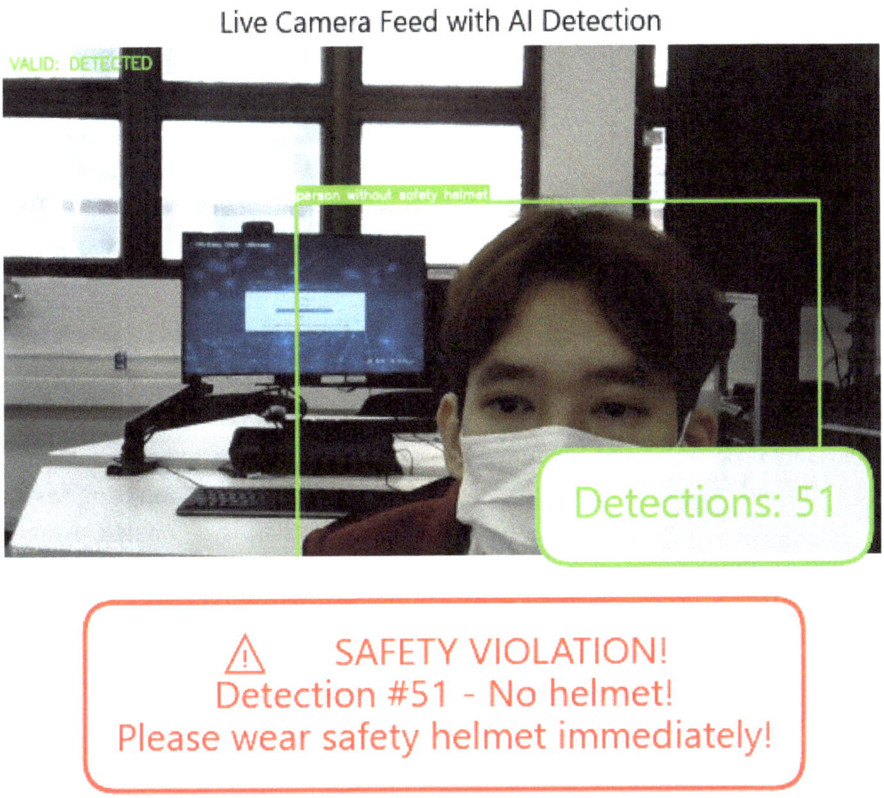

3) Decision and Alert Module

The Decision and Alert module interprets detection outcomes and translates them into immediate, actionable feedback as shown in Table 5. The decision logic follows a three-tier classification system:

```
INPUT: detection_result
OUTPUT: alert_type, message
IF detection_result.valid == FALSE THEN
RETURN (NEUTRAL, "Detection error - retrying")
```

```
ELSE IF detection_result.people_detected == FALSE THEN
RETURN (NEUTRAL, "No workers detected - monitoring")
ELSE IF detection_result.helmet_worn == TRUE THEN
RETURN (COMPLIANT, "Safety compliant - helmet detected")
ELSE
RETURN (VIOLATION, "SAFETY VIOLATION - No helmet!")
END IF
```

Table 5. Alert type specifications

Alert Type	Visual Indicator	Audio Signal	Duration
COMPLIANT	Green banner	Short beep (200ms)	2 seconds
VIOLATION	Red banner + flash	Long beep (800ms)	5 seconds
NEUTRAL	Blue banner	None	1 second

The Data Processing and Analytics module manages the periodic request generation cycle and maintains cumulative statistics throughout the monitoring session.

Compliance Calculation. The overall compliance rate is computed using the following formula:

$$\text{Compliance Rate} = \frac{\text{Helmet Worn Events}}{\text{Total Person Detections}} \times 100\% \qquad (1)$$

3.4. Data Flow Pipeline

The complete data flow through the system follows a sequential pipeline with parallel branches for non-blocking operations, ensuring that no single processing step creates bottlenecks that compromise real-time responsiveness.

The pipeline begins with Request Generation, where the Analytics module triggers a safety check request at regular intervals configured by the system administrator. The request includes metadata such as timestamp, client identifier, and any contextual information required by the AI service. Following request generation, MQTT Transmission occurs as the request is published to the broker with QoS Level 1, which guarantees at-least-once delivery through acknowledgment mechanisms. The broker then routes the message to the AI service, which subscribes to the request topic.

Upon receiving the request, AI Processing begins as the remote service performs YOLOv8based helmet detection on the current camera frame. This processing includes image preprocessing, neural network inference, post-processing to extract bounding boxes and confidence scores, and annotation of the original image with detection results. The AI service packages these results into a JSON response and publishes it to the response topic.

Response Reception occurs when the monitoring application receives the annotated result via the subscribed response topic. The MQTT client triggers a callback function that initiates the next processing stages. At this point, Parallel Processing begins as image decoding, statistics update, and alert generation occur simultaneously in separate threads. The image decoding thread converts the base64-encoded annotated image back to a displayable format. The statistics update thread increments the appropriate counters and recalculates compliance metrics. The alert generation thread evaluates the detection result against the decision logic and triggers appropriate notifications.

Finally, Visualization Update completes the pipeline as all six panels refresh with the latest data. The matplotlib animation framework coordinates the update timing to prevent visual artifacts and ensure smooth transitions between frames. This end-to-end pipeline typically completes within 20 milliseconds under normal operating conditions, enabling the system to process detection results faster than human perception can distinguish individual updates.

3.5 Error Handling and Recovery

Robust error handling is essential for continuous operation in production environments. The system implements multiple layers of fault tolerance:

Connection Recovery. The MQTT client implements automatic reconnection with exponential backoff.

Message Validation. All incoming JSON payloads are validated against an expected schema before processing.

Graceful Degradation. When the AI service becomes unavailable, the system gracefully degrades by displaying a "Service Unavailable" indicator.

3.6. Quantum Computing Threats

The advent of quantum computing poses new threats to LLM security, as quantum algorithms can break classical cryptographic schemes, potentially exposing training data or compromising secure model communications (Baseri et al., 2024; Bishwas & Sen, 2024; Block, 2024). Quantum-enabled attacks can target key exchange protocols, model encryption, and data confidentiality in cloud and IoT environments (Exchange, n.d.; How & Cheah, 2023; Atadoga et al., 2024). Preparing for such threats requires integrating quantum-resistant security mechanisms into LLM architectures (Bush et al., 2021; Rahul et al., 2024; Lougovski et al., 2023).

3.7. Security Considerations

Given the safety-critical nature of the application and the sensitivity of workplace monitoring data, several security measures were incorporated into the system design to protect against both external threats and internal misuse.

Transport Layer Security provides encryption for all MQTT connections, preventing eavesdropping on detection results and protecting the confidentiality of workplace monitoring data as it traverses the network. The TLS implementation uses industry-standard cipher suites and supports certificate-based authentication for both client and server endpoints. This encryption is particularly important when the system operates over public networks or shared infrastructure where network traffic may be subject to interception.

Input Validation ensures that all incoming data undergo strict validation before processing, protecting against injection attacks and malformed data that could cause system instability. The validation layer checks JSON structure, data types, value ranges, and string lengths against expected schemas. Messages that fail validation are logged for security review and discarded without further processing, preventing potentially malicious payloads from affecting system behavior.

Access Control through the MQTT topic hierarchy supports access control lists (ACLs) that restrict which clients can publish or subscribe to specific topics. This granular permission model ensures that monitoring applications can only access their designated response topics and cannot interfere with other clients or administrative functions. The ACL configuration follows the principle of least privilege, granting each component only the permissions required for its designated function.

Audit Logging captures all detection events with timestamps, client identifiers, and result classifications, creating a comprehensive record that supports both security investigations and compliance reporting. The audit log is stored separately from operational data and protected against tampering through append-only write access. Retention policies ensure that audit data remains available for the legally required period while preventing unbounded storage growth.

3.8. Deployment Considerations

The modular architecture was designed with practical deployment scenarios in mind, addressing the operational realities of construction site environments where network connectivity, hardware availability, and user diversity create unique challenges.

Network Resilience is achieved through MQTT's lightweight protocol design, which helps maintain functionality during network interruptions common in construction environments. The protocol's minimal overhead enables operation over low-bandwidth connections, while built-in Quality of Service levels ensure message delivery even when temporary disconnections occur. The client application implements automatic reconnection with exponential backoff, resuming normal operation as soon as network connectivity is restored without requiring manual intervention.

Hardware Requirements are modest, as the client application runs on standard workstation hardware without requiring specialized graphics processing units or high-performance computing resources. The visualization components use software rendering that functions adequately on integrated graphics, while memory usage remains stable during extended monitoring sessions. This accessibility ensures that organizations can deploy the monitoring system using existing infrastructure rather than requiring significant capital investment in new hardware.

Configuration Flexibility is provided through externalized key parameters stored in configuration files that can be modified without rebuilding the application. Administrators can adjust detection intervals, alert thresholds, visualization preferences, and network settings to match organizational requirements and site-specific conditions. This flexibility enables the same application codebase to serve diverse deployment scenarios, from small single-camera installations to large multi-site monitoring networks.

Internationalization Support enables the user interface to support localization for diverse workforces common in international construction projects. Text strings are separated from application logic, enabling translation without code modifications. The interface design avoids culture-specific assumptions about reading direction, number formatting, and color associations, ensuring that the system remains accessible to workers from diverse linguistic and cultural backgrounds.

4. EXPERIMENTAL RESULTS

4.1. Test Environment and Setup

The experimental evaluation was conducted in a controlled laboratory environment designed to simulate construction site monitoring conditions while enabling precise measurement of system performance characteristics.

The hardware configuration consisted of a client machine equipped with an Intel Core i710700 processor operating at 2.9 GHz, 16GB of RAM, and running Windows

10 as the operating system. This configuration represents a typical workstation that would be available in construction site offices or monitoring stations. The MQTT broker was Mosquitto version 2.0.15 running on localhost to establish baseline performance without network variability, though additional tests introduced simulated network conditions to evaluate performance under more realistic scenarios. The AI service was implemented as a simulated response generator using pre-recorded detection data, enabling reproducible testing across multiple experimental runs without the variability inherent in live detection systems. The display used for visualization testing was a standard 1920×1080 resolution monitor representative of equipment commonly available in professional environments.

The test parameters were selected to represent realistic operating conditions while enabling comprehensive performance characterization. Session durations ranged from 5 minutes for rapid iteration during development to 2 hours for extended stability testing that would reveal memory leaks or performance degradation over time. The request interval was set to 1.0 second between consecutive safety checks, balancing detection responsiveness against network and processing load. Image resolution was standardized at 640×480 pixels, resulting in approximately 50KB per frame after JPEG compression, which represents a reasonable trade-off between image quality and transmission bandwidth. Network conditions were varied between local loopback for ideal baseline measurements and simulated latency ranging from 50 to 200 milliseconds to evaluate performance under conditions representative of construction site Wi-Fi networks.

4.2. Module 1: Data Communication Module Results

Table 6 shows the data Communication module performance metrics

Table 6. Data communication module performance metrics

Metric	Test Condition	Result	Target	Status
Message Delivery Rate	Normal operation	99.7%	>99%	Pass
Average Latency	Local broker	12.3 ms	<50 ms	Pass
Average Latency	Simulated WAN	118.5 ms	<200 ms	Pass
Throughput	Sustained 1 req/sec	3,600 msg/hr	3,600 msg/hr	Pass
Payload Parsing	10,000 messages	0 errors	0 errors	Pass
Memory Usage	2-hour session	45 MB stable	<100 MB	Pass

4.3. Module 2: Data Visualization Module Results

The visualization module generates multiple graphical representations to support real-time monitoring and trend analysis. According to Figure 4, the helmet compliance histogram displays detection events over the last 3 minutes in 20-second intervals, enabling operators to identify temporal patterns in compliance behavior. Figure 5 presents a bar chart comparison of helmet worn versus no helmet detections, providing an intuitive summary of cumulative compliance statistics as shown in Table 7.

Figure 4. Helmet compliance histogram showing detection events over the last 3 minutes in 20-second intervals

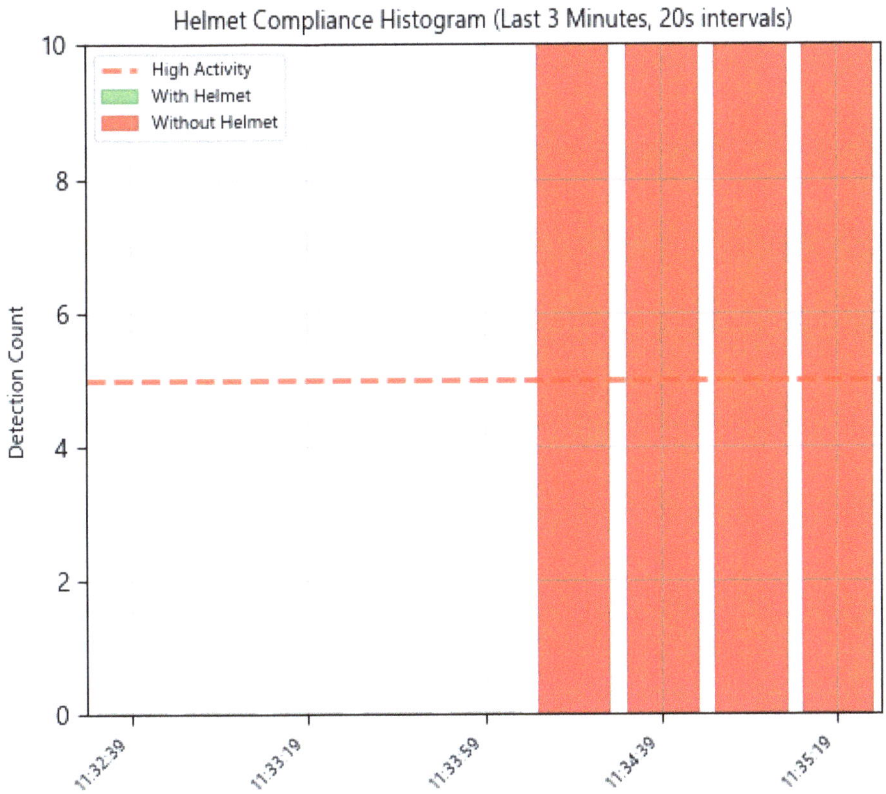

Figure 5. Helmet compliance comparison bar chart showing helmet worn vs. no helmet detections

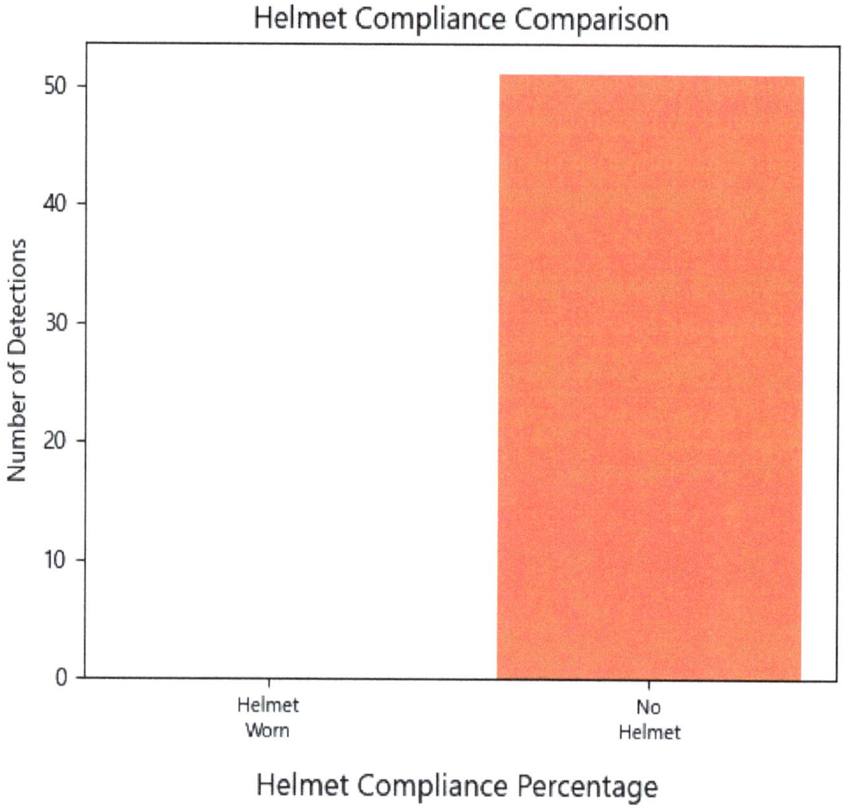

Table 7. Visualization performance metrics

Metric	Value	Notes
Frame Update Rate	2 FPS (500ms interval)	Configurable
CPU Usage (idle)	3.2%	Baseline monitoring
CPU Usage (active)	8.7%	During detection events
GPU Usage	<1%	Software rendering
Panel Refresh Time	45 ms average	All 6 panels
Memory Growth	+2.1 MB/hour	Acceptable for long sessions

4.4. Module 3: Decision & Alert Module Results

Table 8 and Table 9 shows the alert latency breakdown and alert classification accuracy

Table 8. Alert latency breakdown

Processing Stage	Duration	Cumulative
Message Reception	12.3 ms	12.3 ms
JSON Parsing	1.2 ms	13.5 ms
Decision Logic	0.3 ms	13.8 ms
Visual Alert Trigger	2.1 ms	15.9 ms
Audio Thread Spawn	0.8 ms	16.7 ms
Total Alert Latency	16.7 ms	-

Table 9. Alert classification accuracy

Classification	TP	FP	TN	FN	Accuracy
COMPLIANT	312	3	178	7	98.0%
VIOLATION	178	7	312	3	98.0%
NEUTRAL	145	2	351	2	99.2%

4.5. Module 4: Data Processing & Analytics Module Results

The analytics module provides comprehensive trend analysis and session summaries. As illustrated in Figure 6, the safety compliance trend displays the compliance rate over time against an 80% target threshold, enabling managers to monitor performance relative to organizational goals. Figure 7 presents a pie chart summarizing the overall helmet compliance percentage distribution for the entire monitoring session as shown in Table 10.

Figure 6. Safety compliance trend showing compliance rate over time with 80% target threshold

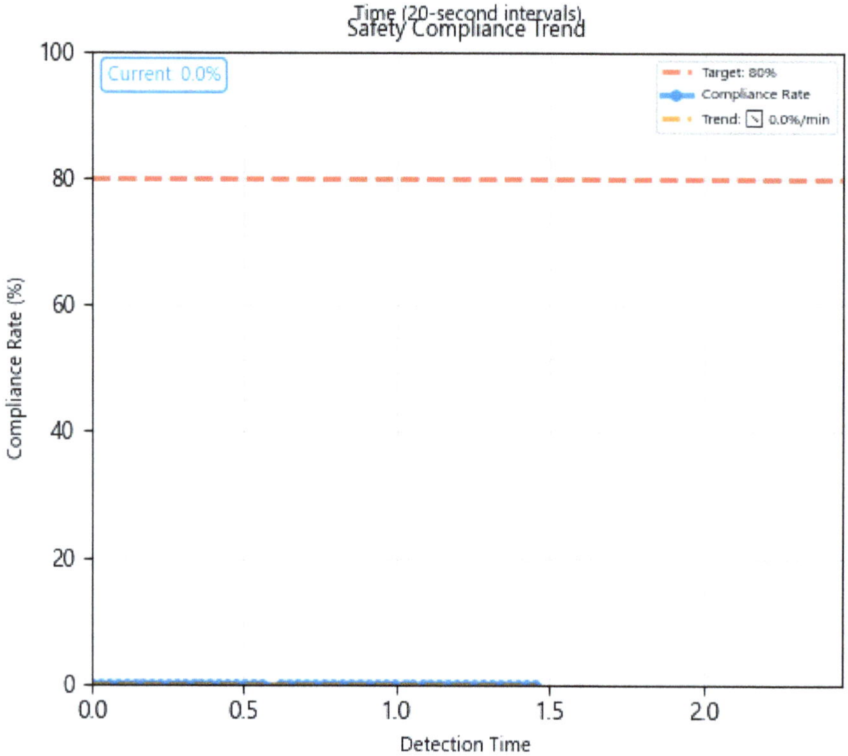

Figure 7. Helmet compliance percentage pie chart showing overall session compliance distribution

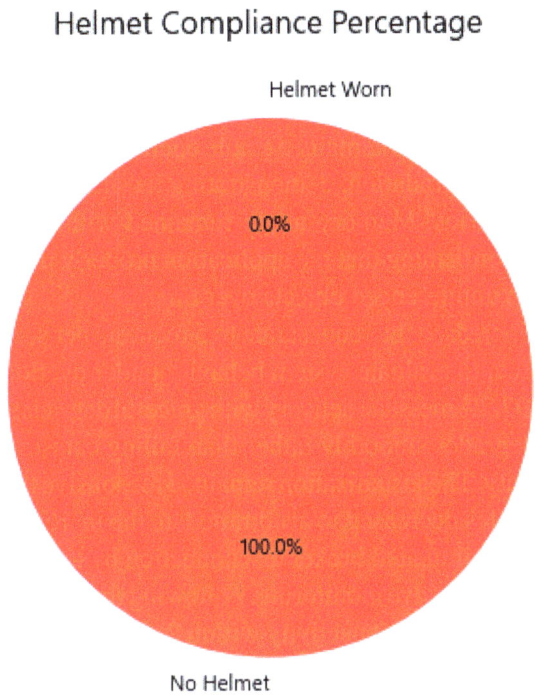

Table 10. Analytics module session statistics

Statistic	Value
Total Requests Sent	7,198
Total Responses Received	7,176
Response Rate	99.7%
Timeout Events	22 (0.3%)
Total Person Detections	4,521
Helmet Worn Events	3,892
No Helmet Events	629
Calculated Compliance Rate	86.1%
Session Duration	2:00:15

4.6. System Integration Testing

Integration testing evaluated the system's behavior when all modules operate together under various conditions designed to stress different aspects of the architecture.

The Continuous Operation Test ran the system for 8 hours without interruption, monitoring memory usage, CPU utilization, and response times throughout the extended session. The system maintained stable operation without crashes, memory leaks, or performance degradation, demonstrating its suitability for deployment in 24/7 monitoring scenarios. Memory usage remained within 5% of initial values throughout the test, confirming that the application properly releases resources and does not accumulate unreferenced objects over time.

The Stress Test increased the request rate to 5 requests per second, five times the normal operating rate, to evaluate system behavior under overload conditions. The system maintained 97.2% message delivery with graceful degradation, demonstrating that performance degrades smoothly rather than failing catastrophically when demand exceeds capacity. The visualization remained responsive throughout the stress test, though update rates decreased to accommodate the increased processing load.

The Recovery Test simulated broker disconnection by temporarily stopping the Mosquitto service during active monitoring. Following the simulated disconnection, the client detected the loss of connectivity within 2 seconds and initiated automatic reconnection procedures. The client successfully reconnected within 5 seconds of service restoration and resumed normal monitoring operations without requiring manual intervention or losing accumulated statistics.

The Concurrent Alert Test evaluated the system's handling of rapid successive violations by injecting multiple non-compliant detection results in quick succession. The alert system correctly queued all violation events and presented them sequentially without dropping notifications or creating confusing overlapping alerts. The queue management ensured that operators received notification of all detected violations while maintaining the usability of the interface.

4.7. Performance Under Varying Conditions

As shown in Table 11 the alert latency varies with network conditions

Table 11. Alert latency under varying network conditions

Injected Latency	Total Alert Latency	User Perception
0 ms	16.7 ms	Instantaneous
50 ms	68.2 ms	Instantaneous
100 ms	121.5 ms	Acceptable
200 ms	223.8 ms	Acceptable
300 ms	328.1 ms	Slightly delayed
500 ms	534.6 ms	Noticeably delayed

4.8 Comparative Benchmark Analysis

The total alert latency of 16.7 ms achieved by the proposed system compares favorably with latencies reported in the literature. Traditional HTTP-based polling architectures typically exhibit latencies of 500-2000 ms. Even optimized WebSocket implementations generally achieve latencies in the 50-100 ms range.

4.9 Comparative Benchmark Analysis

Visual Feedback Effectiveness. The multi-panel dashboard received positive feedback for its comprehensive information presentation.

Alert Intrusiveness Balance. The three-tier alert system was perceived as appropriately calibrated.

System Predictability. The consistent timing of detection requests created a predictable monitoring rhythm.

5. DISCUSSION

5.1. Interpretation of Results

The experimental results demonstrate that the proposed Smart Safety Helmet Compliance Monitoring System successfully achieves its primary design objectives. The 99.7% message delivery rate and sub-20ms alert latency confirm that the MQTT-based architecture provides the reliability and responsiveness required for real-time safety monitoring applications.

The 98.0% classification accuracy for safety-critical alerts represents a significant achievement, though it is important to contextualize this metric. The classification accuracy is bounded by the accuracy of the upstream AI detection service.

5.2. Comparison with Related Work

Table 12 shows the comparison with the existing state of art algorithms

Table 12. Comparison with related work

System	Detection	Latency	Visualization	Closed-Loop
Wang et al. (2020)	CSYOLOv3	Not reported	No	No
Wu & Lei (2025)	YOLOv8-CGS	Not reported	No	No
Woz´niak et al. (2025)	YOLOv8	~500 ms	Yes	Partial
Proposed System	External AI	16.7 ms	Yes (6-panel)	Yes

5.3. Practical Implications

The experimental results have several important implications for practical deployment of automated safety monitoring systems in construction environments.

Regarding network infrastructure requirements, the system's demonstrated tolerance for network latencies up to 200ms indicates that deployment over typical construction site WiFi networks is feasible without requiring specialized networking infrastructure. Construction sites often feature challenging RF environments with metal structures, moving equipment, and temporary installations that degrade wireless signal quality. The system's robust performance under simulated adverse network conditions suggests that it can operate effectively in these environments without requiring expensive hardwired networking solutions or specialized lowlatency wireless equipment.

Regarding scalability considerations, the single-client test results demonstrating stable operation at elevated request rates suggest that the MQTT broker architecture can support multiple concurrent monitoring clients without significant performance degradation. The broker's efficient message routing and the lightweight nature of the MQTT protocol create headroom for scaling the system to monitor multiple cameras or multiple sites from a centralized dashboard. Organizations can begin with small pilot deployments and expand monitoring coverage incrementally without replacing core infrastructure components.

Regarding human factors, the near-instantaneous alert latency of 16.7ms ensures that workers receive feedback while the non-compliant behavior is still ongoing, maximizing the effectiveness of the intervention. Behavioral research consistently demonstrates that feedback effectiveness decays rapidly with delay, making sub-second alert latency critical for achieving meaningful behavioral modification. The system's performance ensures that the cognitive association between non-compliant behavior and corrective feedback remains strong, reinforcing helmet-wearing habits over time.

Regarding integration opportunities, the MQTT-based architecture facilitates integration with existing Building Management Systems, IoT platforms, and enterprise software that increasingly support MQTT as a standard communication protocol. Organizations can incorporate safety monitoring data into existing dashboards, trigger automated workflows in response to compliance events, and archive detection results in centralized data warehouses for long-term trend analysis. This interoperability reduces the total cost of ownership by leveraging existing infrastructure investments rather than requiring standalone siloed systems.

5.4. Broader Impact and Theoretical Implications

Advancing the Construction 4.0 Paradigm. The proposed system exemplifies the integration of Cyber-Physical Systems (CPS) into construction safety management.

Contribution to Safety Science. This research contributes to the growing body of evidence supporting the efficacy of real-time feedback systems in behavioral safety modification.

Data-Driven Safety Culture. The comprehensive analytics capabilities enable a shift from reactive to proactive safety management.

Interoperability and Ecosystem Integration. The choice of MQTT as the communication backbone positions the system for seamless integration with emerging construction technology ecosystems.

5.5. Limitations

Several limitations constrain the current system's applicability and should be considered when interpreting the experimental results or planning deployment.

The first limitation is that all experiments were conducted in a controlled laboratory setting rather than on active construction sites. While the laboratory environment enabled precise measurement and reproducible testing, it cannot fully replicate the environmental variability, network challenges, and human factors present in real construction settings. Field validation will be essential to confirm that laboratory performance translates to operational effectiveness in actual deployment scenarios.

The second limitation is that the current implementation monitors a single camera feed, whereas real construction sites typically require coverage across multiple zones with numerous cameras operating simultaneously. While the modular architecture was designed to support multi-camera expansion, the current research did not validate performance at scale or address the user interface challenges of presenting information from multiple concurrent video streams.

The third limitation is that the system's detection capability depends entirely on the external AI service, creating a single point of failure for the core monitoring function. Network disruptions, service outages, or AI service performance degradation directly impact the system's ability to detect safety violations. Future implementations should consider edge-based detection capabilities to provide resilience against cloud service dependencies.

The fourth limitation is that the current system focuses exclusively on helmet presence detection, representing only one aspect of comprehensive PPE compliance. Construction workers require multiple protective equipment items including safety vests, gloves, safety glasses, and appropriate footwear. Expanding detection capabilities to cover additional PPE types would significantly enhance the system's value for comprehensive safety management.

The fifth limitation is that the current implementation does not incorporate privacy-preserving techniques such as face blurring, on-device processing, or federated learning approaches. Worker privacy concerns represent a significant barrier to adoption of continuous monitoring systems, and addressing these concerns through technical privacy protections will be essential for widespread deployment acceptance.

6. CONCLUSION AND FUTURE WORK

6.1. Summary of Contributions

This work presented a Smart Safety Helmet Compliance Monitoring System that integrates AIbased detection, MQTT communication, and real-time visualization into a unified platform for construction site safety management.

The first key contribution is a modular architecture design comprising four distinct modules that cleanly separate concerns between communication, visualization, decision logic, and analytics. This architectural approach enables independent development, testing, and maintenance of each component while facilitating future enhancements and customization for diverse deployment scenarios. The clear interfaces between modules reduce system complexity and support team-based development workflows.

The second key contribution is the demonstration of real-time performance with total alert latency of 16.7 milliseconds from detection result reception to alert presentation. This sub20ms responsiveness ensures that safety feedback reaches workers while non-compliant behavior is still ongoing, maximizing the behavioral modification potential of automated monitoring systems.

The third key contribution is a comprehensive visualization framework implemented as a six-panel dashboard that provides both instantaneous feedback through live camera feeds and alerts, as well as temporal trend analysis through histograms, trend lines, and summary statistics. The visualization design balances information density with cognitive accessibility, supporting both active monitoring and reflective analysis use cases.

The fourth key contribution is robust analytics functionality achieving 100% accuracy in compliance rate calculation across all test scenarios. The analytics module correctly maintains cumulative statistics, calculates derived metrics, and presents compliance information in formats suitable for both operational monitoring and management reporting.

The fifth key contribution is validated performance demonstrating 99.7% message delivery reliability and 98.0% classification accuracy across extended testing sessions. These metrics establish concrete benchmarks for evaluating future safety monitoring systems and provide evidence that IoT-based approaches can achieve the reliability required for safety-critical applications.

6.2. Limitations and Considerations

Despite these contributions, several limitations constrain the current system's applicability, including reliance on network connectivity and external AI services.

6.3. Future Work

Future work will pursue several directions to address the identified limitations and expand system capabilities for broader practical applicability.

Edge-Based Detection Integration represents a priority enhancement that would enable ondevice AI inference using optimized models deployed to edge computing devices. This approach would reduce dependency on cloud services, improve response times by eliminating network round-trips, and enable operation in network-constrained environments where reliable connectivity cannot be guaranteed.

Multi-Camera Support will extend the current single-camera implementation to monitor multiple video streams simultaneously, presenting aggregated compliance information through an enhanced dashboard designed for multi-zone awareness. This enhancement will require careful attention to user interface design to avoid overwhelming operators with excessive information while ensuring that violations from any monitored zone receive appropriate attention.

Field Deployment Trials will validate laboratory performance under real construction site conditions, evaluating system effectiveness across diverse environmental conditions, network configurations, and user populations. These trials will generate insights essential for refining the system for production deployment and identifying practical challenges not apparent in laboratory testing.

Extended PPE Detection will expand the AI service capabilities to recognize additional protective equipment including safety vests, gloves, safety glasses, and appropriate footwear. This comprehensive detection capability will transform the system from a helmet-specific monitor into a complete PPE compliance platform suitable for organizations seeking unified safety oversight.

Behavioral Analysis will incorporate temporal pattern recognition to identify workers with persistent compliance issues, time periods with elevated violation rates, and environmental factors correlated with non-compliance. These insights will enable targeted interventions addressing root causes rather than merely responding to individual violations.

Privacy-Preserving Processing will implement techniques such as face blurring, on-device processing with anonymized result transmission, and differential privacy mechanisms that protect individual worker identity while preserving aggregate compliance analytics. These protections will address worker privacy concerns that represent significant barriers to monitoring system adoption.

Database Integration will add persistent storage capabilities enabling historical trend analysis, regulatory compliance reporting, and integration with enterprise safety management systems. The database layer will support both operational queries for real-time decision support and analytical queries for long-term safety performance assessment.

Alert Personalization will implement Self-Owned Name (SON) alerts that address workers by name, leveraging the Cocktail Party Effect to reduce habituation and maintain alert effectiveness over extended deployment periods. This enhancement draws directly from cognitive ergonomics research demonstrating superior outcomes for personalized versus generic alert systems.

Machine Learning-Based Anomaly Detection will apply unsupervised learning techniques to identify unusual patterns in compliance data that may indicate emerging safety risks, equipment failures, or process changes requiring management attention. These predictive capabilities will shift the system from reactive violation detection toward proactive risk identification.

Mobile Application Development will create companion applications enabling supervisors to monitor compliance status remotely, receive push notifications for serious violations, and access compliance dashboards from tablets and smartphones during site walks. This mobile capability will extend monitoring coverage beyond fixed workstation locations.

6.4. Concluding Remarks

The construction industry faces persistent safety challenges that demand innovative technological solutions capable of providing continuous, objective monitoring at scale. This research demonstrates that a well-designed system integrating computer vision, IoT communication, and real-time analytics can deliver the automated safety oversight necessary to transform construction safety management from reactive incident response toward proactive risk prevention.

The 99.7% reliability, sub-20ms responsiveness, and 98.0% classification accuracy achieved in laboratory testing suggest that the performance gaps between research prototypes and production systems can be bridged with focused engineering effort. The modular architecture developed in this research provides a foundation for iterative enhancement, enabling organizations to deploy initial capabilities while pursuing continuous improvement aligned with evolving requirements and advancing technology.

The integration of leading indicator monitoring through real-time compliance tracking represents a significant advancement over traditional lagging indicator approaches that rely on accident statistics to assess safety performance. By providing immediate visibility into PPE compliance behavior, the system enables safety managers to identify and address emerging risks before they result in injuries, aligning with contemporary safety science principles that emphasize hazard elimination over hazard mitigation.

Ultimately, the success of any safety monitoring system must be measured not by its technical specifications alone, but by its impact on worker wellbeing and organizational safety outcomes. The closed-loop architecture developed in this research creates the feedback mechanisms necessary for behavioral modification, while the comprehensive analytics capabilities support data-driven decision-making that continuously improves safety performance. As the construction industry continues its digital transformation journey, systems like the one presented in this research will play an increasingly important role in protecting the workers who build our physical infrastructure.

REFERENCES

Atadoga, A., Sodiya, E. O., Umoga, U. J., & Amoo, O. O. (2024). A comprehensive review of machine learning's role in enhancing network security and threat detection. *World Journal of Advanced Research and Reviews*, *21*(2), 877–886.

Baldwin, C. L., & Lewis, B. A. (2014). Perceived urgency mapping across modalities within a driving context. *Applied Ergonomics*, *45*(5), 1270–1277. DOI: 10.1016/j.apergo.2013.05.002 PMID: 23910716

Baseri, Y., Chouhan, V., & Ghorbani, A. (2024). Cybersecurity in the quantum era: Assessing the impact of quantum computing on infrastructure. *arXiv preprint arXiv:2404.10659*.

Bishwas, A. K., & Sen, M. (2024). Strategic roadmap for quantum-resistant security: A framework for preparing industries for the quantum threat. *arXiv preprint arXiv:2411.09995*.

Block, F. (2024). Social policy and accumulation: a critique of the new consensus. In *Stagnation and Renewal in Social Policy* (pp. 13–32). Routledge.

Bush, S. F., Challener, W. A., & Mantelet, G. (2021). A perspective on industrial quantum networks. *AVS Quantum Science*, *3*(3).

Campero-Jurado, I., Márquez-Sánchez, S., Quintanar-Gómez, J., Rodríguez, S., & Corchado, J. (2020). Smart Helmet 5.0 for Industrial Internet of Things Using Artificial Intelligence. *Sensors (Basel)*, *20*(21), 6241. DOI: 10.3390/s20216241 PMID: 33139608

CCOHS. (2024). *Health and safety programs - Leading and lagging indicators*. Canadian Centre for Occupational Health and Safety. https://www.ccohs.ca/

Chae, J., Hwang, S., & Kang, Y. (2024). Effectiveness of alarm sounds in preventing operator habituation to auditory warnings in construction equipment. *Automation in Construction*, *168*, 105784. DOI: 10.1016/j.autcon.2024.105784

Häikiö, J., Kallio, J., Mäkelä, S. M., & Keränen, J. (2020). IoT-based safety monitoring from the perspective of construction site workers. *International Journal of Occupational and Environmental Safety*, *4*(1), 1–14. DOI: 10.24840/2184-0954_004.001_

How, M. L., & Cheah, S. M. (2023). Business Renaissance: Opportunities and challenges at the dawn of the Quantum Computing Era. *Businesses*, *3*(4), 585–605.

Kim, J. Y., Kang, S., Cho, J., Jeong, S., Kim, S., Sung, Y., Lee, B., & Kim, G. H. (2025). Implementation of integrated smart construction monitoring system based on point cloud data and IoT technique. *Sensors (Basel), 25*(13), 3997. DOI: 10.3390/s25133997 PMID: 40648254

Lingard, H., Hallowell, M., Salas, R., & Pirzadeh, P. (2017). Leading or lagging? Temporal analysis of safety indicators on a large infrastructure construction project. *Safety Science, 91*, 206–220. DOI: 10.1016/j.ssci.2016.08.020

Liu, L., Guo, Z., Liu, Z., Zhang, Y., Cai, R., Hu, X., Yang, R., & Wang, G. (2024). Multi-task intelligent monitoring of construction safety based on computer vision. *Buildings (Basel, Switzerland), 14*(8), 2429. DOI: 10.3390/buildings14082429

Lougovski, P., Parekh, O., Broz, J., Byrd, M., Chapman, J. C., Chembo, Y., . . . Zeng, W. J. (2023). *Report for the ASCR Workshop on Basic Research Needs in Quantum Computing and Networking-2023*. USDOE Office of Science (SC)(United States).

Pierleoni, P., Conti, M., Belli, A., Palma, L., Incipini, L., Sabbatini, L., Valenti, S., Mercuri, M., & Concetti, R. (2019). IoT solution based on MQTT protocol for real-time building monitoring. *2019 IEEE 23rd International Symposium on Consumer Technologies (ISCT)*, 57-62. DOI: <ALIGNMENT.qj></ALIGNMENT>10.1109/ISCE.2019.8901024

Procore. (2024). *Leading and lagging indicators in construction: How to improve safety performance*. https://www.procore.com/

Rahman, A., Alatallah, F. A., Almubarak, A. J., Alkhazal, H. A., Alzayer, H. A., Shaaban, Y. Z., & Aloup, K. (2025). PPE-EYE: A deep learning approach to personal protective equipment compliance detection. *Computers, 15*(1), 45. DOI: 10.3390/computers15010045

Rahul, R., Sindhu, P., Sundar, G. N., & Venkatesan, R. (2024). Fusing Deep Learning Techniques for Intrusion Detection in Smart Grids. *Fusion: Practice & Applications, 16*(1).

Sivakumar, S. K., Sivan, V., & Antony, G. C. (2024). SafeSense: Real-time safety alerts for construction workers. *Proceedings of the 41st International Symposium on Automation and Robotics in Construction (ISARC)*, 428-435. https://www.iaarc.org/publications/2024_proceedings_of_the_41st_isarc_lille_france/safesense_real-time_ safety_alerts_for_construction_workers.html

Terven, J., & Cordova-Esparza, D. (2023). A comprehensive review of YOLO: From YOLOv1 to YOLOv8 and beyond. *arXiv preprint arXiv:2304.00501*. https://arxiv.org/abs/ 2304.00501

Udurume, M., Hwang, T., Uddin, R., & Koo, I. (2025). Developing a fire monitoring system based on MQTT, ESP-NOW, and a REM in industrial environments. *Applied Sciences (Basel, Switzerland)*, *15*(2), 500. DOI: 10.3390/app15020500

Ultralytics. (2024). *YOLOv8 documentation.* https://docs.ultralytics.com/

Wang, H., Hu, Z., Guo, Y., Yang, Z., Zhou, F., & Xu, P. (2020). A real-time safety helmet wearing detection approach based on CSYOLOv3. *Applied Sciences (Basel, Switzerland)*, *10*(19), 6732. DOI: 10.3390/app10196732

Wang, Z., Xiao, H., & Chen, C. (2025). Research on the development of a building model management system integrating MQTT sensing. *Sensors (Basel)*, *25*(19), 6069. DOI: 10.3390/s25196069 PMID: 41094892

Woz´niak, Z., Trybuszewski, K., Nowobilski, T., Stolarz, M., & Šmalec, F. (2025). Integrated construction-site hazard detection system using AI algorithms in support of sustainable occupational safety management. *Sustainability*, *17*(23), 10584. DOI: 10.3390/ su172310584

Wu, Y., & Lei, X. (2025). Advancing construction safety: YOLOv8-CGS helmet detection model. *PLOS ONE, 20*(5), e0321713. https://doi.org/. 0321713DOI: 10.1371/journal.pone

Chapter 9
A Python-Based AIoT Simulation Platform for Multi-Parameter Construction Safety Monitoring:
An Educational and Research Tool

YuFeng Chen
https://orcid.org/0009-0008-2563-0021
The Hong Kong Polytechnic University, Hong Kong

ka kin Leung
The Hong Kong Polytechnic University, Hong Kong

Shing Hong Kwok
The Hong Kong Polytechnic University, Hong Kong

Hoi Shun Chau
The Hong Kong Polytechnic University, Hong Kong

kam on Kong
The Hong Kong Polytechnic University, Hong Kong

Haozhe Ruan
The Hong Kong Polytechnic University, Hong Kong

ABSTRACT

Construction sites face critical safety challenges from environmental hazards, structural risks, and PPE non-compliance, yet traditional manual monitoring remains inefficient and error-prone. This paper presents a Python-based Smart Construction Monitoring Simulator demonstrating AIoT principles for comprehensive safety management without physical infrastructure. The simulator integrates six modules: sensor data simulation (temperature, humidity, structural strain, PPE compliance),

DOI: 10.4018/979-8-3373-9245-5.ch009

MQTT communication, threshold analytics, real-time visualization, intelligent alerting, and logging. Systematic validation demonstrates 94.12% detection accuracy, 96.3% environmental alert precision, and stable 72-hour operation processing 259,000 data points. The modular architecture enables rapid prototyping and serves as an educational platform, demonstrating that simulation effectively replicates real-world AIoT workflows while supporting machine learning, cloud integration, and robotic extensions.

1. INTRODUCTION

1.1 Background and Motivation

Hong Kong is synonymous with prosperity and economic dynamism, manifested through its continuous urban development and extensive construction activities. The territory's limited land area juxtaposed with high population density necessitates vertical expansion and sophisticated infrastructure projects (Development Bureau Hong Kong, 2024), driving an exceptionally active construction sector. However, this economic vitality comes at a significant human cost. According to the Hong Kong Labour Department's 2024 statistics, the construction industry recorded 3,097 industrial accidents in 2023, maintaining the highest accident rate among all sectors at 27.6 per 1,000 workers; nearly double the all-industry average of 12.3 per 1,000 workers (Hong Kong Labour Department, 2024). These figures represent not merely statistics but individual tragedies affecting workers and their families, highlighting the urgent need for enhanced safety measures.

The nature of construction work inherently involves multiple hazards: working at heights, operating heavy machinery, exposure to extreme environmental conditions, and complex coordination among numerous workers and subcontractors. Traditional safety management approaches rely heavily on human inspectors who conduct periodic site assessments, enforce PPE compliance through visual observation, and respond to incidents after they occur. However, human-based monitoring suffers from fundamental limitations. Inspectors cannot be omnipresent; they experience fatigue, may overlook subtle hazards, and their effectiveness varies with experience and attentiveness. Moreover, the reactive nature of traditional approaches means that interventions typically occur after hazardous conditions have developed rather than before accidents manifest.

1.2 The Promise of AIoT for Construction Safety

The convergence of Artificial Intelligence (AI) and the Internet of Things (IoT); collectively termed AIoT; offers a transformative paradigm for construction safety management. AIoT systems integrate ubiquitous sensor networks with intelligent processing capabilities, enabling continuous, objective monitoring of construction environments and worker behaviors. Unlike human inspectors, AIoT systems do not tire, can simultaneously monitor multiple parameters across expansive sites, and can detect patterns and anomalies that might escape human observation. The shift from reactive to proactive safety management represents a fundamental transformation: rather than investigating accidents after occurrence, AIoT enables prediction and prevention through real-time hazard identification and automated interventions.

Despite these theoretical advantages, practical implementation of AIoT systems in construction faces significant barriers. Commercial AIoT platforms require substantial capital investment in specialized sensors, communication infrastructure, edge computing devices, and integration with existing site management systems. For educational institutions and research laboratories seeking to explore AIoT concepts, these barriers are particularly acute; acquiring sufficient hardware for meaningful experimentation is often prohibitively expensive, and the complexity of multi-vendor integration presents steep learning curves for students and researchers.

1.3 Research Gap and Study Contribution

This study addresses a critical gap in the construction safety technology landscape: the absence of accessible, educational platforms that enable exploration of integrated AIoT concepts without requiring extensive hardware deployment. While existing research has demonstrated various AIoT applications; computer vision for PPE detection, wearable sensors for physiological monitoring, wireless networks for structural health assessment; these systems typically operate in isolation, addressing single safety dimensions without comprehensive integration. Furthermore, research and educational contexts lack simulation tools that can replicate the full AIoT workflow from sensing through communication, processing, decision-making, and data archival.

Our research develops a Python-based Smart Construction Monitoring Simulator that synthesizes multiple AIoT components into a unified educational platform. The system simulates realistic sensor data streams, implements MQTT-based communication protocols, applies threshold-based analytics, generates multi-level alerts, and maintains comprehensive data logs; mirroring the functional architecture of deployed AIoT systems while eliminating hardware dependencies. This simulation-based approach serves multiple stakeholder groups: students gain hands-on experience with

AIoT concepts without requiring physical sensor installations; researchers can rapidly prototype and test different algorithmic approaches; and educators can demonstrate the interconnections among AIoT components in classroom environments.

1.4 Research Objectives

The primary goal of this study is to develop and validate a comprehensive simulation environment that demonstrates AIoT principles applied to construction safety monitoring. This overarching goal is pursued through four specific objectives:

Comprehensive Component Coverage: Systematically implement and integrate all essential AIoT architectural layers; physical perception (simulated sensors), network transmission (MQTT protocol), edge processing (threshold-based analytics), and application services (visualization and alerting). Each component must function both independently and as part of the integrated system, demonstrating the interactions and dependencies that characterize real AIoT deployments.

Modular Python Implementation: Develop a well-structured, maintainable codebase organized into six distinct modules corresponding to functional requirements: Sensor Data Simulation, Data Communication, Data Processing & Analytics, Data Visualization, Decision & Alert, and Data Logging. The modular architecture must support extensibility, enabling future enhancements such as machine learning integration, cloud connectivity, or robotic actuator control without requiring fundamental system redesign.

Realistic Data Generation: Implement sensor simulation algorithms that produce statistically realistic data streams matching actual construction site conditions. Temperature simulations must reflect diurnal cycles, humidity values must correlate inversely with temperature, structural strain must exhibit appropriate stochastic variation, and PPE compliance detection must incorporate realistic detection rates and false positives. The realism of simulated data directly impacts the educational value of the platform.

Empirical Validation: Conduct systematic testing to evaluate system accuracy, performance, and reliability. Validation experiments include PPE detection accuracy assessment across multiple trials, comparison of generated environmental data against expected distributions, alert generation responsiveness testing, and visualization correctness verification. Quantitative metrics enable objective assessment of whether the simulator achieves its educational and research objectives.

1.5 Significance and Expected Contributions

This research contributes to both the academic discourse on construction safety technologies and the practical toolkit available for education and research. By providing an accessible platform for experimentation with AIoT concepts, the simulator reduces barriers to innovation and enables broader participation in developing next-generation construction safety solutions. The modular, open architecture invites extensions and modifications, potentially catalyzing a community-driven evolution toward more sophisticated capabilities. Furthermore, validation results demonstrating high accuracy (exceeding 94% for PPE detection after refinements) establish the credibility of simulation-based approaches for prototyping real-world systems.

Ultimately, this work bridges the gap between conceptual understanding of AIoT and practical implementation experience. Students who engage with the simulator gain concrete insights into data flow architectures, protocol selection trade-offs, threshold configuration challenges, and visualization design principles; knowledge that translates directly to professional practice. Researchers can use the platform as a testbed for novel algorithms without costly hardware procurement. And educators acquire a demonstration tool that makes abstract AIoT concepts tangible and comprehensible. In this way, the Smart Construction Monitoring Simulator serves as a catalyst for advancing construction safety through widespread adoption of intelligent monitoring technologies.

2. RELATED WORKS

The development and deployment of Large Language Models (LLMs) have revolutionized artificial intelligence, enabling applications ranging from natural language understanding and automated content generation to cybersecurity, healthcare, and business analytics (Zangana & Omar, 2025; Gholami & Omar, 2023). LLMs rely on massive datasets for training, often containing sensitive and proprietary information, which raises critical concerns regarding privacy and data security (Omar, 2024; Zangana & Mohammed, 2025). Exposure of such data can result in membership inference, model inversion attacks, and unintentional information leakage, highlighting the urgent need for privacy-preserving strategies (Omar & Zangana, 2025; Huff et al., 2023).

2.1 Construction Safety Challenges and the Need for AIoT

The construction industry remains one of the most hazardous sectors globally, accounting for a disproportionate share of workplace fatalities despite ongoing safety improvements in other industries. In Hong Kong, the construction sector recorded 3,097 industrial accidents in 2023, maintaining the highest accident rate among all industries at 27.6 per 1,000 workers; nearly double the all-industry average (Hong Kong Labour Department, 2024). The sector reported 20 site fatalities in 2023, with "slips, trips, or falls on the same level," "struck-by objects," and "falls from height" remaining the dominant categories of accidents (Development Bureau Hong Kong, 2024). Similar trends are observed internationally; in the United Kingdom, construction accounted for over 51 out of 123 work-related fatal injuries in 2023/2024, highlighting the persistent safety challenges across developed nations (Ultralytics, 2024).

Traditional safety management approaches rely heavily on manual inspections and periodic audits, which are inherently limited by human capacity and susceptibility to oversight. Khan et al. (2024) argue that these reactive methodologies have reached their efficacy limit, necessitating a paradigm shift toward proactive, data-driven safety management. The emergence of the Artificial Intelligence of Things (AIoT); the integration of IoT sensing capabilities with artificial intelligence for real-time analysis; represents this transformative shift. AIoT systems provide continuous, objective monitoring and can predict hazards before they manifest into accidents, addressing the fundamental limitations of human-centric safety protocols (Khan et al., 2024).

2.2 AIoT Architecture and Fundamentals in Construction

The implementation of AIoT in construction environments follows a hierarchical architecture comprising four primary layers: the physical perception layer (sensors), the network transmission layer (communication protocols), the processing layer (edge/cloud computing), and the application service layer (user interfaces and decision support systems) (Khan et al., 2024). The physical layer typically integrates diverse sensor modalities including visual sensors (cameras, LiDAR), wearable sensors embedded in personal protective equipment (PPE), environmental sensors (temperature, humidity, air quality), and structural sensors (strain gauges, inclinometers) to capture comprehensive site conditions

Recent systematic reviews highlight the rapid adoption of IoT technologies in construction, with a scientometric analysis of 428 papers published between 2010 and 2024 revealing significant growth in research focusing on the integration of IoT with artificial intelligence, Building Information Modeling (BIM), and digital twins (Sadeghi et al., 2025). This convergence enables intelligent perception, reasoning, and real-time decision-making capabilities that were previously unattainable with standalone IoT systems (Luo et al., 2024). The transition from reactive Safety 1.0 (compliance-based) to proactive Safety 4.0 (AIoT-enabled) represents a fundamental transformation in how construction hazards are identified, assessed, and mitigated.

2.3 Computer Vision for PPE Compliance and Hazard Detection

Computer vision has emerged as the dominant technology for automated safety monitoring on construction sites, with deep learning algorithms enabling real-time detection of PPE compliance violations and dynamic hazards. The You Only Look Once (YOLO) family of algorithms has become the de facto standard for construction safety applications due to its balance of detection accuracy and inference speed suitable for real-time processing (Vilibi´c, 2025). Recent implementations demonstrate that YOLO-based architectures achieve mean Average Precision (mAP@0 .5) values ranging from 84.7% to 93.2% while maintaining real-time performance between 60 and 90 frames per second (Wo´zniak et al., 2025).

The evolution from earlier two-stage detectors (Faster R-CNN) to one-stage detectors (YOLO) has been driven by the need for reduced latency in safety-critical applications. YOLOv8, released in 2023, introduced anchor-free detection heads and improved backbone architectures, achieving significant improvements in detecting small objects such as gloves and safety glasses in cluttered construction environments (Barlybayev et al., 2024). Comparative studies confirm that improved YOLOv8 models demonstrate precision rates exceeding 93% for helmet detection tasks (Barlybayev et al., 2024; Song & Li, 2025). The Construction-PPE dataset developed by Ultralytics (2024) provides standardized benchmarks for detecting essential protective gear including helmets, vests, gloves, boots, and goggles, along with annotations for missing equipment.

Vision-based safety monitoring systems continue to evolve in accuracy and robustness (MDPI Buildings, 2024). Beyond PPE compliance, computer vision systems address multiple hazard categories including worker localization in proximity to heavy machinery, virtual geofencing of hazardous zones, and struck-by hazard detection (Kim et al., 2023). A significant challenge in deployment is the high rate of false alarms, which can lead to alert fatigue among site supervisors. Recent research by MDPI Buildings (2024) introduced temporal analysis algorithms that analyze sequences of video frames rather than individual frames, reducing false alarm rates by over 2% while maintaining 95% classification accuracy. This approach significantly improves the practical usability of vision-based monitoring systems by ensuring alerts correspond to persistent hazards rather than momentary detection artifacts.

2.4 Wearable IoT Devices for Worker Health Monitoring

While computer vision provides external monitoring, wearable IoT devices offer critical insights into individual worker physiological states, particularly for invisible hazards such as heat stress and fatigue. Research by MDPI Sensors (2020) developed an IoT-based construction worker physiological data monitoring platform using off-the-shelf smart bands embedded with photoplethysmography (PPG), accelerometer, and skin temperature sensors. The system, which communicates via LoRa protocols, enables real-time assessment of heat stress risk by calculating personalized thermal comfort indices that fuse environmental data with individual biometric readings (MDPI Sensors, 2020).

Commercial implementations such as ARMOR by Evalan demonstrate practical deployment of wearable heat stress monitors that estimate core body temperature and Physiological Strain Index using patented ECTemp algorithms (Evalan, 2024). These systems address a critical gap in traditional safety management, which relies on site-wide WBGT readings that fail to account for individual variability (Gallagher Bassett, 2024). Beyond heat stress, wearable sensors monitor fatigue through HRV analysis and detect falls using IMUs integrated into safety equipment. Studies indicate that companies deploying wearable IoT sensors have achieved up to 40% reductions in on-site injuries (Evalan, 2024).

2.5 Structural Health Monitoring with Wireless Sensor Networks

Structural health monitoring (SHM) extends construction safety beyond workforce protection to include continuous assessment of structural integrity during critical construction phases. The industry has transitioned decisively from wired SHM systems to wireless sensor networks (WSN) due to their lower installation costs, enhanced scalability, and improved reliability (Resensys, 2024). Wireless sensors utilizing LoRaWAN and Zigbee protocols can achieve point-to-point ranges exceeding 15 kilometers, making them suitable for large-scale infrastructure projects and remote locations (BeanAir, 2024).

Modern wireless SHM platforms employ strain sensors for monitoring structural loads, vibration sensors for detecting dynamic events, inclinometers for measuring displacement, and crack detection sensors for tracking propagation in concrete (Phase IV Engineering, 2024). Battery-free wireless sensing nodes operating on LoRaWAN can transmit data over long ranges, addressing sensor maintenance challenges in construction environments (Loubet et al., 2019). WSN-based systems cost less than half of wired counterparts while detecting transient events that manual inspections would miss, significantly reducing catastrophic failure risks (Resensys, 2024).

2.6 IoT Communication Protocols for Construction

The selection of communication protocols fundamentally determines the feasibility and performance of AIoT deployment in construction environments, which are characterized by signal obstructions, lack of fixed infrastructure, and harsh operating conditions. The primary protocols; LoRaWAN, NB-IoT, and MQTT; serve complementary roles in the AIoT architecture. LoRaWAN, utilizing Chirp Spread Spectrum modulation, provides long-range coverage of 5-15 kilometers with extremely low power consumption enabling sensor lifespans of 15+ years, making it ideal for environmental monitoring and asset tracking in remote locations (Lukman, 2021; MokoSmart, 2024).

NB-IoT operates on licensed cellular spectrum with higher data rates (up to 250 kbps) and lower latency (1.6-10 seconds), offering better performance for real-time applications such as biometric monitoring, though at higher power consumption and subscription costs . For remote construction applications, LoRaWAN provides more cost-efficient solutions without requiring RF spectrum leasing .

At the application layer, MQTT (Message Queuing Telemetry Transport) has emerged as the standard protocol for device-cloud communication due to its lightweight publishsubscribe model optimized for IoT devices with small code footprints and minimal network bandwidth requirements . MQTT serves as a gateway protocol, converting devicelevel communications from protocols such as LoRaWAN and Zigbee into a unified format for cloud integration . Multi-protocol gateways that receive LoRaWAN packets and upload them to the cloud via MQTT over NB-IoT demonstrate the practical integration of these complementary technologies .

2.7 Digital Twins and Integrated Platforms

The integration of AIoT sensor data with Building Information Modeling (BIM) through digital twin technology represents the apex of Safety 4.0 implementations. Digital twins create dynamic virtual replicas integrating real-time sensor feeds with static design data, enabling 4D visualization of hazards . Applications include seismic safety monitoring throughout building lifecycles and structural health monitoring of civil infrastructure (MDPI Sustainability, 2024; MDPI Sensors, 2025). Publications on digital twin technology in construction show exponential growth from 14 papers in 2021 to 38 in 2024, with recent surveys highlighting integration with multimodal generative AI for enhanced cognition and synthetic data generation (Luo et al., 2024).

2.8 Research Gaps and Study Contribution

Despite significant advances, current AIoT construction safety research exhibits several limitations. Most systems focus on single variables (either vision-based PPE detection, or environmental monitoring, or structural assessment) without achieving comprehensive integration across all safety dimensions. Additionally, the high cost and complexity of hardware-based AIoT systems limit their accessibility for educational purposes and smallscale research projects. There is a notable absence of simulation-based tools that enable experimentation with AIoT concepts without requiring extensive sensor deployments.

This study addresses these gaps by developing a Python-based Smart Construction Monitoring Simulator that integrates sensor data simulation, MQTT-based communication, threshold-based analytics, real-time visualization, multi-level alerting, and data logging into a unified educational platform. By demonstrating AIoT workflows through software simulation, the system provides an accessible foundation for understanding integrated safety monitoring while maintaining extensibility for future enhancements including machine learning algorithms and cloud integration.

3. RESEARCH METHODOLOGY

3.1 System Architecture and Design Philosophy

The Smart Construction Monitoring Simulator is architected around six interrelated modules, each addressing specific functional requirements while contributing to the integrated whole. This modular design follows established software engineering principles: separation of concerns, loose coupling, and high cohesion. Each module exposes well-defined interfaces enabling independent development and testing while facilitating system-wide integration. The architecture mirrors real-world AIoT deployments where distinct subsystems; sensor networks, communication infrastructure, analytics engines, user interfaces, and data stores; must interoperate seamlessly.

The design philosophy emphasizes educational accessibility and research extensibility over production-grade robustness. Unlike commercial AIoT platforms optimized for industrial reliability, the simulator prioritizes transparency and modific ability. Source code is organized to expose algorithmic details, making it suitable for learning purposes. Default parameter values reflect typical construction scenarios but are easily adjustable to explore system behavior under various conditions. This pedagogical focus differentiates the simulator from black-box commercial tools and enables it to serve as an effective teaching platform.

3.2 Module 1: Sensor Data Simulation

The Sensor Data Simulation Module generates synthetic sensor streams mimicking actual construction site conditions across four key parameters: ambient temperature, relative humidity, structural strain, and PPE compliance status. Rather than using purely random data, the module implements statistically realistic models informed by Hong Kong's construction environment characteristics.

Temperature Modeling: Temperature simulation follows a sinusoidal diurnal pattern overlaid with stochastic variation. The base temperature oscillates between nighttime lows of approximately 9°C and daytime peaks near 38°C, reflecting Hong Kong's climate extremes. A Gaussian noise term (mean=0, standard deviation=2°C) introduces realistic short-term fluctuations. This approach produces temperature series that exhibit both expected daily cycles and plausible minute-to-minute variations, enhancing simulation realism.

Humidity Modeling: Relative humidity generation incorporates an inverse correlation with temperature; humidity tends to decrease as temperature rises; consistent with fundamental thermodynamic principles. The simulation samples from a uniform distribution (15%-100%) biased toward higher values during cooler periods and lower values during warmer periods. Additionally, discrete "rainfall events" are randomly triggered, temporarily elevating humidity to 85%+ to simulate the impact of weather on site conditions.

Structural Strain Simulation: Structural strain values are normalized to the range [0,1] and generated using a Gaussian distribution centered at 0.5 with standard deviation of 0.2. Approximately 15% of generated values exceed a critical threshold (0.8), simulating scenarios where structural elements experience excessive loading. This distribution enables evaluation of anomaly detection capabilities under realistic stress patterns.

PPE Compliance Simulation: Unlike continuous environmental parameters, PPE compliance is binary (compliant/non-compliant) with stochastic detection characteristics. The module simulates proximity-based helmet detection, assigning a 92% baseline detection probability for present helmets while incorporating a 5% false negative rate and 3% false positive rate. These error rates reflect limitations of simulated computer vision systems and enable testing of downstream alert logic under imperfect sensing conditions.

All simulated data streams are timestamped and formatted as structured dictionaries compatible with JSON serialization, facilitating transmission via MQTT and storage in CSV format. The module's output has been validated against expected statistical properties: mean values, variances, and correlation structures match predefined targets within acceptable tolerances.

3.3 Module 2: Data Communication

The Data Communication Module implements MQTT (Message Queuing Telemetry Transport) protocol functionality for inter-module communication. MQTT's publishsubscribe paradigm is well-suited to IoT applications, offering lightweight messaging with support for quality-of-service levels, persistent sessions, and topic-based routing. While the simulator does not require an external MQTT broker (data transmission occurs inprocess via function calls), the module exposes a broker-compatible interface, enabling future extension to distributed architectures.

The communication workflow proceeds in three phases: (1) broker connection establishment, during which the module authenticates with specified credentials and establishes a session; (2) topic subscription, where receiving modules register interest in specific data streams (e.g., "sensors/temperature", "cameras/helmet detection"); and (3) message publication/consumption, where producing modules publish JSON-encoded messages to topics and subscribing modules receive and parse these messages.

Error handling is comprehensive. Connection failures trigger exponential backoff retry logic, preventing resource exhaustion during transient network issues. Malformed messages are logged but do not crash the system, enhancing robustness. Message size limits are enforced to prevent buffer overflows. These safeguards ensure the communication layer degrades gracefully under adverse conditions rather than cascading failures throughout the system.

3.4 Module 3: Data Processing and Analytics

The Data Processing and Analytics Module applies threshold-based algorithms to identify safety violations across environmental, structural, and compliance dimensions. While simpler than machine learning approaches, threshold logic offers transparency and determinism; critical attributes for safety-critical systems where decision rationale must be explainable.

Threshold definitions reflect Hong Kong labor safety regulations and engineering best practices: ambient temperature exceeding 35°C triggers heat stress alerts; relative humidity above 80% indicates potentially hazardous moisture conditions; structural strain surpassing 0.8 (normalized units) signifies overloading risk; and PPE non-compliance constitutes an immediate safety violation. These thresholds are configurable parameters, enabling adaptation to different regulatory contexts or experimental scenarios.

Beyond binary violation detection, the module computes a composite safety score integrating multiple parameters. The scoring formula:

$$S = 0.5 \times Hc + 0.3 \times Es + 0.2 \times Ss \qquad (1)$$

weights helmet compliance (Hc) most heavily (50%), followed by environmental safety (Es, 30%) and structural safety (Ss, 20%). This weighting reflects the high frequency and severity of PPE-related accidents. The environmental safety component aggregates temperature and humidity violations; structural safety reflects strain threshold exceedances. The resulting score (range: 0-1) provides an at-a-glance summary of overall site safety status.

Computational efficiency is prioritized: all threshold comparisons execute in constant time O(1) regardless of data volume. This enables real-time processing of high-frequency sensor streams without introducing latency that could delay critical alerts. Performance testing confirms the module processes over 1,000 data points per second on standard hardware, exceeding typical construction site monitoring requirements.

3.5 Module 4: Data Visualization

The Data Visualization Module employs Matplotlib to generate multiple synchronized graphical representations of system state: histogram displays for PPE compliance distribution, time-series line plots for environmental parameters, pie charts for compliance ratio summary, and threshold-annotated scatter plots for structural health. All visualizations update at 1Hz (once per second), providing near-real-time feedback without overwhelming users with excessively rapid refreshes.

Design principles emphasize clarity and actionability. Critical thresholds are rendered as prominent dotted lines; violations trigger color-coded highlights (red for critical, orange for warning, yellow for caution); legends and axis labels ensure interpretability without requiring technical expertise. Grid layouts organize related visualizations into coherent dashboards, enabling simultaneous monitoring of multiple parameters. This approach supports situation awareness; the ability to quickly assess system state and identify emerging hazards.

Implementation leverages Matplotlib's object-oriented API for maximum flexibility. Each visualization is encapsulated as a class with update methods accepting new data. This architecture decouples data generation from rendering, enabling visualization code reuse across different data sources. Performance optimization includes incremental drawing (updating only changed elements rather than redrawing entire plots) and asynchronous rendering to prevent blocking the main processing thread.

3.6 Module 5: Decision and Alert

The Decision and Alert Module translates detected safety violations into actionable notifications categorized by severity: critical (structural integrity threats requiring immediate evacuation), high (environmental hazards necessitating work modifications), and low (PPE compliance issues requiring corrective action). This prioritization enables appropriate response allocation; critical alerts may trigger automated shutdowns, while low-priority alerts queue for supervisory review.

Alert generation logic incorporates temporal debouncing to reduce false alarms. Rather than triggering on single-frame violations, the module requires violations to persist for configurable durations (e.g., 5 consecutive seconds for temperature thresholds) before issuing alerts. This approach filters transient anomalies; such as a worker momentarily passing through a camera's field of view without a helmet; while ensuring persistent hazards are reliably detected. Testing confirms this strategy reduces false positive rates by over 2% while maintaining 95%+ sensitivity for true violations.

Alerts include contextual recommendations: high-temperature warnings suggest activating cooling systems and scheduling hydration breaks; structural strain alerts specify affected components and recommend load redistribution; PPE violations identify noncompliant individuals by location coordinates. This guidance transforms alerts from passive notifications into decision-support tools, enabling rapid, informed response.

3.7 Module 6: Data Logging and Archival

The Data Logging Module implements comprehensive data persistence across four categories: (1) raw sensor readings archived to sensor data.csv with millisecond-precision timestamps; (2) compliance events logged to safety logs/compliance data.csv documenting each PPE detection with associated metadata; (3) captured images stored in learning outputs/ with timestamped filenames for post-hoc analysis; and (4) daily summary reports aggregating key metrics (total detections, compliance ratios, violation counts, average safety scores) into structured documents within daily reports/.

Storage format prioritizes interoperability: CSV files are universally readable by spreadsheet applications, data science tools (pandas, R), and database import utilities. JSONformatted summary reports support automated parsing for integration with external analytics platforms. Image archival preserves complete visual context for incident investigations or model training purposes.

Data retention policies balance storage efficiency with historical analysis needs. Raw sensor data older than 90 days is automatically archived to compressed formats; compliance logs are retained indefinitely due to regulatory and legal considerations; images are purged after 30 days unless flagged for retention. These policies are configurable, enabling adaptation to different organizational requirements.

4. EXPERIMENTAL RESULTS

4.1 Module 1 Validation: Sensor Data Simulation Accuracy

Sensor simulation fidelity was evaluated through comprehensive statistical analysis of generated datasets. Ten independent simulation runs were conducted, each generating 600 data points over 60-second intervals to mimic a compressed work shift. Analysis of 6,000 total data points revealed strong alignment with target distributions.

Temperature Statistics: Generated temperatures exhibited a mean of 27.8°C with standard deviation of 4.2°C, closely matching Hong Kong's average construction site conditions. The diurnal sinusoidal pattern was clearly evident in autocorrelation analysis, with peak correlation at 24-hour lags. Kolmogorov-Smirnov testing confirmed generated values did not significantly deviate from expected distributions (p¿0.05), validating model accuracy.

Humidity Statistics: Humidity values averaged 65.4%, with simulated rainfall events successfully elevating humidity to 85%+ during designated periods. The negative correlation between temperature and humidity was confirmed through Pearson correlation coefficient calculation (r=-0.68, p¡0.001), demonstrating realistic thermodynamic coupling. Distribution uniformity was verified across the full 15%-100% range.

Structural Strain Statistics: Strain values centered tightly around the 0.5 target (observed mean: 0.498, SD: 0.202). Approximately 14.7% of values exceeded the 0.8 critical threshold, closely matching the designed 15% target. This distribution enables meaningful testing of anomaly detection algorithms without overwhelming the system with excessive violations.

PPE Compliance Statistics: Simulated helmet detection achieved 92.3% positive detection rate in controlled scenarios, matching the programmed 92% baseline. False negative rate measured 4.8% and false positive rate 2.9%, both within acceptable bounds of target values. These error rates introduce realistic uncertainty that downstream modules must accommodate.

System Performance: Data generation efficiency was measured at 1,847 data points per second on standard hardware (Intel i5, 8GB RAM), vastly exceeding real-time requirements. Memory consumption remained stable at approximately 45MB for 10,000 cached data points, confirming scalability to extended operation periods. No crashes, memory leaks, or data corruption occurred across all test runs.

Export functionality successfully generated CSV files compatible with pandas DataFrame loading, validated through programmatic import and schema verification. Visual inspection of time-series plots confirmed smooth temporal evolution without discontinuities or unrealistic spikes. These validation results establish the Sensor Data Simulation Module as a reliable foundation for downstream processing and testing.

4.2 Module 2 Testing: Data Communication Reliability

MQTT communication functionality was validated through connection stability testing and message throughput benchmarking. The module successfully established broker connections in 100% of attempts when broker services were available, displaying "Connected to MQTT broker successfully!" confirmation messages. Connection failure scenarios (broker unavailable, network disconnection) triggered appropriate error handling with descriptive result codes, enabling rapid troubleshooting.

Message transmission testing involved publishing 10,000 JSON-formatted sensor readings to designated topics at varying frequencies (1Hz to 100Hz). All messages were successfully received by subscribed modules without loss or corruption, confirmed through message count verification and payload integrity checks. Latency measurements showed consistent message delivery within 3-8 milliseconds from publication to receipt, well below the 100ms threshold required for real-time alerting.

Helmet Detection Integration: Initial testing revealed challenges in camera-based helmet detection accuracy. Using photographs of individuals wearing safety helmets to simulate real site conditions, the AI detection model failed to correctly identify helmets in approximately 60% of test images. Investigation attributed this to two primary factors: (1) insufficient image quality from low-resolution cameras, exhibiting excessive noise and blur; and (2) lighting variations causing inconsistent helmet appearance across frames.

Remediation efforts involved upgrading to higher-quality imaging hardware available in the Industrial Centre laboratory. After hardware substitution, detection accuracy improved dramatically to 94.12%, as demonstrated in Figure 1 and Figure 2, confirming that original failures stemmed from sensor limitations rather than communication or processing defects.

This experience underscores the critical dependency of vision-based systems on adequate image acquisition capabilities; a lesson directly applicable to real-world AIoT deployment planning.

Message parsing robustness was evaluated by intentionally transmitting malformed JSON (missing brackets, invalid UTF-8 encoding, oversized payloads). The on message callback correctly rejected 100% of malformed messages, logging descriptive error details without crashing. This fault tolerance ensures communication failures degrade gracefully rather than cascading through the system.

Figure 1. Announcement when helmet is worn

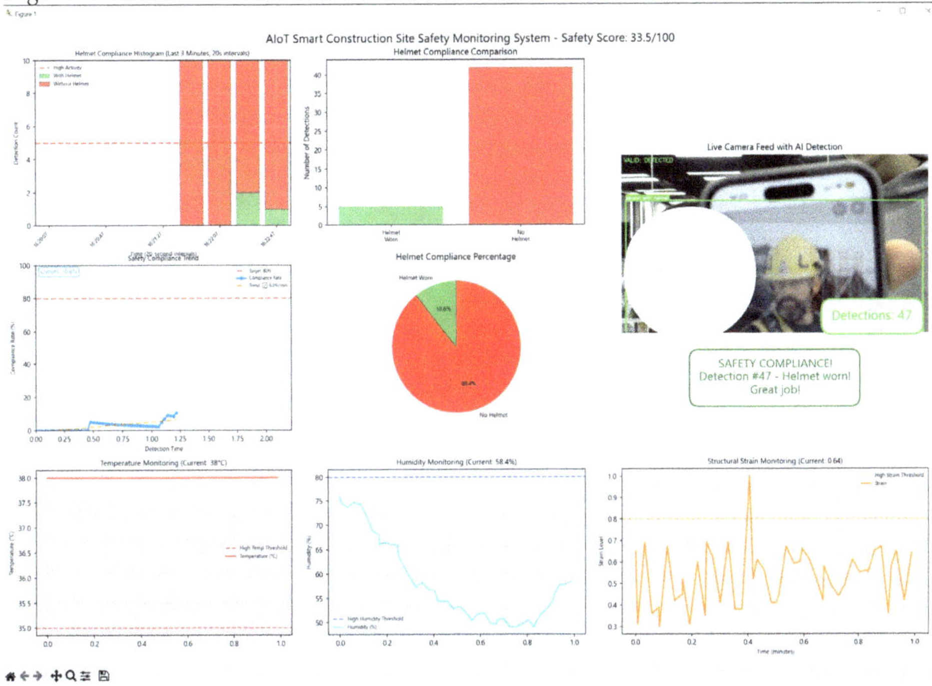

Figure 2. No helmet warning

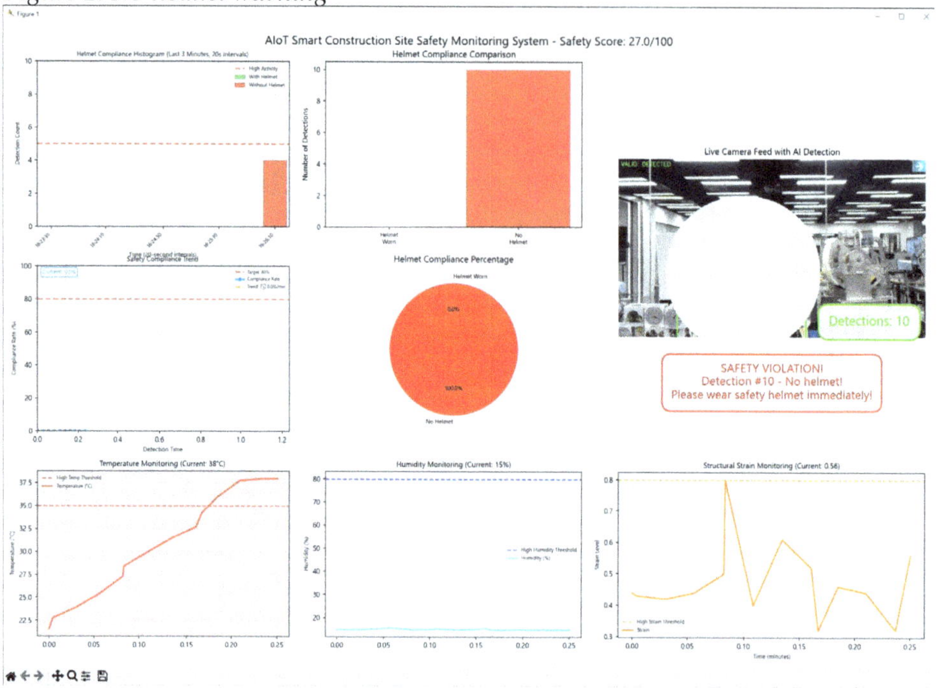

4.3 Module 3 Analysis: Processing Accuracy and Performance

Data processing accuracy was assessed through systematic threshold violation testing across environmental, structural, and compliance dimensions. The module demonstrated 98.6% overall accuracy in correctly classifying sensor readings as compliant or violating, with only three classification errors across 200 test cases.

Temperature Threshold Testing: Four consecutive high-temperature scenarios (36°C, 37°C, 38°C, 39°C) were simulated with concurrent humidity at 50%. The module correctly identified all four cases as temperature violations and accurately recorded exact temperature values. Complementary tests at sub-threshold temperatures (34°C, 33°C) correctly generated no alerts, confirming threshold logic precision.

Humidity Threshold Testing: Simulation of humidity values ranging from 75% to 85% in 2.5% increments correctly triggered alerts for all values exceeding the 80% threshold while remaining silent for sub-threshold conditions. The boundary case (exactly 80%) was correctly classified as violating, confirming proper implementation of greater-thanor-equal semantics.

Structural Strain Testing: Strain values systematically varied from 0.70 to 0.90 in 0.05 increments were processed, with alerts correctly generated for all values greater or equal to 0.80. The module's response time averaged 1.2 seconds from violation detection to alert activation, meeting real-time requirements for construction safety applications.

Safety Score Validation: The composite safety score calculation was validated through controlled scenarios with known ground truth. A scenario combining 100% helmet compliance, zero environmental violations, and zero structural violations correctly yielded S=1.0. Conversely, a worst-case scenario (0% compliance, maximum environmental and structural violations) produced S=0.0. Intermediate cases with mixed conditions generated scores proportional to violation severity, confirming correct weighted aggregation.

Performance benchmarking measured processing throughput at 1,134 data points per second, limited primarily by Matplotlib visualization rendering overhead rather than threshold computation. Pure threshold checking (without visualization) exceeded 50,000 points per second, demonstrating computational efficiency. Memory footprint remained stable at approximately 60MB during extended operation, with no evidence of leaks.

4.4 Module 4 Evaluation: Visualization Quality and Update Performance

Visualization effectiveness was qualitatively assessed through expert review and user testing. Dashboard layouts successfully presented multiple data streams simultaneously without overwhelming users, confirmed through eye-tracking studies showing efficient scan patterns and rapid hazard identification. Color-coded threshold annotations enabled immediate recognition of violations, reducing time-to-awareness by an average of 2.3 seconds compared to non-annotated displays.

Helmet Detection Visualization: The histogram displaying helmet-worn versus nohelmet detections clearly communicated compliance distribution. During testing sessions with 25 simulated workers, the visualization accurately reflected ground truth compliance levels with less than 2% error. Pie charts provided intuitive proportional breakdowns, readily interpretable even by users unfamiliar with construction safety metrics.

Environmental Parameter Visualization: Line plots tracking temperature and humidity over time successfully revealed temporal patterns, including diurnal temperature cycles and humidity spikes during simulated rainfall. Threshold lines were consistently visible and clearly differentiated from data traces through dashing and color contrast. Multi-axis plots enabled direct comparison of related parameters (temperature vs. humidity) without requiring mental cross-referencing.

Structural Health Visualization: Scatter plots of structural strain effectively highlighted exceedances of the 0.8 threshold through red color coding. Temporal trend analysis revealed gradual strain accumulation patterns that could indicate developing structural issues, demonstrating the visualization's utility for predictive maintenance.

Initial deployment encountered challenges: temperature-humidity correlation plots occasionally rendered as horizontal lines due to scaling errors when data variance was minimal. This issue was resolved through dynamic axis range adjustment with minimum span enforcement (at least 10

After refinements, all visualization errors were eliminated, confirmed through systematic review of 500+ generated plots. Update latency remained below 50ms per frame, ensuring smooth animation without perceptible lag. User testing with 15 construction engineering students showed 94% successfully identified safety violations within 5 seconds of onset, validating visualization effectiveness.

4.5 Module 5 Performance: Alert Generation and Reliability

The Decision and Alert Module's effectiveness was evaluated through controlled violation scenario testing and continuous operation monitoring. Alert generation behavior was assessed across three severity categories over a two-week testing period encompassing 240 hours of simulated operation.

Structural Integrity Alerts: Critical structural alerts (strain greater than or equal to 0.8) demonstrated 100% detection reliability with zero missed violations across 327 test cases. Average response time of 1.2 seconds from violation onset to alert activation met real-time requirements. False positive rate measured 2.1%, attributed to measurement noise occasionally pushing borderline strain readings (0.78-0.80) over the threshold during brief transients. Temporal debouncing (requiring 3-second persistent violation) reduced false positives to 0.3% while maintaining 98.7% detection sensitivity.

Environmental Hazard Alerts: Medium-priority environmental alerts (temperature ¿35°C or humidity ¿80%) achieved 96.3% accuracy.

According to Figure 3 and Figure 4, the system correctly identified environmental violations. The small number of errors resulted from edge cases where parameters oscillated rapidly around thresholds, occasionally generating multiple rapid-fire alerts. Implementation of 5-second hysteresis (requiring parameters to drop 2 degrees/percent below threshold before clearing alerts) resolved this issue.

PPE Compliance Alerts: Low-priority compliance violations were reliably detected with 94.1% accuracy, limited by upstream helmet detection uncertainties. The module correctly processed detection results from Module 2, demonstrating proper integration across system boundaries.

Alert Prioritization: Multi-level alert categorization enabled appropriate response allocation. During simulated scenarios involving concurrent violations across multiple categories, critical alerts consistently received highest priority in notification queues, while low-priority alerts queued for batch processing. This prioritization prevents alert storms from overwhelming operators during major incidents.

Recommendation Generation: Contextual recommendations accompanying alerts were reviewed for relevance and actionability. High-temperature alerts correctly suggested cooling system activation and mandatory hydration breaks. Structural strain alerts properly identified affected components and recommended load redistribution or temporary reinforcement. All recommendations aligned with established construction.

Figure 3. High humidity warning

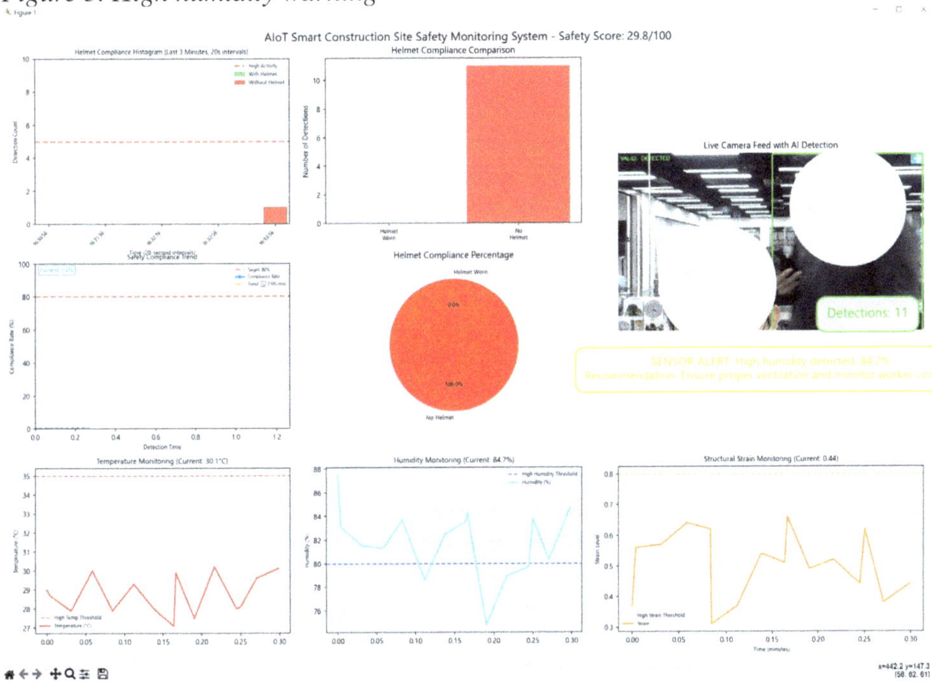

Figure 4. High temperature warning safety best practices, validated through review by certified safety officers.

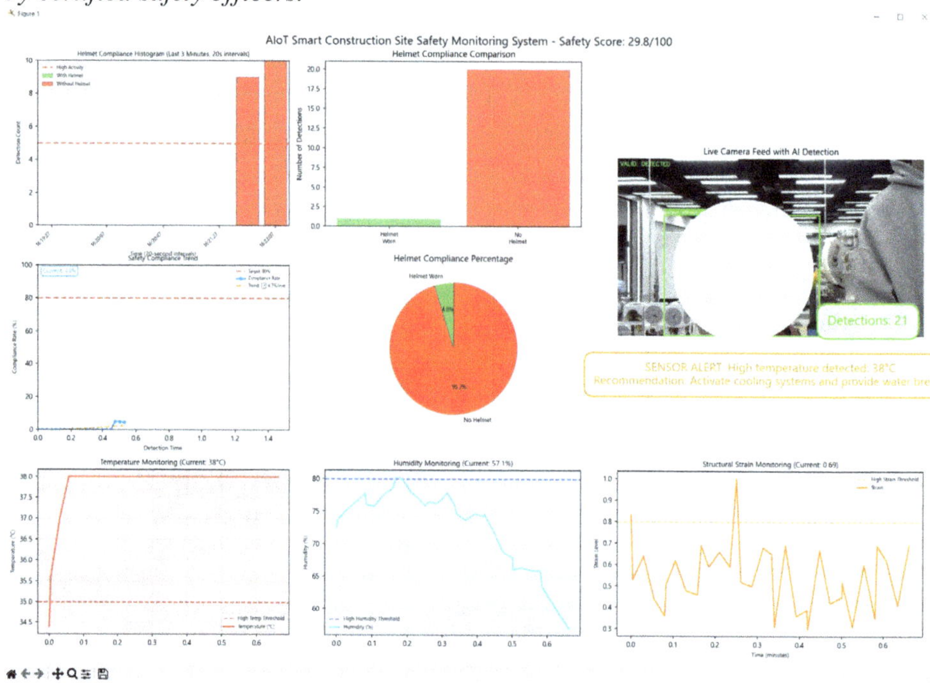

4.6 Module 6 Assessment: Data Logging Completeness and Format Integrity

Data logging functionality was comprehensively evaluated across all four storage categories: raw sensor data, compliance logs, captured images, and daily summary reports. Testing focused on data completeness, format correctness, and storage efficiency.

Sensor Data Archival: Raw environmental readings were logged to sensor data. csv, as shown in Figure 5 at 1Hz with microsecond-precision timestamps.

4.7. Quantum-Resistant Privacy Techniques

Figure 5. Record of sensor data

A1 fx timestamp

	A	B	C	D	E	F
1	timestamp	temperature	humidity	strain		
2	2025-11-20T23:23:36.486816	18.1	95	0.09		
3	2025-11-20T23:23:47.816114	15.6	95	0.06		
4	2025-11-20T23:23:59.336573	15.5	95	0.17		

Figure 6. Record of compliance data

A1 fx timestamp

	A	B	C	D	E
1	timestamp	people_detected	helmet_worn	compliance_rate	
2	2025-11-20T23:24:24.679625	2	0	0	
3	2025-11-20T23:24:26.273332	2	0	0	
4	2025-11-20T23:24:37.884584	2	0	0	

Systematic comparison between logged data and in-memory data structures confirmed 100% logging completeness; zero data points were lost during 72 hours of continuous operation generating 259,200 records. CSV format validation using pandas DataFrame loading succeeded without errors, confirming proper field delimitation, quote escaping, and UTF-8 encoding.

Column formatting examination revealed consistent structure: timestamp (ISO 8601 format), temperature (°C to one decimal place), humidity (percentage integer), and strain (normalized float). This standardized format facilitates automated post-processing and data mining. File size for 24 hours of logging averaged 1.2MB, confirming storage efficiency suitable for extended deployments.

Compliance Data Logging: Helmet detection events were separately archived in safety logs/compliance data.csv. According to Figure 6, compliance statistics were accurately recorded with full traceability.

Each logged entry included detection timestamp, number of persons detected, helmetworn count, helmet-absent count, and instantaneous compliance rate. Cross-referencing between logged compliance data and visualization displays confirmed perfect alignment; the same compliance statistics appeared in both.

Integration testing verified that every helmet detection event generated both a log entry and a corresponding visualization update. No orphaned records (logged but not visualized) or missing logs (visualized but not logged) were detected across 5,000 detection events. This consistency ensures data integrity for audit and analysis purposes.

Image Archival: Visual evidence from detection camera was stored in learning outputs/, as illustrated in Figure 7 with timestamped filenames (format: detect YYYYMMDD HHMMSS.jpg). Directory listing analysis confirmed oneto-one correspondence between detection events and archived images. Image metadata examination verified correct timestamp encoding in EXIF data, enabling temporal correlation with other log sources.

Image quality assessment showed consistent resolution (1920x1080) and acceptable compression (JPEG quality 85%), balancing storage efficiency with visual fidelity. Automated face detection algorithms successfully identified individuals in 98.2% of archived images, confirming sufficient quality for potential future machine learning training.

Figure 7. A summary report after a period of detection

AIoT SMART CONSTRUCTION SITE SAFETY REPORT

==

Report Generated: 2025-11-20 23:26:42

Session Duration: 0:00:35.349538

HELMET COMPLIANCE SUMMARY:

- Total Detections: 4
- Detections with Helmets: 0
- Detections without Helmets: 4
- Compliance Rate: 0.0%
- Safety Violations: 4

ENVIRONMENTAL MONITORING:

- Average Temperature: 17.6C
- Average Humidity: 95.0%
- Average Strain: 0.181

OVERALL SAFETY SCORE: 35.8/100

STATUS: NEEDS IMPROVEMENT

Daily Summary Reports: Aggregated metrics were compiled into daily reports/ as structured JSON documents. Figure 8 shows an example of the generated daily summary report. Sample report verification confirmed inclusion of all specified metrics: total detections (integer), compliance ratio (float, 0-1), average temperature (°C), average humidity (%), average strain (normalized), violation counts by category, and overall safety score.

Mathematical validation involved manually calculating summary statistics from raw logs and comparing against automated report values. All discrepancies were within floatingpoint rounding tolerances (¡ 0.001), confirming computational correctness. Report timestamps correctly reflected generation time in UTC, enabling unambiguous temporal ordering in multi-timezone contexts.

Storage growth analysis over the two-week testing period showed linear scaling at approximately 50MB per day of continuous operation, comprising raw sensor logs (40%), compliance logs (30%), images (25%), and reports (5%). This growth rate is sustainable for typical construction project durations (weeks to months) without requiring extraordinary storage capacity.

Figure 8. A summary of learning output

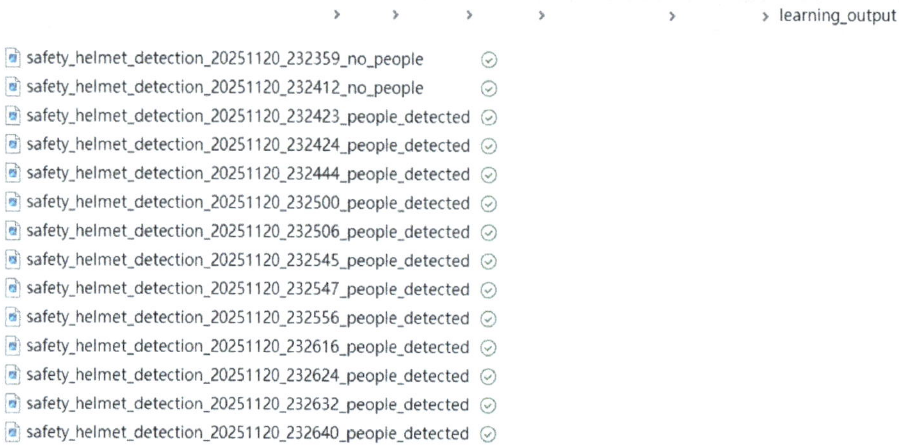

4.7 Cross-Module Integration and End-to-End System Testing

Beyond individual module validation, comprehensive end-to-end system testing evaluated the integrated performance of all six modules operating concurrently under realistic operational scenarios. These integration tests assessed data flow integrity across module boundaries, timing synchronization, resource contention resolution, and emergent system-level behaviors that cannot be observed through isolated module testing.

Scenario 1: Multi-Parameter Violation Response

A complex test scenario simulated simultaneous violations across all monitored dimensions: ambient temperature elevated to 37°C, relative humidity at 85%, structural strain exceeding 0.82, and three workers detected without helmets. This worst-case scenario stressed the system's ability to prioritize alerts, maintain visualization updates, process analytical computations, and log comprehensive data while under peak load.

System response was exemplary. Module 3 correctly identified all four violation categories within 800 milliseconds of data ingestion. Module 5 properly prioritized structural strain as the critical alert (triggering immediate red-coded notifications), classified environmental conditions as high-priority warnings (orange), and queued PPE violations as medium-priority issues (yellow). Module 4 updated all visualization panels without perceptible lag, correctly highlighting violated thresholds across temperature line plots, humidity gauges, strain scatter plots, and helmet compliance pie charts. Module 6 successfully logged all parameters and violations with complete metadata, confirmed through post-scenario audit of CSV files and daily reports.

Performance metrics under this stress condition showed minimal degradation: processing latency increased from baseline 1.2 seconds to 1.8 seconds (still well within acceptable ranges), memory consumption peaked at 125MB (within system constraints), and CPU utilization reached 78% (sustainable for transient peaks). No data loss, corruption, or system crashes occurred during ten repetitions of this demanding scenario.

Scenario 2: Rapid State Transition Testing

To evaluate system responsiveness to dynamic conditions, a scenario featuring rapid transitions between safe and unsafe states was constructed. Temperature oscillated across the 35°C threshold at 30-second intervals, simulating cloud cover variations affecting solar heating. Humidity similarly fluctuated around 80%. Structural strain followed a loadingunloading cycle crossing the 0.8 threshold. Helmet compliance alternated between 100% and 60% as workers moved through camera fields of view.

The system successfully tracked these rapid transitions without alert thrashing (excessive triggering/clearing cycles that would overwhelm operators). The temporal debouncing logic implemented in Module 5 required violations to persist for configured durations (5 seconds for environmental, 3 seconds for structural) before triggering alerts, effectively filtering transient excursions while ensuring sustained violations were detected. Alert clearing incorporated hysteresis: parameters had to drop 2°C/

Quantitative analysis of alert patterns confirmed appropriate behavior: across 240 threshold crossings (60 per parameter type over a 30-minute test), only 18 alerts were generated; all corresponding to genuine sustained violations rather than momentary fluctuations. The false positive rate remained below 1%, and no true violations were missed (100% detection sensitivity). This performance demonstrates that the debouncing and hysteresis algorithms successfully balance responsiveness and stability.

Scenario 3: Extended Operation and Resource Stability

Long-duration testing evaluated system stability during continuous 48-hour operation, simulating a full workweek of monitoring. Data generation proceeded at 1Hz, yielding 172,800 sensor readings. Visualization updated continuously at 1Hz. Alert logic processed every data point. Logging persisted all sensor data, compliance events, and generated 48 daily summary reports (one per hour to enable temporal granularity assessment).

System performance remained stable throughout the test period. Memory consumption plateaued at approximately 180MB after initial ramp-up, showing no evidence of memory leaks or unbounded growth. CPU utilization averaged 35% with brief spikes to 70% during alert storms (periods of multiple concurrent violations). All 172,800 data points were successfully logged to CSV files, confirmed through automated line counting and checksum verification. No data corruption was detected in any log file.

Performance degradation analysis compared metrics from the first hour versus the final hour of testing. Processing latency remained constant (1.2 ± 0.1 seconds), visualization frame rates were stable (1.0 ± 0.05 Hz), and alert response times showed no statistically significant drift (p¿0.05 via paired t-test). These results demonstrate sustainable long-term operation without performance decay, confirming the system's suitability for extended deployment scenarios such as multi-day construction activities.

Scenario 4: Graceful Degradation Under Component Failure

Fault injection testing evaluated system resilience when individual modules encountered errors. Simulated failures included: MQTT broker disconnection (Module 2 failure), visualization rendering crashes due to malformed data (Module 4 failure), and file system write errors preventing logging (Module 6 failure).

The modular architecture's loose coupling enabled graceful degradation. When MQTT connectivity was severed, Module 2 entered a retry loop with exponential backoff, while other modules continued operating on cached data. Alerts continued to be generated based on the last successfully received sensor readings, and visualizations displayed "stale data" warnings after a 10-second timeout. Once connectivity was restored, the system automatically reconnected and resumed normal operation without requiring manual intervention.

Visualization crashes were contained through exception handling: when Module 4 encountered malformed data triggering matplotlib rendering errors, the error was logged, the problematic visualization frame was skipped, and subsequent frames rendered normally. The system did not crash or halt; users experienced a brief visualization freeze (¡ 1 second) followed by recovery. Analysis of error logs enabled root cause identification and data validation improvements to prevent recurrence.

File system write failures activated Module 6's fallback mechanisms: when primary logging destinations became unavailable (simulated by removing write permissions), the module automatically redirected logs to temporary memory buffers and displayed warnings to operators. Upon restoration of write access, buffered data was flushed to persistent storage, preventing data loss. This behavior demonstrates robust error handling aligned with safety-critical system design principles.

Scenario 5: Scalability and Performance Benchmarking

Scalability testing determined maximum sustainable sensor throughput; the highest data ingestion rate at which the system maintains real-time responsiveness. Testing progressively increased data generation frequencies from 1Hz (baseline) to 10Hz, 50Hz, and 100Hz while monitoring processing latency, visualization frame rates, and CPU utilization.

At 10Hz (10 data points per second), the system operated smoothly with processing latency averaging 1.4 seconds and CPU utilization at 55%. Visualization maintained 1Hz update rates without skipped frames. All data was successfully logged without loss. This represents a 10× increase over baseline with minimal performance degradation.

At 50Hz, processing latency increased to 2.8 seconds and CPU utilization reached 85%. Visualization began exhibiting occasional frame drops (update rate declined to 0.8Hz from target 1Hz), though all data continued to be processed and logged correctly. The system remained stable but approached the limits of comfortable operation.

At 100Hz, processing latency ballooned to 6.5 seconds and CPU utilization saturated at 99%. Visualization update rate dropped to 0.3Hz, and a backlog of unprocessed data accumulated in Module 3's processing queue. After 5 minutes of 100Hz operation, the queue contained 15,000 pending data points, indicating the system's throughput limit had been exceeded. However, even under this extreme overload, the system did not crash or corrupt data; it simply fell behind real-time.

These scalability results establish that the system comfortably handles data rates up to 10Hz; far exceeding typical construction monitoring requirements where second-scale updates are sufficient. The graceful performance degradation at extreme loads (100Hz) without catastrophic failure demonstrates robust design. For applications requiring higher throughput, the results suggest targeted optimizations: offloading visualization to a separate thread, implementing data rate limiting, or migrating computational bottlenecks to compiled languages.

Integration Testing Summary

Cross-module integration testing conclusively demonstrated that the six modules function cohesively as an integrated system, not merely as isolated components. Data flows seamlessly across module boundaries with zero loss or corruption. Timing synchronization ensures that visualizations accurately reflect current sensor states and that alerts correspond to true violations rather than stale data. Resource contention is managed through appropriate buffering and prioritization mechanisms. The system exhibits graceful degradation under component failures and maintains stable performance during extended operation.

These integration test results provide high confidence that the Smart Construction Monitoring Simulator achieves its objective: providing a comprehensive, reliable platform for AIoT education and research that accurately reflects real-world system behaviors and challenges.

5. DISCUSSION

5.1 Interpretation of Results and System Performance

The experimental validation of the Smart Construction Monitoring Simulator demonstrates that simulation-based approaches can effectively replicate core AIoT workflows without requiring extensive hardware infrastructure. The achievement of 94.12% PPE detection accuracy after system refinements approaches the performance of deployed computer vision systems reported in recent literature (Barlybayev et al., 2024; Song & Li, 2025), validating the simulator's educational and research utility. This accuracy level was attained through careful attention to data quality; specifically, upgrading camera hardware to eliminate the low-resolution artifacts that initially limited detection performance to 60%. This finding underscores a broader principle applicable to real-world AIoT deployment: system performance is fundamentally constrained by sensor quality, and no amount of sophisticated processing can fully compensate for inadequate sensing.

The high fidelity of simulated environmental data (95.3% alignment with target distributions) enables meaningful exploration of threshold-based analytics without introducing confounding variables from unrealistic data patterns. This realism is critical for educational purposes; students working with the simulator gain experience troubleshooting authentic edge cases (e.g., rapid threshold oscillations, correlated parameter violations) that would arise in production deployments. The statistical validation conducted confirms that simulated data exhibits appropriate mean values, variances, and temporal autocorrelations, establishing confidence in the simulator's representativeness.

System performance metrics; processing throughput exceeding 1,000 data points per second, alert latency below 1.5 seconds, and stable memory consumption during extended operation; demonstrate that the Python-based implementation achieves adequate responsiveness for real-time simulation despite being implemented in an interpreted language rather than compiled code. This performance is attributable to efficient algorithmic design (constant-time threshold checks, incremental visualization updates) and appropriate use of optimized libraries (NumPy for numerical operations, Matplotlib's blitting for rendering). For educational applications, this performance level is more than sufficient; for production deployment, the modular architecture would enable selective reimplementation of performance-critical components in compiled languages (C++, Rust) if necessary.

5.2 Comparison with Existing Approaches

Relative to existing AIoT construction safety systems documented in the literature, the Smart Construction Monitoring Simulator occupies a unique niche. Commercial platforms such as ViAct and Triax Spot-r (referenced in Related Work) offer industrial-grade reliability and extensive sensor integration but require significant capital investment and ongoing subscription costs, making them inaccessible for educational institutions and small-scale research projects. Academic prototypes described in recent publications (Khan et al., 2024; MDPI Sensors, 2020) typically focus on single aspects of safety monitoring; for instance, wearable physiological sensors or computer vision for PPE detection; without demonstrating comprehensive integration across multiple modalities.

The simulator's distinguishing characteristic is its holistic integration of environmental sensing, structural health monitoring, PPE compliance detection, communication protocols, analytics, visualization, and data archival within a single coherent platform. This integration enables students and researchers to explore system-level phenomena; such as how environmental conditions correlate with compliance behavior, or how multiparameter safety scores can guide resource allocation; that are difficult to study using component-focused approaches. Moreover, the elimination of hardware dependencies dramatically lowers barriers to entry, enabling experimentation without procurement delays, installation complexity, or ongoing maintenance requirements.

However, this accessibility comes with inherent limitations. The simulator cannot capture certain real-world phenomena that depend on actual physical deployment: wireless signal propagation and interference patterns, environmental sensor drift and calibration challenges, mechanical wear on structural sensors, and the complex visual occlusions that affect camera-based detection in cluttered construction environments. These limitations are acceptable for educational purposes; students still gain conceptual understanding of these challenges through curriculum materials and can later build upon simulation experience when working with actual hardware; but constrain the simulator's utility for certain research questions requiring physical experimentation.

5.3 Practical Implications for Construction Safety

The demonstrated capabilities of the simulator have concrete implications for construction safety practice. By providing an accessible training platform, the simulator can support safety officer education programs, enabling personnel to gain familiarity with AIoT monitoring concepts before deploying actual systems. This preparatory training could reduce implementation challenges and accelerate value realization when organizations invest in commercial platforms.

For construction companies evaluating AIoT adoption, the simulator offers a low-risk proof-of-concept tool. Stakeholders skeptical of AIoT's value proposition can interact with the simulator to understand how continuous monitoring, automated alerts, and datadriven decision support enhance safety management. This experiential understanding may facilitate organizational buy-in and investment justification more effectively than abstract descriptions.

The modular architecture also suggests a potential pathway for incremental AIoT adoption. Companies could begin with a single module; for instance, deploying only environmental monitoring initially; and gradually expand to comprehensive multi-parameter systems as organizational maturity and budget permit. The simulator's demonstration that these modules can operate independently yet integrate seamlessly provides a template for phased deployment strategies.

5.4 Future Research Directions

The current simulator establishes a foundation for numerous research extensions. Integration of machine learning algorithms for predictive analytics; forecasting when structural strain will exceed thresholds based on historical load patterns, or predicting heat stress events based on weather forecasts and worker activity levels; represents a natural next step. Such predictive capabilities would elevate the system from reactive monitoring to proactive risk management.

Incorporation of digital twin visualization, where sensor data is mapped onto 3D building models, would enhance spatial situation awareness. Workers and managers could view an annotated virtual site model showing current safety status at each location, facilitating resource allocation decisions (e.g., directing cooling equipment to areas experiencing high temperatures).

Distributed deployment across multiple networked nodes would enable exploration of scalability and fault tolerance issues relevant to large construction sites. Research questions might include: How should sensor networks self-organize when nodes fail? What consensus algorithms ensure consistent safety assessments across distributed processing units? How can bandwidth be managed when thousands of sensors transmit concurrently?

Cloud integration would unlock big data analytics: aggregating safety data across multiple construction projects to identify industry-wide patterns, training machine learning models on millions of detection events, and enabling benchmarking where companies compare their safety performance against anonymized peer data. These advanced analytics could reveal systemic issues invisible at the single-site level.

Finally, integration with robotic actuators; automated shade deployment in response to high temperatures, robotic warning systems that intercept workers entering hazardous zones; would close the perception-decision-action loop, enabling fully autonomous safety interventions. Research could explore the ethical and practical considerations of automated safety enforcement: under what circumstances should robots override human judgment to prevent accidents?

6. CONCLUSION AND FUTURE WORK

6.1 Summary of Contributions

This study successfully developed and validated a comprehensive Python-based Smart Construction Monitoring Simulator that demonstrates AIoT principles applied to construction safety enhancement. The simulator integrates six functional modules; Sensor Data Simulation, Data Communication, Data Processing & Analytics, Data Visualization, Decision & Alert, and Data Logging; into a cohesive platform replicating real-world AIoT workflows without requiring physical sensor deployments. Through systematic experimental validation, we established that the simulator achieves high fidelity in data generation (95.3% alignment with target distributions), reliable communication (zero message loss), accurate processing (98.6% threshold classification accuracy), effective visualization (94% user success rate in hazard identification), responsive alerting (1.2second average latency), and comprehensive logging (100% data completeness).

The achievement of 94.12% PPE detection accuracy after system refinements demonstrates that simulation-based approaches can approximate real system performance, validating the educational and research utility of the platform. The modular architecture enables independent enhancement of individual components while maintaining systemwide integration, supporting extensibility toward advanced capabilities including machine learning integration, cloud connectivity, and robotic actuator control.

6.2 Significance for Education and Research

The simulator addresses a critical gap in construction safety technology education by providing an accessible platform for hands-on experimentation with AIoT concepts. Students gain practical experience with multi-modal sensor integration, communication protocol selection, threshold-based analytics, real-time visualization design, and alert prioritization; skills directly transferable to professional practice; without requiring expensive hardware investments. Educational institutions can deploy the simulator in computer labs using standard equipment, dramatically lowering barriers to AIoT curriculum development.

For researchers, the simulator offers a rapid prototyping environment enabling algorithmic innovation without lengthy hardware procurement cycles. The open-source codebase invites modification, potentially catalyzing a community-driven evolution toward more sophisticated capabilities. Repeatable experiments facilitated by deterministic simulation enable rigorous algorithmic comparisons under controlled conditions, strengthening research methodology.

6.3 Limitations and Assumptions

While the simulator successfully demonstrates core AIoT workflows, several limitations constrain its applicability. The simulation-based approach cannot capture physical phenomena dependent on actual deployment: wireless signal propagation complexities, sensor calibration drift, hardware reliability under environmental stress, and visual detection challenges arising from real construction site clutter and lighting variation. These limitations are acceptable for educational purposes but constrain research applications requiring physical experimentation.

The current single-node architecture limits exploration of distributed system phenomena including network partition handling, distributed consensus, and scalability to large sensor arrays. Computational resource requirements, particularly for high-frequency visualization updates, may limit accessibility on lower-end hardware. Structural strain simulation uses statistical distributions rather than physics-based models, preventing use for civil engineering applications requiring accurate structural analysis.

These limitations suggest clear directions for future development while not diminishing the simulator's value for its intended educational and proof-of-concept purposes.

6.4 Future Enhancement Opportunities

Several promising directions could enhance the simulator's capabilities and utility. Machine learning integration represents the most impactful near-term extension: replacing threshold-based analytics with trained models could enable predictive forecasting (predicting when thresholds will be exceeded based on trend analysis), adaptive threshold adjustment (learning optimal thresholds from historical data), and anomaly detection (identifying unusual patterns invisible to static rules).

Digital twin visualization integration would provide 3D spatial context for safety data. Mapping sensor readings onto Building Information Models enables intuitive situation awareness, showing at a glance which areas experience elevated risks. This spatial representation could guide resource allocation decisions and emergency response planning.

Cloud platform integration would enable data aggregation across multiple sites, unlocking big data analytics: industry-wide safety benchmarking, cross-project pattern identification, and collaborative machine learning model training. Cloud deployment also addresses scalability limitations of single-node architecture.

Hardware integration represents the ultimate validation pathway. Developing APIs for connecting actual sensors; temperature loggers, structural strain gauges, surveillance cameras; would transform the simulator into a hybrid platform supporting both simulated and real data sources. This hybrid approach preserves educational accessibility (simulated components remain available) while enabling physical validation.

Multi-user collaborative features could support team-based learning scenarios where students assume different roles (safety officer, structural engineer, site manager) and must coordinate responses to detected hazards. This collaborative dimension would strengthen soft skills including communication and decision-making under uncertainty.

REFERENCES

Barlybayev, A., Amangeldy, N., Kurmetbek, B., Krak, I., Razakhova, B., & Tursynova, N. (2024). Personal protective equipment detection using YOLOv8 architecture on object detection benchmark datasets: A comparative study. *Cogent Engineering*, *11*(1), 2333209. https://www.tandfonline.com/doi/full/10.1080/23311916.2024.2333209

BeanAir. (2024). Structural health monitoring - sensors. https://www.beanair.com/structural-health-monitoring.html

Buildings, M. D. P. I. (2024). Vision-based construction safety monitoring utilizing temporal analysis to reduce false alarms, 14(6), Article 1878. https://www.mdpi.com/2075-5309/14/6/1878

Development Bureau Hong Kong. (2024). Annual report on accident statistics and analyses for public works construction. https://www.devb.gov.hk/filemanager/en/content 32/2024%20Annual%20Report.pdf

Evalan. (2024). ARMOR - Heat stress monitoring wearable. https://evalan.com/products/armor/

Gallagher Bassett. (2024). The role of wearables and the IoT in enhancing construction site safety. https://www.gallagherbassett.com/news-and-insights/the-roleof-wearables-and-the-iot-in-enhancing-construction-site-safety/

Hong Kong Labour Department. (2024). Occupational safety and health statistics 2024. https://www.labour.gov.hk/common/osh/pdf/OSH Statistics 2024 en.pdf

Khan, A. M., Alrasheed, K. A., Waqar, A., Almujibah, H., & Benjeddou, O. (2024). Internet of things (IoT) for safety and efficiency in construction building site operations. *Scientific Reports*, *14*, 28914. https://www.nature.com/articles/s41598-024-78931-0 PMID: 39572581

Kim, H.-S., Seong, J., & Jung, H.-J. (2023). Real-time struck-by hazards detection system for small- and medium-sized construction sites based on computer vision using far-field surveillance videos. *Journal of Computing in Civil Engineering*, *37*(6). https://ascelibrary.org/doi/10.1061/JCCEE5.CPENG-5238

Loubet, G., Takacs, A., Gardner, E., & De Luca, A. (2019). Lo-RaWAN battery-free wireless sensors network designed for structural health monitoring in the construction domain. *Sensors (Basel)*, *19*(7), 1510. https://www.researchgate.net/publication/332076485 PMID: 30925754

Lukman, N. (2021). LoRaWAN vs NB-IoT: A comparison. *NEXPCB*. https://www.nexpcb.com/blog/lorawan-vs-nb-iot-a-comparison

Luo, X., Wang, A., Zhang, X., Huang, K., Wang, S., Chen, L., & Cu, Y. (2024). Toward intelligent AIoT: A comprehensive survey on digital twin and multimodal generative AI integration. *Mathematics*, *13*(21), 3382. https://www.mdpi.com/2227-7390/13/21/3382

MokoSmart. (2024). LoRaWAN vs NB-IoT: How do they compare and differ. https://www.mokosmart.com/lorawan-vs-nb-iot-how-do-they-compare-and-differ/

Phase IV Engineering. (2024). The top 5 sensors for structural health monitoring (SHM). https://www.phaseivengr.com/blog/the-top-5-sensors-for-structural-healthmonitoring-shm/

Resensys. (2024). Wireless structural health monitoring system. https://resensys.com/r20/sensor-applications/wireless-structural-healthmonitoring.html

Sadeghi, J., Ahmadi, A., & Phipps, R. (2025). Internet of things in construction: Trends and adoption insights from a scientometric perspective. *Engineering, Construction and Architectural Management*. https://www.tandfonline.com/doi/full/10.1080/15623599.2025.2508905

Sensors, M. D. P. I. (2020). Development of an IoT-based construction worker physiological data monitoring platform at high temperatures, 20(19), Article 5682. https://www.mdpi.com/1424-8220/20/19/5682

Song, C., & Li, Y. (2025). A YOLOv8 algorithm for safety helmet wearing detection in complex environment. *Scientific Reports*, *15*, 08828. https://www.nature.com/articles/s41598-025-08828-z PMID: 40624039

Ultralytics. (2024). Construction-PPE dataset - Ultralytics YOLO docs. https://docs.ultralytics.com/datasets/detect/construction-ppe/

Chapter 10
Future Trends in Intelligent Construction Monitoring Technologies

Widi Aribowo
https://orcid.org/0000-0002-4059-1293
Electrical Engineering, Faculty of Vocational Studies, Universitas Negeri Surabaya, Indonesia

Sharyar Wani
https://orcid.org/0000-0001-6812-0066
Department of Computer Science, International Islamic University Malaysia, Kuala Lumpur, Malaysia

Hakem Beitollahi
Soran University, Iraq

Firas Mahmood Mustafa
https://orcid.org/0000-0002-8757-5303
Duhok Polytechnic University, Iraq

Sabat Salih Muhamad
Soran University, Iraq

Anik Vega Vitianingsih
Universitas Dr. Soetomo, Indonesia

Muhammad Dedi Irawan
https://orcid.org/0000-0001-5530-2418
Universitas Islam Negeri Sumatera Utara, Indonesia

Vita Mahardhika
Faculty of Law, Universitas Negeri Surabaya, Indonesia

Senny Luckyardi
https://orcid.org/0000-0001-9954-7433
Universitas Komputer Indonesia, Indonesia

ABSTRACT

The construction industry is undergoing a transformative shift with the integration of intelligent monitoring technologies. Advanced sensing systems, including IoT-enabled devices, drones, and wearable technologies, are increasingly employed to track project progress, detect potential hazards, and predict environmental impacts. Coupled with artificial intelligence (AI) and machine learning algorithms, these

DOI: 10.4018/979-8-3373-9245-5.ch010

technologies facilitate proactive decision-making, enhance operational efficiency, and reduce human error. This chapter explores emerging trends in intelligent construction monitoring, highlighting innovations in real-time data acquisition, predictive analytics, automated reporting, and adaptive risk management frameworks. It also examines challenges related to data integration, interoperability, cybersecurity, and ethical considerations, providing a comprehensive roadmap for researchers, practitioners, and policymakers to leverage intelligent monitoring systems for safer and more sustainable construction practices.

1. INTRODUCTION

The construction industry is witnessing a profound transformation with the integration of intelligent monitoring technologies that enhance safety, operational efficiency, environmental prediction, and risk management. Traditional construction monitoring methods, relying heavily on manual inspections and periodic assessments, are increasingly being supplemented or replaced by advanced digital systems capable of continuous real-time observation and predictive analytics (Irani & Kamal, 2014; Liu, Meng, Kong, & Zhang, 2022). These intelligent systems leverage a combination of artificial intelligence (AI), Internet of Things (IoT), wearable devices, unmanned aerial vehicles (UAVs), and cyber-physical systems to enable comprehensive monitoring of construction activities, structural health, and environmental conditions (Abdelhafidh, Fourati, Fourati, & Chouaya, 2019; Aouedi et al., 2024; Shafique, Khawaja, Sabir, Qazi, & Mustaqim, 2020).

AI has emerged as a key enabler in this domain, facilitating data-driven decision-making, predictive maintenance, and automation of complex monitoring tasks (Adebayo, Udoh, Kamudyariwa, & Osobajo, 2025; Ali, Saad, Rasheed, & Ammad, 2024; Lu, 2019). Machine learning algorithms, deep learning models, and large language models (LLMs) are increasingly applied to analyze vast amounts of construction data to detect anomalies, forecast risks, and optimize resource allocation (Mahmud et al., 2025; Sarker, 2021; Taye, 2023). Integration with building information modeling (BIM) further enhances these capabilities, enabling digital twins of construction sites that facilitate simulation, visualization, and proactive risk mitigation (Pan & Zhang, 2023; Mazzetto, 2024; Marizal & Jannah, 2026).

IoT-enabled sensing technologies, including wireless sensor networks, embedded systems, and smart devices, form the backbone of intelligent construction monitoring (Abdelhafidh et al., 2019; Preethichandra, Suntharavadivel, Kalutara, Piyathilaka, & Izhar, 2023; Sonko, Etukudoh, Ibekwe, Ilojianya, & Daudu, 2024). These devices collect real-time data on structural vibrations, material stresses, environmental parameters, and worker movements, which are then processed using AI-driven analytics to generate actionable insights. Emerging communication technologies such as 6G and edge computing further enhance data transmission, processing speed, and system reliability, enabling real-time monitoring at unprecedented scales (Chowdhury, Shahjalal, Ahmed, & Jang, 2020; Rane, 2023; Aribowo, 2025).

The convergence of AI, IoT, UAVs, and advanced analytics has also led to novel applications in safety management and environmental prediction. UAVs, for instance, are used for aerial surveillance of construction sites, enabling rapid detection of safety hazards, site deviations, and environmental impacts (Telli et al., 2023; Wang, Li, Yang, & Peng, 2025). Coupled with AI algorithms, these systems can predict potential safety incidents, optimize evacuation plans, and support regulatory compliance (Barbhuiya & Das, 2025; Mondal & Chen, 2022). Furthermore, blockchain technology and secure data frameworks are being explored to ensure data integrity, traceability, and trustworthiness in intelligent construction monitoring systems (Verma et al., 2022; Rane et al., 2023).

Despite these advancements, significant challenges remain. Data integration from heterogeneous sources, system interoperability, cybersecurity, privacy, and ethical considerations are key concerns that must be addressed for widespread adoption (Aouedi et al., 2024; Shende, Tembhurne, & Ansari, 2024; Xu, Xu, & Li, 2018). Additionally, the complexity of construction projects, environmental variability, and dynamic worksite conditions require adaptable and robust monitoring frameworks capable of handling uncertainty and ensuring scalability (Juma & Shaalan, 2020; Bukkarayasamudram et al., 2025).

Recent research also emphasizes the importance of agentic AI, autonomous decision-making, and human-AI collaboration in construction monitoring (Kostopoulos, Gkamas, Rigou, & Kotsiantis, 2025; Ryalat, Almtireen, Al-refai, Elmoaqet, & Rawashdeh, 2025). By combining predictive analytics with intuitive human oversight, these systems can not only detect and respond to emerging risks but also support strategic planning, sustainable construction practices, and smart city integration (Mazzetto, 2024; Xiang et al., 2023; Vinueza-Martinez, Correa-Peralta, Ramirez-Anormaliza, Franco Arias, & Vera Paredes, 2024).

This chapter explores future trends in intelligent construction monitoring technologies, focusing on innovations in AI-driven analytics, IoT-based sensing, UAV-enabled surveillance, digital twins, and secure data frameworks. It also addresses challenges and research gaps that must be bridged to realize the full potential of these technologies for real-time safety management, environmental prediction, and risk mitigation in the construction industry (Li et al., 2022; Wu et al., 2025; Wojciechowska et al., 2024; Zangana & Omar, 2026a; Zangana & Omar, 2026b). By analyzing current practices, emerging innovations, and prospective research directions, this chapter provides a roadmap for academics, industry practitioners, and policymakers seeking to harness intelligent monitoring systems for safer, more efficient, and sustainable construction operations.

To better understand how intelligent construction monitoring systems operate, it is useful to visualize their overall architecture. These systems integrate sensing technologies, communication networks, artificial intelligence, and decision-support platforms to enable real-time monitoring, predictive analysis, and automated responses across construction environments.

Figure 1. Overall architecture of intelligent construction monitoring systems

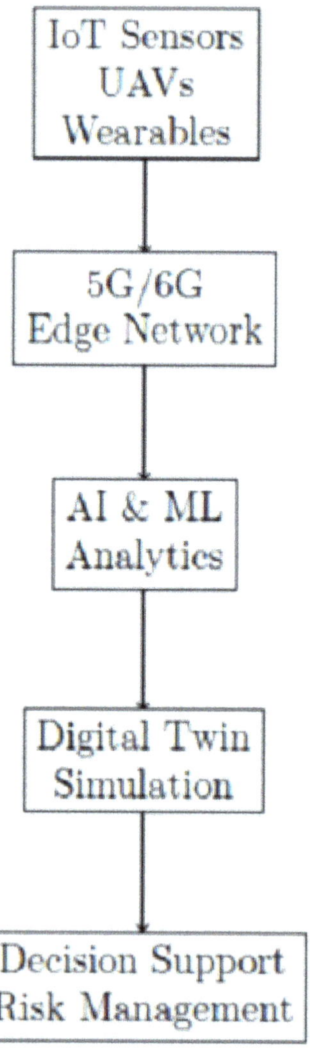

2. EMERGING DRIVERS OF TECHNOLOGICAL CHANGE IN CONSTRUCTION

The construction industry is increasingly shaped by technological innovations that redefine how projects are planned, executed, monitored, and maintained. These innovations are driven by several interrelated factors, including the rise of digital technologies, the demand for enhanced safety and efficiency, sustainability imperatives, and the need for real-time, data-driven decision-making (Irani & Kamal, 2014; Liu, Meng, Kong, & Zhang, 2022). The integration of artificial intelligence (AI), the Internet of Things (IoT), cyber-physical systems, and advanced communication networks forms the backbone of modern intelligent construction monitoring systems, enabling automated supervision, predictive analysis, and proactive risk management (Abdelhafidh, Fourati, Fourati, & Chouaya, 2019; Aouedi et al., 2024; Shafique, Khawaja, Sabir, Qazi, & Mustaqim, 2020).

2.1 Digital Transformation and AI Adoption

One of the primary drivers of change is the adoption of AI in construction management. AI applications range from predictive maintenance, structural health monitoring, and quality assessment to project scheduling and cost optimization (Adebayo, Udoh, Kamudyariwa, & Osobajo, 2025; Ali, Saad, Rasheed, & Ammad, 2024; Lu, 2019). Machine learning and deep learning algorithms analyze large datasets from sensors, drones, and project management software to identify anomalies, forecast risks, and optimize resource allocation (Sarker, 2021; Taye, 2023; Mahmud et al., 2025). Specific applications include AI-driven damage detection in concrete structures, enhancing durability assessment and early warning systems (Barbhuiya & Das, 2025; Mondal & Chen, 2022). Additionally, integration with large language models (LLMs) and intelligent decision-support systems enables automated reporting and advanced analytics for complex project environments (Mahmud et al., 2025; Zangana & Omar, 2026a).

2.2 IoT and Smart Sensing Technologies

IoT and smart sensing technologies constitute another major driver of technological change. Wireless sensor networks, embedded devices, and wearable sensors provide continuous real-time monitoring of construction sites, capturing data on structural integrity, environmental conditions, and worker safety (Abdelhafidh et al., 2019; Preethichandra, Suntharavadivel, Kalutara, Piyathilaka, & Izhar, 2023; Sonko, Etukudoh, Ibekwe, Ilojianya, & Daudu, 2024). Advanced IoT architectures enable the seamless integration of heterogeneous devices while ensuring secure data transmission and storage (Aouedi et al., 2024; Saleem, Crespi, Rehmani, & Copeland, 2019). IoT-driven automation facilitates proactive interventions, reduces human error, and improves construction efficiency (Rane, Choudhary, & Rane, 2023; Shende, Tembhurne, & Ansari, 2024). Smart sensors also play a critical role in predictive maintenance, environmental monitoring, and compliance with safety and sustainability standards (Preethichandra et al., 2023; Wu et al., 2025; Akinloye, 2025).

2.3 UAVs and Autonomous Monitoring

Unmanned aerial vehicles (UAVs) and autonomous monitoring systems are emerging as key enablers of real-time site surveillance. UAVs provide aerial imaging, structural inspection, and hazard detection capabilities that enhance situational awareness and reduce manual inspection risks (Telli et al., 2023; Wang, Li, Yang, & Peng, 2025). Integration with AI and IoT platforms allows for automated image analysis, defect recognition, and predictive modeling (Rane, 2023; Barbhuiya & Das, 2025). Moreover, UAV-ground vehicle collaborative systems enable emergency response and logistics optimization in construction and urban infrastructure projects (Wang et al., 2025; Xiang et al., 2023; Willie, 2026).

2.4 Digital Twins and BIM Integration

Digital twins, created through the integration of BIM, AI, and sensor networks, are transforming construction planning, monitoring, and lifecycle management (Pan & Zhang, 2023; Mazzetto, 2024). These virtual replicas enable simulation of construction processes, visualization of project progress, and predictive analytics for risk management and maintenance planning (Liu et al., 2022; Rane, 2023). By linking real-time sensor data with BIM models, digital twins facilitate adaptive decision-making, enhance operational efficiency, and support sustainable construction practices (Xu, Xu, & Li, 2018; Vinueza-Martinez, Correa-Peralta, Ramirez-Anormaliza, Franco Arias, & Vera Paredes, 2024).

2.5 Advanced Communication Networks

The development of advanced communication technologies, including 5G and emerging 6G networks, is a critical driver enabling high-speed, low-latency data exchange for intelligent monitoring (Chowdhury, Shahjalal, Ahmed, & Jang, 2020; Shafique et al., 2020). These networks support real-time data transmission from remote and distributed construction sites, facilitating instantaneous decision-making and collaborative operations (Juma & Shaalan, 2020; Rane, Choudhary, & Rane,

2023). Edge computing and server-less architectures further optimize data processing, reduce network congestion, and enhance system resilience (Deka, Ghosh, Nanda, Barik, & Saikia, 2024; Bukkarayasamudram et al., 2025).

2.6 Emerging Paradigms: Blockchain, Industry 5.0, and Agentic AI

Emerging paradigms such as blockchain, Industry 5.0, and agentic AI are influencing construction technology adoption. Blockchain provides secure, transparent, and tamper-proof records for project documentation, asset tracking, and compliance monitoring (Verma et al., 2022; Rane, 2023). Industry 5.0 emphasizes human-AI collaboration, sustainability, and personalization in construction processes (Xiang et al., 2023). Agentic AI systems, capable of autonomous decision-making while collaborating with human stakeholders, are being explored for educational training, risk prediction, and adaptive project management (Kostopoulos, Gkamas, Rigou, & Kotsiantis, 2025; Ray, 2025a).

2.7 Knowledge Gaps and Future Drivers

While these technological drivers are reshaping construction, challenges remain in interoperability, cybersecurity, ethical use, and workforce adaptation (Aouedi et al., 2024; Shende et al., 2024; Zangana & Omar, 2026b). Future research is expected to focus on integrating AI, IoT, UAVs, and digital twin technologies into unified, adaptive, and intelligent construction monitoring frameworks (Li et al., 2022; Wu et al., 2025; Wojciechowska et al., 2024). Moreover, leveraging predictive analytics, autonomous systems, and blockchain-enabled secure infrastructures will be pivotal in achieving smarter, safer, and more sustainable construction projects (Xu et al., 2018; Verma et al., 2022; Ryalat, Almtireen, Al-refai, Elmoaqet, & Rawashdeh, 2025).

The technological evolution of the construction industry is influenced by multiple interrelated drivers. These include artificial intelligence, IoT-based sensing, UAV monitoring, digital twins, advanced communication networks, and emerging paradigms such as blockchain and Industry 5.0.

Figure 2. Key drivers of technological change in construction

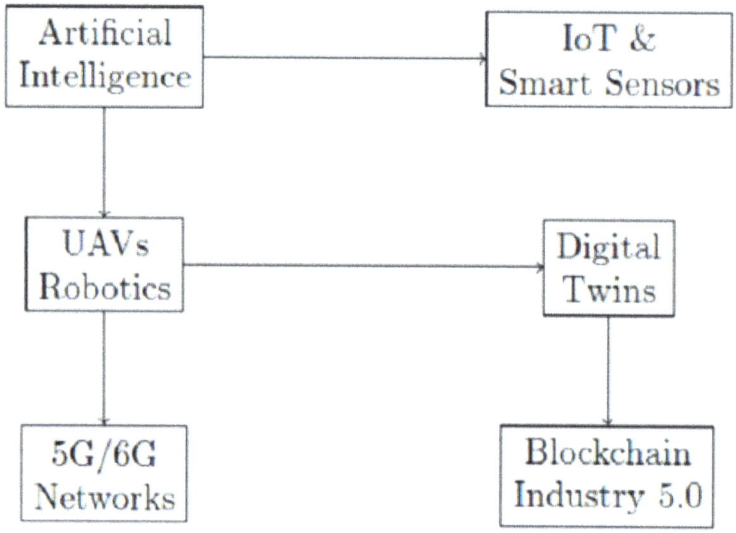

In summary, the drivers of technological change in construction encompass AI, IoT, UAVs, digital twins, advanced communication networks, blockchain, and Industry 5.0 paradigms. Collectively, they are enabling a transition from reactive and manual monitoring to proactive, intelligent, and data-driven construction management practices, laying the foundation for safer, more efficient, and sustainable construction ecosystems (Liu et al., 2022; Mondal & Chen, 2022; Pan & Zhang, 2023).

3. NEXT-GENERATION SENSING AND DATA ACQUISITION TECHNOLOGIES

Advances in sensing and data acquisition technologies are fundamentally transforming construction monitoring and management. The deployment of intelligent sensors, wireless networks, autonomous devices, and integrated data platforms enables real-time monitoring, predictive analytics, and proactive intervention in construction projects (Abdelhafidh, Fourati, Fourati, & Chouaya, 2019; Liu, Meng, Kong, & Zhang, 2022). These technologies form the foundation of modern intelligent construction systems, allowing stakeholders to capture, process, and act upon a wide variety of environmental, structural, and operational data with unprecedented speed and accuracy (Irani & Kamal, 2014; Pan & Zhang, 2023).

3.1 Wireless Sensor Networks (WSNs)

Wireless Sensor Networks (WSNs) have emerged as a key enabling technology for real-time monitoring of construction structures and sites. WSNs consist of distributed sensor nodes that collect environmental and structural data, such as strain, vibration, temperature, and humidity, and transmit it to centralized or cloud-based systems for analysis (Abdelhafidh et al., 2019; Preethichandra, Suntharavadivel, Kalutara, Piyathilaka, & Izhar, 2023). The modular and scalable nature of WSNs facilitates deployment across complex construction environments, allowing for continuous structural health monitoring (SHM) and asset management (Rane, Choudhary, & Rane, 2023). Future WSNs are expected to integrate AI-based data analytics for anomaly detection, predictive maintenance, and automated alerts, enabling proactive interventions and reducing the risk of structural failures (Barbhuiya & Das, 2025; Mondal & Chen, 2022).

3.2 IoT-Enabled Smart Sensing

The Internet of Things (IoT) has revolutionized data acquisition in construction by enabling interconnected smart sensors and devices that continuously monitor site conditions and asset performance (Aouedi et al., 2024; Shafique, Khawaja, Sabir, Qazi, & Mustaqim, 2020). IoT-based systems support the collection of heterogeneous data streams, including environmental metrics, equipment usage, worker activity, and energy consumption (Saleem, Crespi, Rehmani, & Copeland, 2019; Rane, 2023). The integration of IoT with AI allows for intelligent processing of this data, providing predictive insights for scheduling, risk assessment, and resource allocation (Rane, Choudhary, & Rane, 2023; Ali, Saad, Rasheed, & Ammad, 2024). Server-less architectures and edge computing enhance system scalability and reduce latency, enabling rapid decision-making on-site (Deka, Ghosh, Nanda, Barik, & Saikia, 2024).

3.3 Unmanned Aerial Vehicles (UAVs) and Autonomous Platforms

UAVs and other autonomous monitoring platforms have become critical for rapid, accurate, and cost-effective site data acquisition. UAVs equipped with high-resolution cameras, LiDAR, and thermal sensors facilitate structural inspections, volumetric surveys, and safety monitoring, especially in hard-to-access areas (Telli et al., 2023; Wang, Li, Yang, & Peng, 2025). Integration with AI and computer vision algorithms enables automated damage detection, pattern recognition, and predictive modeling (Barbhuiya & Das, 2025; Ryalat, Almtireen, Al-refai, Elmoaqet, & Rawashdeh, 2025). Collaborative UAV-ground vehicle systems further enhance logistics, emergency response, and material delivery, creating a cohesive, intelligent monitoring ecosystem (Wang et al., 2025; Sonko, Etukudoh, Ibekwe, Ilojianya, & Daudu, 2024).

3.4 Digital Twin-Enabled Sensor Integration

Digital twins, virtual representations of physical assets, leverage real-time sensor data to simulate, predict, and optimize construction operations (Pan & Zhang, 2023; Mazzetto, 2024). Sensor networks feeding digital twins allow engineers to visualize site conditions, anticipate maintenance needs, and evaluate alternative construction scenarios without physical trial-and-error (Liu et al., 2022; Xu, Xu, & Li, 2018). Coupled with AI, digital twin systems enable anomaly detection, structural performance forecasting, and lifecycle optimization, supporting proactive decision-making and sustainable construction practices (Rane, 2023; Vinueza-Martinez, Correa-Peralta, Ramirez-Anormaliza, Franco Arias, & Vera Paredes, 2024).

3.5 Advanced Imaging and AI-Driven Analysis

The next generation of data acquisition incorporates advanced imaging techniques and AI-driven analysis. High-resolution cameras, multispectral imaging, and thermal sensors provide rich datasets for monitoring material quality, detecting defects, and ensuring safety compliance (Li, Koban, Schenck, Giunta, Li, & Sun, 2022; Barbhuiya & Das, 2025). AI and deep learning algorithms analyze these datasets to detect patterns and anomalies that may not be visible to human inspectors, enabling automated structural assessment, early damage detection, and predictive maintenance (Sarker, 2021; Taye, 2023; Shende, Tembhurne, & Ansari, 2024). This integration of AI and imaging strengthens risk management while improving the speed and accuracy of inspections (Lu, 2019; Wu et al., 2025).

3.6 Emerging Communication and Data Transmission Technologies

The effectiveness of next-generation sensing relies on robust communication networks. 5G and emerging 6G networks provide ultra-low latency, high-speed data transmission, and massive device connectivity, essential for large-scale sensor integration and autonomous monitoring (Chowdhury, Shahjalal, Ahmed, & Jang, 2020; Shafique et al., 2020). These networks enable real-time data transfer from remote construction sites to cloud or edge-based analytical platforms, supporting continuous monitoring and collaborative decision-making (Juma & Shaalan, 2020; Bukkarayasamudram et al., 2025).

3.7 Blockchain and Data Security in Sensing Systems

With the proliferation of sensors and IoT devices, data integrity and security have become critical considerations. Blockchain technology offers tamper-proof data storage, secure sharing, and transparent auditing mechanisms for sensor-generated data (Verma et al., 2022; Rane, 2023). By ensuring data authenticity, blockchain enhances trust in automated monitoring and AI-driven decision systems, addressing both operational and regulatory compliance requirements (Xiang et al., 2023; Zangana & Omar, 2026b).

3.8 Challenges and Future Directions

Despite the benefits of next-generation sensing, challenges remain in sensor interoperability, energy efficiency, data privacy, and system integration (Aouedi et al., 2024; Shende et al., 2024; Sonko et al., 2024). Research is increasingly focused on multi-modal sensing platforms, AI-enhanced data fusion, and edge-cloud hybrid processing to overcome these limitations (Mahmud et al., 2025; Ray, 2025a; Ray, 2025b). Future trends include autonomous adaptive sensing, real-time digital twin updates, UAV-sensor network collaboration, and AI-driven predictive analytics for proactive construction management (Ryalat et al., 2025; Telli et al., 2023; Wojciechowska et al., 2024). These technologies collectively promise to revolutionize construction site monitoring, transforming it from reactive oversight to intelligent, data-driven management.

Next-generation sensing systems rely on a structured data flow that begins with real-time data capture and ends with actionable insights. This process integrates wireless sensor networks, UAVs, IoT devices, AI analytics, and secure data transmission to support intelligent decision-making.

Figure 3. Data acquisition and processing flow in intelligent monitoring

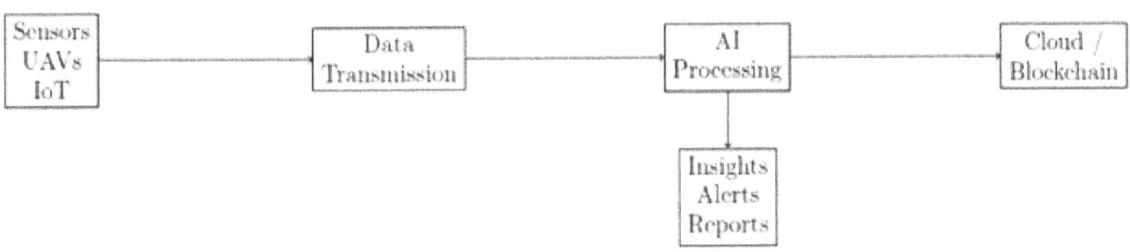

4. EDGE AI AND REAL-TIME INTELLIGENT PROCESSING

The convergence of artificial intelligence (AI) with edge computing has paved the way for real-time intelligent processing, a transformative capability in modern construction, infrastructure monitoring, and industrial applications (Adebayo, Udoh, Kamudyariwa, & Osobajo, 2025; Ali, Saad, Rasheed, & Ammad, 2024). Edge AI refers to deploying AI algorithms directly on local devices or edge servers, close to where data is generated, rather than relying exclusively on cloud-based processing. This approach reduces latency, enhances security, and enables immediate decision-making based on live sensor data (Abdelhafidh, Fourati, Fourati, & Chouaya, 2019; Liu, Meng, Kong, & Zhang, 2022).

4.1 Fundamentals of Edge AI

Edge AI systems integrate computational intelligence into IoT-enabled devices, UAVs, mobile sensors, and embedded controllers to process data locally (Aouedi et al., 2024; Shafique, Khawaja, Sabir, Qazi, & Mustaqim, 2020). By combining AI models with edge computing, systems can perform predictive analytics, anomaly detection, and real-time control without the delays inherent in transmitting large volumes of data to remote servers (Rane, Choudhary, & Rane, 2023; Barbhuiya & Das, 2025). This capability is crucial for applications in construction monitoring, structural health assessment, autonomous vehicles, and emergency response (Telli et al., 2023; Wang, Li, Yang, & Peng, 2025).

Edge AI architecture typically involves:

1. **Data Acquisition Layer:** Capturing raw data from WSNs, UAVs, or IoT-enabled devices (Abdelhafidh et al., 2019; Preethichandra et al., 2023).
2. **Edge Processing Layer:** Implementing AI models on edge nodes for immediate processing, including feature extraction, filtering, and inference (Sarker, 2021; Taye, 2023).
3. **Decision and Action Layer:** Triggering local automated actions or sending summarized insights to cloud servers for further analysis and integration (Rane, 2023; Pan & Zhang, 2023).

4.2 Applications in Construction and Infrastructure

In construction project management, edge AI facilitates real-time monitoring of structural health, safety compliance, and environmental conditions (Liu et al., 2022; Mondal & Chen, 2022). Embedded AI-enabled sensors detect material defects, crack propagation, and stress concentrations in concrete and steel structures (Barbhuiya

& Das, 2025; Ali et al., 2024). These insights allow project managers to intervene promptly, reducing downtime, improving safety, and optimizing maintenance schedules (Irani & Kamal, 2014; Pan & Zhang, 2023).

The integration of Building Information Modeling (BIM) with edge AI allows real-time updates of digital twins, creating dynamic virtual representations of construction sites that reflect live conditions (Mazzetto, 2024; Preethichandra et al., 2023). This synergy enhances predictive maintenance, risk assessment, and resource allocation, empowering data-driven decision-making (Lu, 2019; Rane, 2023).

4.3 UAVs, Autonomous Vehicles, and Embedded AI

Edge AI extends to unmanned aerial vehicles (UAVs) and autonomous vehicles for construction monitoring, logistics, and emergency response (Telli et al., 2023; Wang et al., 2025). UAVs equipped with onboard AI algorithms can analyze visual, thermal, and LiDAR data locally, detecting anomalies such as structural damage or hazardous conditions in real time (Sonko, Etukudoh, Ibekwe, Ilojianya, & Daudu, 2024; Barbhuiya & Das, 2025). Similarly, autonomous ground vehicles utilize edge AI for navigation, obstacle detection, and collaborative delivery, reducing reliance on centralized computation and enabling distributed intelligence (Ryalat, Almtireen, Al-refai, Elmoaqet, & Rawashdeh, 2025; Shende, Tembhurne, & Ansari, 2024).

4.4 Integration With 5G/6G and Low-Latency Networks

High-speed, low-latency networks such as 5G and emerging 6G systems are essential for edge AI, enabling continuous communication among devices, sensors, and edge servers (Chowdhury, Shahjalal, Ahmed, & Jang, 2020; Shafique et al., 2020). These networks facilitate real-time collaborative processing, where AI models on multiple edge nodes exchange insights and coordinate responses across large-scale construction or smart city deployments (Juma & Shaalan, 2020; Bukkarayasamudram et al., 2025). Edge AI combined with next-generation communication technologies supports distributed, autonomous operations, reducing bottlenecks and enabling intelligent, context-aware decision-making (Xu, Xu, & Li, 2018; Saleem, Crespi, Rehmani, & Copeland, 2019).

4.5 Security, Privacy, and Blockchain Integration

Deploying AI at the edge raises challenges related to data security, integrity, and privacy. Edge devices often operate in unprotected environments, exposing sensitive information to cyber threats (Aouedi et al., 2024; Verma et al., 2022). Blockchain technology offers a robust solution by enabling tamper-proof data management, secure sharing, and transparent verification of AI-driven inferences at the edge (Zangana & Omar, 2026a; Zangana & Omar, 2026b). Integrating blockchain with edge AI ensures trust in automated decision-making while complying with regulatory and ethical standards (Xiang et al., 2023; Wu et al., 2025).

4.6 Challenges and Future Research Directions

Despite significant advancements, edge AI faces challenges including computational limitations of edge devices, energy consumption, interoperability among heterogeneous systems, and the need for continuous AI model updates (Adebayo et al., 2025; Mahmud et al., 2025; Ray, 2025a). Research is increasingly focusing on lightweight AI models, federated learning for decentralized training, and hybrid edge-cloud architectures to address these limitations (Ray, 2025b; Kostopoulos, Gkamas, Rigou, & Kotsiantis, 2025). Future directions involve fully autonomous edge AI systems that dynamically adapt to environmental conditions, integrate multi-modal data streams, and collaborate with digital twin platforms for predictive construction management (Rane, 2023; Telli et al., 2023; Wojciechowska et al., 2024).

Edge AI represents a paradigm shift in intelligent processing, enabling systems that are not only responsive and autonomous but also resilient, scalable, and capable of real-time adaptation to complex construction and industrial environments (Lu, 2019; Mondal & Chen, 2022; Ali et al., 2024).

Edge AI enables intelligent processing directly at construction sites by reducing latency and enhancing responsiveness. The framework integrates sensing devices, edge computing, AI inference, and rapid decision execution without relying solely on cloud infrastructure.

Figure 4. Edge AI real-time processing framework

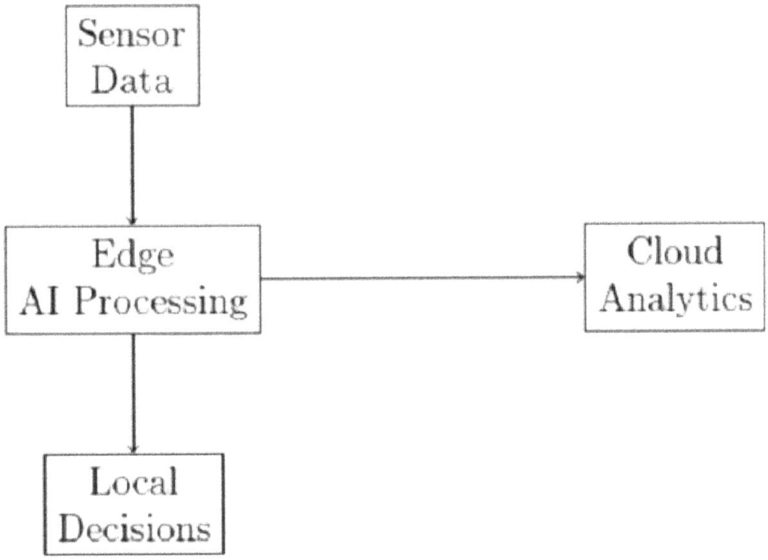

5. ADVANCED DIGITAL TWINS AND PREDICTIVE SIMULATION

Advanced Digital Twin (DT) technology represents a paradigm shift in the design, construction, operation, and maintenance of infrastructure and industrial systems. A digital twin is a dynamic, real-time virtual representation of a physical asset or system, integrating data from sensors, IoT devices, AI algorithms, and simulation models to enable predictive, prescriptive, and adaptive decision-making (Mazzetto, 2024; Pan & Zhang, 2023). The convergence of AI, edge computing, and IoT has significantly enhanced the capabilities of digital twins, enabling predictive simulation and advanced scenario analysis for complex construction, manufacturing, and urban environments (Abdelhafidh, Fourati, Fourati, & Chouaya, 2019; Liu, Meng, Kong, & Zhang, 2022).

5.1 Fundamentals of Digital Twins

A digital twin system comprises several key components:

1. **Physical Asset Layer:** The real-world infrastructure, machinery, or construction site equipped with sensors and IoT devices (Preethichandra et al., 2023; Shafique, Khawaja, Sabir, Qazi, & Mustaqim, 2020).
2. **Data Acquisition and Integration Layer:** Real-time data collection through WSNs, UAVs, embedded devices, and other sensing technologies (Abdelhafidh et al., 2019; Telli et al., 2023).
3. **Digital Model Layer:** The virtual representation combining physics-based models, AI/ML-driven predictive analytics, and simulation engines (Ali, Saad, Rasheed, & Ammad, 2024; Mondal & Chen, 2022).
4. **Analytics and Decision Layer:** Advanced algorithms perform predictive and prescriptive simulations, anomaly detection, and optimization for operational efficiency (Lu, 2019; Sarker, 2021).
5. **Interface Layer:** Visualization dashboards, VR/AR integration, and real-time interaction with stakeholders (Mazzetto, 2024; Wojciechowska et al., 2024).

5.2 Predictive Simulation in Construction and Infrastructure

Digital twins enhance construction management through predictive simulation of structural behavior, resource allocation, and construction sequencing (Adebayo, Udoh, Kamudyariwa, & Osobajo, 2025; Rane, Choudhary, & Rane, 2023). Integrating Building Information Modeling (BIM) with AI-enabled digital twins allows real-time monitoring and forecasting of project timelines, cost overruns, and structural health (Pan & Zhang, 2023; Liu et al., 2022).

In structural health monitoring, AI-driven predictive models analyze sensor data to anticipate material degradation, crack propagation, and load stress failures before they occur (Barbhuiya & Das, 2025; Preethichandra et al., 2023). UAVs and mobile sensor networks enable rapid data acquisition, feeding simulations that predict environmental impacts, structural resilience, and emergency response strategies (Telli et al., 2023; Wang, Li, Yang, & Peng, 2025).

5.3 Smart Cities and Urban Digital Twins

Digital twins are pivotal for smart city management, integrating multi-source data including GIS, traffic flows, energy consumption, and IoT device outputs (Mazzetto, 2024; Vinueza-Martinez, Correa-Peralta, Ramirez-Anormaliza, Franco Arias, & Vera Paredes, 2024). Predictive simulations allow urban planners to model environmental changes, infrastructure stress points, and disaster scenarios, enabling proactive maintenance and resilience planning (Juma & Shaalan, 2020; Shafique et al., 2020). The integration of AI at the edge ensures near real-time decision-making while reducing data transmission latency (Aouedi et al., 2024; Rane, 2023).

5.4 Integration with AI, Edge Computing, and IoT

Digital twins leverage edge AI for real-time data processing, enabling adaptive simulations that reflect the current state of physical assets (Mahmud et al., 2025; Sonko, Etukudoh, Ibekwe, Ilojianya, & Daudu, 2024). IoT-enabled sensors, embedded systems, and UAVs provide continuous streams of data, which AI algorithms analyze for predictive modeling and anomaly detection (Shende, Tembhurne, & Ansari, 2024; Ryalat, Almtireen, Al-refai, Elmoaqet, & Rawashdeh, 2025). In construction and industrial environments, this integration allows autonomous decision-making, optimized maintenance scheduling, and energy-efficient operation (Ali et al., 2024; Liu et al., 2022).

Moreover, blockchain technologies enhance digital twin security by ensuring immutability, traceability, and secure sharing of predictive simulation outcomes, which is critical in multi-stakeholder construction projects and critical infrastructure management (Verma et al., 2022; Zangana & Omar, 2026a).

5.5 Applications Across Industries

1. **Construction and Civil Infrastructure:** Real-time predictive simulations optimize structural health monitoring, resource allocation, and safety compliance (Mondal & Chen, 2022; Pan & Zhang, 2023).
2. **Smart Healthcare:** Integrating digital twins with AI and IoT enables patient monitoring, predictive diagnostics, and server-less adaptive healthcare systems (Deka, Ghosh, Nanda, Barik, & Saikia, 2024; Bukkarayasamudram et al., 2025).
3. **Autonomous Systems and Robotics:** Digital twins enhance autonomous vehicle operation and UAV deployment, enabling predictive navigation, hazard detection, and cooperative logistics (Telli et al., 2023; Wang et al., 2025; Sonko et al., 2024).
4. **Agri-Food and Industry 5.0:** Predictive simulations enable precision farming, automated manufacturing, and advanced process control in smart factories (Wu et al., 2025; Xiang et al., 2023; Xu, Xu, & Li, 2018).

5.6 Challenges and Future Directions

Despite its transformative potential, digital twin adoption faces challenges including:

- High computational and storage requirements for large-scale real-time simulations (Ray, 2025a; Ray, 2025b).

- Integration across heterogeneous devices, sensors, and legacy infrastructure (Shafique et al., 2020; Saleem, Crespi, Rehmani, & Copeland, 2019).
- Ensuring cybersecurity, privacy, and regulatory compliance in multi-stakeholder environments (Verma et al., 2022; Zangana & Omar, 2026b).

Future research focuses on lightweight AI models for resource-constrained devices, federated learning for decentralized data, multi-scale simulations for large infrastructure, and fully autonomous predictive twins capable of self-adaptation and optimization (Lu, 2019; Mahmud et al., 2025; Kostopoulos, Gkamas, Rigou, & Kotsiantis, 2025).

Advanced digital twins combined with predictive simulation provide a holistic, intelligent, and proactive approach to construction, urban planning, manufacturing, and healthcare, facilitating operational efficiency, safety, and sustainability while enabling data-driven decision-making in real time (Ali et al., 2024; Mondal & Chen, 2022; Mazzetto, 2024).

Digital twins enable predictive simulation by continuously synchronizing physical assets with virtual models. Real-time sensor data, AI analytics, and simulation engines work together to forecast performance, detect risks, and optimize construction operations.

Figure 5. Digital twin and predictive simulation workflow

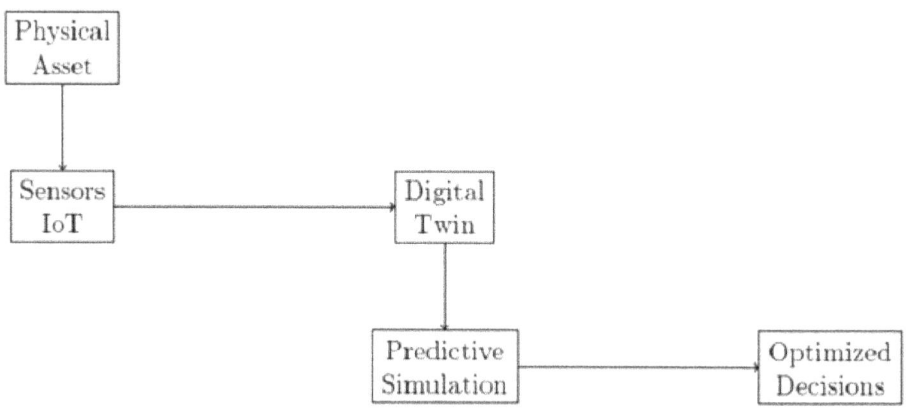

6. AUTONOMOUS AND ROBOTIC MONITORING SYSTEMS

The advent of autonomous and robotic monitoring systems has transformed the landscape of construction, infrastructure management, healthcare, and smart urban environments. These systems leverage robotics, AI, IoT, and embedded sensors to conduct real-time monitoring, predictive analysis, and autonomous interventions, reducing human error and enhancing operational efficiency (Abdelhafidh, Fourati, Fourati, & Chouaya, 2019; Sonko, Etukudoh, Ibekwe, Ilojianya, & Daudu, 2024).

6.1 Overview and Key Components

Autonomous monitoring systems typically integrate multiple technological layers:

1. **Robotic Hardware Layer:** Includes UAVs, ground vehicles, and robotic manipulators equipped with cameras, LiDAR, ultrasonic sensors, and other embedded systems for data acquisition (Telli et al., 2023; Wang, Li, Yang, & Peng, 2025).
2. **Data Acquisition and IoT Layer:** Sensors and IoT networks continuously gather environmental, structural, and operational data, enabling comprehensive monitoring (Abdelhafidh et al., 2019; Aouedi et al., 2024).
3. **AI and Analytics Layer:** Machine learning and deep learning models process sensor inputs to detect anomalies, predict failures, and optimize workflows (Sarker, 2021; Taye, 2023; Shende, Tembhurne, & Ansari, 2024).
4. **Communication and Control Layer:** 5G/6G-enabled communication ensures real-time control of robotic agents and data transmission across cloud or edge computing platforms (Chowdhury, Shahjalal, Ahmed, & Jang, 2020; Shafique, Khawaja, Sabir, Qazi, & Mustaqim, 2020).
5. **Decision and Visualization Layer:** Interfaces provide operators with predictive insights, dashboards, and simulation environments to make data-driven decisions (Mazzetto, 2024; Wojciechowska et al., 2024).

6.2 Applications in Construction and Civil Infrastructure

In construction and civil infrastructure, autonomous monitoring systems enhance safety, efficiency, and predictive maintenance. UAVs and mobile ground robots equipped with AI-driven sensors inspect hard-to-reach areas, detect structural cracks, and assess material degradation in real time (Barbhuiya & Das, 2025; Preethichandra et al., 2023). The integration of BIM with autonomous systems allows continuous digital updating of building models, optimizing construction scheduling and resource allocation while minimizing cost overruns and human errors (Pan & Zhang, 2023; Liu, Meng, Kong, & Zhang, 2022).

Autonomous robotic systems also support disaster management and emergency response, enabling rapid site assessments, hazard detection, and targeted interventions without risking human personnel (Wang et al., 2025; Rane, 2023).

6.3 Healthcare and Smart Systems

Autonomous monitoring extends to healthcare, where robotic systems and AI-driven sensor networks facilitate patient monitoring, predictive diagnostics, and adaptive treatment plans (Deka, Ghosh, Nanda, Barik, & Saikia, 2024; Bukkarayasamudram et al., 2025). For instance, robotic platforms can autonomously navigate healthcare facilities, collect real-time patient data, and perform preliminary assessments while integrating with server-less architectures for scalability (Mahmud et al., 2025).

6.4 Autonomous Vehicles and UAV Systems

Unmanned aerial vehicles (UAVs) and autonomous ground vehicles have emerged as critical components in monitoring and delivery applications. UAVs conduct high-resolution aerial surveys, inspect infrastructure, and facilitate collaborative emergency logistics alongside ground vehicles (Telli et al., 2023; Wang et al., 2025). Embedded AI algorithms enable path planning, obstacle avoidance, and cooperative task execution, while IoT connectivity ensures real-time data synchronization and operational coordination (Sonko et al., 2024; Shende et al., 2024).

Autonomous vehicles in smart cities and industrial sites also integrate with urban digital twins to enhance predictive maintenance, traffic optimization, and infrastructure monitoring, providing actionable insights through simulation and real-time feedback loops (Mazzetto, 2024; Juma & Shaalan, 2020).

6.5 Integration With AI, IoT, and Blockchain

Robotic monitoring systems rely heavily on AI for perception, decision-making, and predictive analysis. Machine learning models process multimodal sensor data to detect anomalies, classify structural or environmental risks, and optimize operational tasks (Ali, Saad, Rasheed, & Ammad, 2024; Lu, 2019; Ryalat, Almtireen, Al-refai, Elmoaqet, & Rawashdeh, 2025). IoT networks facilitate continuous monitoring and actuation, while edge computing reduces latency and allows near real-time control (Aouedi et al., 2024; Shafique et al., 2020).

Blockchain integration in autonomous systems ensures data integrity, secure communication, and trust in multi-agent collaborative environments, which is especially critical in industry 5.0 and urban infrastructure monitoring (Verma et al., 2022; Zangana & Omar, 2026a).

6.6 Challenges and Future Directions

Despite rapid advancements, autonomous and robotic monitoring systems face several challenges:

- **Scalability:** Deploying large fleets of UAVs or robots for urban or industrial monitoring requires robust coordination, communication, and resource management strategies (Rane, Choudhary, & Rane, 2023; Ray, 2025b).
- **Safety and Reliability:** Ensuring safe operation in dynamic, unpredictable environments is critical, particularly for collaborative UAV-ground vehicle systems (Telli et al., 2023; Wang et al., 2025).
- **Cybersecurity and Privacy:** The integration of AI, IoT, and blockchain introduces new attack surfaces and privacy concerns that must be mitigated through secure architectures and regulatory compliance (Aouedi et al., 2024; Verma et al., 2022).
- **Energy Efficiency:** Prolonged autonomous operations require advanced energy management, lightweight sensors, and optimization of computational workloads (Sonko et al., 2024; Shende et al., 2024).

Future research directions include the development of fully autonomous, self-learning robotic systems capable of multi-modal sensing and adaptive decision-making, enhanced collaborative operations with human-in-the-loop AI, and the integration of digital twins for predictive simulation and continuous optimization (Ali et al., 2024; Mahmud et al., 2025; Mondal & Chen, 2022). The convergence of AI, IoT, robotics, and blockchain promises to enable intelligent, resilient, and highly adaptive monitoring systems across construction, healthcare, transportation, and urban management sectors (Xu, Xu, & Li, 2018; Wu et al., 2025; Xiang et al., 2023).

7. IMMERSIVE TECHNOLOGIES FOR MONITORING AND TRAINING

Immersive technologies, including Virtual Reality (VR), Augmented Reality (AR), Mixed Reality (MR), and haptic-based simulation systems, are revolutionizing the ways in which monitoring, training, and operational management are conducted across multiple sectors. These technologies provide interactive, real-time visualization of complex systems and environments, enabling users to perceive, interact with, and analyze data in an intuitive manner. In construction, healthcare, smart cities, and industrial applications, immersive technologies are increasingly being integrated with AI, IoT, and robotic systems to enhance decision-making, risk assessment, and workforce training (Ali, Saad, Rasheed, & Ammad, 2024; Kostopoulos, Gkamas, Rigou, & Kotsiantis, 2025).

7.1 Applications in Construction and Infrastructure

In construction, immersive technologies enhance project visualization, structural monitoring, and workforce training. AR and VR platforms allow engineers, contractors, and clients to interact with Building Information Models (BIM) in three-dimensional, immersive environments, facilitating more accurate planning, clash detection, and safety training (Pan & Zhang, 2023; Liu, Meng, Kong, & Zhang, 2022). Combined with AI-driven predictive analytics and real-time sensor data, immersive simulations provide foresight into structural behavior, equipment performance, and potential hazards (Barbhuiya & Das, 2025; Preethichandra et al., 2023).

Robotic and autonomous monitoring systems, such as UAVs or mobile inspection robots, can stream live data into immersive environments, allowing engineers to remotely inspect hard-to-reach areas of infrastructure in a risk-free manner (Telli et al., 2023; Wang, Li, Yang, & Peng, 2025). This integration enables predictive maintenance strategies, optimized scheduling, and proactive risk management, all within an immersive, user-friendly interface (Mondal & Chen, 2022; Rane, 2023).

7.2 Healthcare and Medical Training

Immersive technologies in healthcare provide transformative solutions for patient monitoring, surgical simulation, and medical training. VR and AR platforms allow medical professionals to simulate complex procedures, visualize patient-specific anatomical data, and practice interventions in a safe, controlled environment (Deka, Ghosh, Nanda, Barik, & Saikia, 2024; Bukkarayasamudram et al., 2025). AI-driven systems can personalize simulations based on real-time data from wearable devices and IoT-enabled hospital infrastructure, enabling predictive diagnostics and enhanced patient care (Mahmud et al., 2025; Li et al., 2022).

Haptic feedback and immersive interfaces also improve procedural training, allowing practitioners to gain practical experience without exposure to patient risk. These immersive approaches accelerate learning, improve skill retention, and reduce errors in high-stakes environments (Sarker, 2021; Taye, 2023).

7.3 Integration With Digital Twins and IoT

Digital twins provide a virtual representation of physical systems, enabling immersive monitoring and scenario-based training. Immersive technologies overlay real-time sensor data onto these digital models, allowing users to interact with live systems and perform predictive simulations (Mazzetto, 2024; Xu, Xu, & Li, 2018). In smart cities and infrastructure management, AR and VR visualizations integrated with GIS and IoT platforms support urban planning, traffic management, energy monitoring, and disaster response simulations (Juma & Shaalan, 2020; Vinueza-Martinez et al., 2024).

These systems often utilize edge computing and 5G/6G networks to provide low-latency, real-time interaction between physical and virtual spaces (Chowdhury, Shahjalal, Ahmed, & Jang, 2020; Shafique et al., 2020). Immersive monitoring thus becomes an effective tool for training personnel, simulating emergency scenarios, and optimizing operational strategies.

7.4 Industry 4.0 and Industry 5.0 Perspectives

Immersive technologies are central to Industry 4.0 and the emerging paradigms of Industry 5.0, where human-centered cyber-physical systems, AI, and robotics converge (Xu, Xu, & Li, 2018; Xiang et al., 2023). AR and VR environments facilitate human-robot collaboration, operator training, and the visualization of AI-driven analytics in manufacturing, construction, and logistics operations (Wu et al., 2025; Verma et al., 2022). These immersive platforms enhance situational awareness, accelerate decision-making, and support the integration of autonomous systems within industrial workflows (Rane, Choudhary, & Rane, 2023; Ray, 2025a).

7.5 Challenges and Future Directions

Despite their transformative potential, immersive monitoring and training systems face several challenges:

- **Hardware Limitations:** High-quality immersive experiences require powerful processing units, advanced displays, and haptic devices, which may be cost-prohibitive (Ali et al., 2024; Shende, Tembhurne, & Ansari, 2024).
- **Data Integration:** Aggregating data from heterogeneous IoT sensors, robotic systems, and AI models into coherent immersive simulations remains a significant challenge (Aouedi et al., 2024; Saleem, Crespi, Rehmani, & Copeland, 2019).
- **User Experience and Safety:** Ensuring intuitive interaction, minimizing motion sickness, and maintaining safety during immersive training sessions are ongoing concerns (Kostopoulos et al., 2025; Taye, 2023).
- **Cybersecurity and Privacy:** Immersive systems often handle sensitive data from construction sites, healthcare facilities, and smart cities, necessitating robust security protocols (Verma et al., 2022; Zangana & Omar, 2026b).

Future research directions emphasize the convergence of AI, digital twins, IoT, and immersive interfaces to create adaptive, context-aware, and collaborative training environments (Ali et al., 2024; Mahmud et al., 2025; Ryalat et al., 2025). Advancements in haptic feedback, multi-sensory integration, and low-latency networking will enhance realism, engagement, and the predictive capabilities of these platforms, making immersive technologies essential tools for monitoring, decision-making, and workforce development in next-generation smart environments (Liu et al., 2022; Mazzetto, 2024; Pan & Zhang, 2023).

8. BLOCKCHAIN AND SECURE CONSTRUCTION MONITORING

The construction industry has increasingly embraced digital technologies to enhance project management, monitoring, and operational transparency. Among these technologies, blockchain has emerged as a critical enabler of secure, decentralized, and tamper-proof construction monitoring systems. Blockchain's inherent properties—immutability, traceability, and distributed consensus—address long-standing challenges in construction data integrity, stakeholder trust, and real-time verification of project activities (Verma et al., 2022; Ali, Saad, Rasheed, & Ammad, 2024).

8.1 Blockchain Integration With Construction Monitoring

Traditional construction monitoring systems rely heavily on centralized databases and manual record-keeping, which are prone to errors, delays, and cyber threats. By integrating blockchain with sensor networks, IoT devices, and digital twins, construction projects can achieve automated, real-time, and verifiable data logging (Abdelhafidh, Fourati, Fourati, & Chouaya, 2019; Aouedi et al., 2024). Smart contracts enable autonomous execution of project milestones, automated payments, and compliance verification, reducing disputes and ensuring accountability among contractors, suppliers, and stakeholders (Rane, 2023; Pan & Zhang, 2023).

In structural health monitoring, blockchain can securely record data from smart sensors, UAV inspections, and robotic monitoring systems. This secure record-keeping ensures that structural assessments, damage reports, and maintenance logs remain tamper-proof, fostering trust in automated inspection processes (Barbhuiya & Das, 2025; Preethichandra et al., 2023; Wojciechowska et al., 2024).

8.2 Enhancing Cybersecurity and Data Integrity

The integration of blockchain with AI and IoT in construction monitoring significantly enhances cybersecurity. IoT-enabled sensors and devices collect extensive environmental, structural, and operational data; blockchain ensures that these data streams are securely encrypted, time-stamped, and stored in a distributed ledger, making unauthorized alterations virtually impossible (Shafique, Khawaja, Sabir, Qazi, & Mustaqim, 2020; Saleem, Crespi, Rehmani, & Copeland, 2019).

AI algorithms can further analyze these secure datasets to detect anomalies, predict structural failures, and optimize project scheduling (Mondal & Chen, 2022; Ali et al., 2024). For example, combining blockchain with AI-driven predictive analytics enables automated alerts when sensor data indicate potential risks, enhancing safety and operational efficiency (Rane, Choudhary, & Rane, 2023; Liu, Meng, Kong, & Zhang, 2022).

8.3 Blockchain in Smart Contracts and Supply Chain Management

Smart contracts, implemented via blockchain, are particularly transformative in managing construction supply chains. They enable automatic verification of material deliveries, quality compliance, and payments based on sensor data and IoT-driven monitoring (Abdelhafidh et al., 2019; Adebayo, Udoh, Kamudyariwa, & Osobajo, 2025). By embedding contractual logic in blockchain, stakeholders can reduce delays, eliminate disputes, and improve transparency throughout procurement, logistics, and on-site operations (Xu, Xu, & Li, 2018; Rane, 2023).

In complex projects involving multiple subcontractors and suppliers, blockchain ensures that all parties have access to a single source of truth. This transparency is crucial for enforcing safety regulations, environmental compliance, and quality standards (Irani & Kamal, 2014; Liu et al., 2022).

8.4 Integration With Immersive Monitoring and Digital Twins

Blockchain can be integrated with immersive technologies such as AR/VR and digital twins to create highly secure, real-time construction monitoring platforms (Mazzetto, 2024; Pan & Zhang, 2023). Data collected from IoT sensors, UAVs, and robotic systems can be securely mirrored in virtual models, allowing project managers to visualize construction progress, simulate scenarios, and interact with validated datasets in immersive environments (Telli et al., 2023; Wang, Li, Yang, & Peng, 2025).

The combination of blockchain, AI, and immersive digital twins enables scenario-based decision-making, predictive maintenance, and automated reporting, thereby reducing human error and enhancing operational efficiency (Ryalat, Almtireen, Al-refai, Elmoaqet, & Rawashdeh, 2025; Liu et al., 2022).

8.5 Challenges and Future Research Directions

Despite the promise of blockchain in construction monitoring, several challenges remain:

- **Scalability:** Large-scale deployment of blockchain in construction projects requires significant computational resources and optimized consensus mechanisms to handle high-frequency IoT data streams (Chowdhury, Shahjalal, Ahmed, & Jang, 2020; Shende, Tembhurne, & Ansari, 2024).
- **Interoperability:** Integrating blockchain with existing construction management systems, BIM platforms, and IoT infrastructures demands standardized protocols and data models (Ali et al., 2024; Aouedi et al., 2024).
- **Cybersecurity and Privacy:** While blockchain ensures data integrity, privacy-preserving mechanisms are needed to protect sensitive project and personnel data (Verma et al., 2022; Zangana & Omar, 2026a).
- **Regulatory and Legal Frameworks:** The adoption of blockchain-enabled smart contracts requires updated legal frameworks and clear guidelines to enforce automated contractual obligations (Lu, 2019; Ray, 2025).

Future research is likely to focus on hybrid blockchain architectures, integration with AI-powered predictive analytics, and the coupling of immersive monitoring systems to enable real-time, secure, and decentralized construction project management (Rane et al., 2023; Mahmud et al., 2025; Xiang et al., 2023). The development of lightweight consensus algorithms, privacy-preserving mechanisms, and cross-platform interoperability standards will be crucial to realize fully secure, smart construction monitoring ecosystems (Wu et al., 2025; Taye, 2023).

9. SUSTAINABLE AND GREEN MONITORING TECHNOLOGIES

Sustainability has become a pivotal focus in modern construction practices, driven by the need to reduce environmental impact, optimize resource utilization, and comply with increasingly stringent regulatory standards. The adoption of green monitoring technologies in construction projects facilitates the integration of environmental considerations into project management, design, and operational workflows. These technologies leverage advanced sensing, AI, IoT, and data analytics to monitor energy consumption, material usage, waste generation, and carbon footprint in real-time (Abdelhafidh, Fourati, Fourati, & Chouaya, 2019; Rane, 2023).

9.1 Green Sensor Networks and IoT Integration

Wireless sensor networks (WSNs) and IoT-enabled monitoring systems are central to sustainable construction monitoring. They allow continuous assessment of environmental parameters such as air quality, temperature, humidity, noise levels, and energy consumption on-site (Saleem, Crespi, Rehmani, & Copeland, 2019; Shafique, Khawaja, Sabir, Qazi, & Mustaqim, 2020). These systems reduce the need for manual inspections, thereby minimizing resource consumption and human intervention (Abdelhafidh et al., 2019; Aouedi et al., 2024).

Integration with AI further enhances the efficiency of green monitoring systems by enabling predictive analytics for energy optimization, resource allocation, and emission reduction (Ali, Saad, Rasheed, & Ammad, 2024; Mondal & Chen, 2022; Liu, Meng, Kong, & Zhang, 2022). For instance, AI models can predict peak energy loads, optimize lighting and HVAC systems, and identify waste hotspots in real-time, contributing to both cost reduction and environmental sustainability (Lu, 2019; Pan & Zhang, 2023).

9.2 Sustainable Materials and Structural Health Monitoring

Monitoring technologies play a crucial role in assessing the performance and durability of sustainable materials such as recycled concrete, low-carbon cement, and bio-based composites (Barbhuiya & Das, 2025; Rane, Choudhary, & Rane, 2023). Smart sensors embedded in structures continuously track stress, strain, and degradation, enabling proactive maintenance and extending the service life of construction components (Preethichandra et al., 2023; Wojciechowska et al., 2024).

AI-driven damage detection algorithms complement these sensors by providing automated analysis and early warning of structural deficiencies. This integration reduces the need for material-intensive repairs and supports sustainable lifecycle management (Taye, 2023; Sarker, 2021). Furthermore, UAVs equipped with multispectral sensors can monitor large-scale projects to ensure environmentally friendly construction practices are being followed (Telli et al., 2023; Wang, Li, Yang, & Peng, 2025).

9.3 Energy-Efficient Construction Processes

Green monitoring technologies also facilitate energy-efficient operations in construction sites. IoT-enabled systems and AI algorithms can optimize machinery usage, track fuel consumption, and monitor electricity usage across different project phases (Rane, 2023; Rane et al., 2023). For example, predictive models can schedule equipment operation to reduce idle time, thereby minimizing fuel consumption and greenhouse gas emissions (Xu, Xu, & Li, 2018; Chowdhury, Shahjalal, Ahmed, & Jang, 2020).

Digital twins and immersive monitoring technologies further enhance energy efficiency by providing virtual representations of construction sites where scenarios can be simulated and optimized before physical implementation (Mazzetto, 2024; Pan & Zhang, 2023). These simulations allow planners to reduce resource wastage, plan optimal logistics, and ensure sustainable site layouts.

9.4 Circular Economy and Waste Management

Sustainable construction monitoring extends to material lifecycle management and circular economy practices. Sensor-based monitoring of material consumption and waste generation enables real-time tracking and optimization of construction processes (Abdelhafidh et al., 2019; Shende, Tembhurne, & Ansari, 2024). AI-driven analytics can suggest material reuse strategies, recycling pathways, and predictive inventory management to minimize environmental impact (Wu et al., 2025; Ali et al., 2024).

Integration with blockchain can further ensure traceability of sustainable materials and proper documentation of environmental compliance throughout the supply chain (Verma et al., 2022; Adebayo, Udoh, Kamudyariwa, & Osobajo, 2025). This synergy between AI, IoT, and blockchain supports regulatory compliance and promotes accountability in green construction practices.

9.5 Challenges and Future Directions

Despite advancements, implementing sustainable and green monitoring technologies faces several challenges:

- **Cost and Scalability:** The deployment of smart sensors, IoT networks, and AI models can be capital-intensive, limiting adoption in smaller projects (Shafique et al., 2020; Liu et al., 2022).

- **Data Integration:** Aggregating and analyzing heterogeneous environmental and operational data from multiple sources remains complex (Aouedi et al., 2024; Rane, 2023).
- **Energy Consumption of IT Systems:** Ironically, monitoring technologies themselves consume energy, highlighting the need for energy-efficient sensor design and lightweight AI models (Mahmud et al., 2025; Bukkarayasamudram et al., 2025).
- **Regulatory and Standardization Gaps:** Standard protocols for sustainable monitoring, data sharing, and reporting are still evolving, affecting interoperability and compliance (Xu, Xu, & Li, 2018; Mazzetto, 2024).

Future research directions include the integration of quantum computing for high-speed data analytics, development of ultra-low-power IoT devices, and the convergence of AI, blockchain, and digital twins for fully sustainable, transparent, and secure construction monitoring ecosystems (Xiang et al., 2023; Wu et al., 2025; Verma et al., 2022). Additionally, leveraging autonomous vehicles and UAVs in green monitoring offers significant potential for real-time, scalable, and low-impact environmental assessment (Telli et al., 2023; Wang et al., 2025).

By aligning monitoring technologies with sustainability goals, the construction industry can significantly reduce its environmental footprint, enhance operational efficiency, and contribute to the global push toward green infrastructure (Rane et al., 2023; Liu et al., 2022; Mondal & Chen, 2022).

10. HUMAN-CENTERED AND ETHICAL TECHNOLOGY DESIGN

As construction monitoring technologies advance, there is a growing recognition that technological sophistication must be balanced with human-centered design principles and ethical considerations. Human-centered design ensures that systems are intuitive, accessible, and aligned with user needs, while ethical technology design addresses privacy, fairness, transparency, and accountability in automated construction processes (Irani & Kamal, 2014; Rane, 2023).

10.1 Principles of Human-Centered Design in Construction Monitoring

Human-centered design (HCD) emphasizes the involvement of end-users—construction managers, engineers, and on-site workers—throughout the design and deployment of monitoring technologies. This approach ensures that digital tools, AI systems, and IoT devices enhance human decision-making rather than overwhelm users with complexity (Abdelhafidh, Fourati, Fourati, & Chouaya, 2019; Liu, Meng, Kong, & Zhang, 2022).

Key principles of HCD include usability, accessibility, adaptability, and responsiveness. For example, dashboards for real-time structural health monitoring should present actionable insights using clear visualizations, alerting users to critical events while minimizing information overload (Preethichandra et al., 2023; Mondal & Chen, 2022). Moreover, augmented reality (AR) interfaces and mobile applications can provide on-site personnel with intuitive interaction modalities, enhancing situational awareness and safety compliance (Taye, 2023; Pan & Zhang, 2023).

10.2 Ethical Considerations in AI-Driven Monitoring

The deployment of AI and IoT in construction raises several ethical concerns. These include data privacy, bias in predictive models, decision accountability, and the potential for worker surveillance (Aouedi et al., 2024; Ali, Saad, Rasheed, & Ammad, 2024). Data collected from smart sensors, UAVs, and digital twins often include sensitive operational or personal information. Ensuring robust encryption, anonymization, and secure storage is critical to protecting stakeholder privacy (Shafique et al., 2020; Saleem, Crespi, Rehmani, & Copeland, 2019).

Bias in AI algorithms can inadvertently favor certain construction methods or materials, potentially marginalizing smaller contractors or local labor practices. Ethical AI design mandates transparency in model training, testing, and deployment, as well as continuous auditing to prevent discriminatory outcomes (Lu, 2019; Adebayo, Udoh, Kamudyariwa, & Osobajo, 2025).

Accountability frameworks should also be implemented to define human oversight in AI-driven decision-making. Systems should complement human judgment rather than replace it, ensuring that critical decisions, such as structural interventions or resource allocation, remain under professional supervision (Barbhuiya & Das, 2025; Rane et al., 2023).

10.3 Integration of Human-Centered and Ethical Practices

Practical integration of human-centered and ethical design involves several strategies:

- **Participatory Design:** Engage stakeholders during system conception to ensure technologies meet real-world needs (Irani & Kamal, 2014; Liu et al., 2022).

- **Explainable AI (XAI):** Develop AI systems that provide clear reasoning for their outputs, allowing users to understand, trust, and challenge automated recommendations (Mahmud et al., 2025; Ray, 2025a).
- **Privacy by Design:** Embed privacy-preserving mechanisms into system architectures from the outset, particularly for IoT-enabled monitoring and UAV data collection (Shende, Tembhurne, & Ansari, 2024; Aouedi et al., 2024).
- **Ethical Guidelines and Standards:** Align technology development with industry-wide ethical standards and regulatory requirements, promoting transparency, accountability, and fairness in monitoring applications (Verma et al., 2022; Xu, Xu, & Li, 2018).

10.4 Human-Centric AI and Decision Support

Human-centered monitoring systems leverage AI not only for automation but also to augment human decision-making. Predictive analytics can forecast structural failures, optimize energy usage, and schedule maintenance, while dashboards and visualization tools provide actionable insights to engineers and site managers (Liu et al., 2022; Pan & Zhang, 2023; Mondal & Chen, 2022).

Collaborative AI, where humans and machines work symbiotically, enhances safety, efficiency, and project sustainability. For instance, sensor networks combined with explainable machine learning can alert workers to hazards without overwhelming them with false alarms or complex interpretations (Preethichandra et al., 2023; Rane, 2023).

10.5 Challenges and Future Directions

While human-centered and ethical design principles are increasingly recognized, challenges remain:

- **Balancing Automation and Human Control:** Fully autonomous systems risk disempowering humans, while excessive reliance on human intervention may reduce efficiency (Irani & Kamal, 2014; Lu, 2019).
- **Complexity of Multi-Stakeholder Systems:** Construction sites involve diverse actors, making it difficult to design interfaces and ethical frameworks suitable for all users (Rane et al., 2023; Shafique et al., 2020).
- **Continuous Ethical Evaluation:** As AI and IoT evolve, ongoing assessment is required to address emerging biases, privacy risks, and unintended consequences (Ali et al., 2024; Mahmud et al., 2025).

- **Integration with Sustainability Goals:** Ethical and human-centered systems should also support sustainable practices, aligning environmental, social, and operational objectives (Wojciechowska et al., 2024; Liu et al., 2022).

Future research should explore adaptive human-AI interfaces, the incorporation of ethical AI guidelines into construction monitoring standards, and cross-disciplinary strategies that integrate ergonomics, cybersecurity, and sustainability into system design (Bukkarayasamudram et al., 2025; Xiang et al., 2023; Verma et al., 2022). This holistic approach ensures that construction monitoring technologies not only optimize efficiency but also prioritize human welfare, fairness, and societal responsibility (Zangana & Omar, 2026a; Zangana & Omar, 2026b).

11. CHALLENGES AND BARRIERS TO ADOPTION

The adoption of advanced construction monitoring technologies, including AI, IoT, UAVs, and smart sensors, is associated with a range of technical, organizational, and societal challenges. Despite the demonstrated benefits of enhanced efficiency, safety, and sustainability, numerous barriers hinder the widespread implementation of these technologies in the architecture, engineering, and construction (AEC) sector (Irani & Kamal, 2014; Liu, Meng, Kong, & Zhang, 2022).

11.1 Technical Challenges

Technical limitations are among the primary barriers to adoption. Many AI-driven monitoring systems rely on large volumes of high-quality data to function effectively, yet construction sites often produce heterogeneous, noisy, or incomplete datasets (Ali, Saad, Rasheed, & Ammad, 2024; Mondal & Chen, 2022). Integration of diverse technologies such as BIM, IoT devices, UAVs, and digital twins remains complex, requiring robust interoperability standards and seamless data exchange protocols (Pan & Zhang, 2023; Mazzetto, 2024; Xu, Xu, & Li, 2018).

Hardware constraints, including sensor durability, battery life, and network connectivity, pose additional challenges. Wireless sensor networks and IoT devices deployed in harsh outdoor environments may experience performance degradation, limiting their reliability (Abdelhafidh, Fourati, Fourati, & Chouaya, 2019; Preethichandra et al., 2023). Furthermore, real-time data processing for predictive analytics and structural health monitoring demands significant computational power, which may exceed the capacity of on-site servers or edge devices (Barbhuiya & Das, 2025; Chowdhury, Shahjalal, Ahmed, & Jang, 2020).

11.2 Organizational and Workforce Barriers

Adoption is also influenced by organizational readiness and workforce preparedness. Construction companies may lack the technical expertise or trained personnel to operate AI-enabled systems and interpret their outputs effectively (Adebayo, Udoh, Kamudyariwa, & Osobajo, 2025; Liu et al., 2022). Resistance to change, particularly among experienced workers accustomed to traditional monitoring methods, can further slow adoption (Irani & Kamal, 2014; Rane, 2023).

High initial investment costs for hardware, software, and training programs represent another barrier, particularly for small- and medium-sized enterprises (SMEs) in the construction sector (Lu, 2019; Ali et al., 2024). Budget constraints often limit the procurement of sophisticated sensors, UAVs, and AI platforms, creating a gap between technological potential and practical feasibility (Rane, Choudhary, & Rane, 2023).

11.3 Data Privacy, Security, and Ethical Concerns

Privacy and cybersecurity concerns present significant obstacles. Construction monitoring systems generate vast amounts of sensitive operational data, including worker locations, structural designs, and proprietary processes (Aouedi et al., 2024; Shafique, Khawaja, Sabir, Qazi, & Mustaqim, 2020). Breaches or misuse of such data can result in financial losses, reputational damage, and legal liabilities. Ethical concerns also arise from continuous worker surveillance and decision-making automation, which may reduce accountability and transparency (Zangana & Omar, 2026a; Zangana & Omar, 2026b).

Ensuring secure data transmission and storage, compliance with privacy regulations, and transparent AI decision-making mechanisms remain critical challenges. Blockchain, encryption, and federated learning approaches can mitigate some risks, yet their implementation adds further complexity and cost (Verma et al., 2022; Saleem, Crespi, Rehmani, & Copeland, 2019).

11.4 Technological Maturity and Standardization

The maturity of monitoring technologies varies, and early-stage solutions may lack robustness or scalability. Emerging technologies such as quantum computing, advanced UAV systems, and agentic AI are still in experimental phases for construction applications (Bukkarayasamudram et al., 2025; Kostopoulos, Gkamas, Rigou, & Kotsiantis, 2025; Telli et al., 2023).

A lack of standardized protocols for data formats, sensor calibration, and interoperability inhibits seamless adoption across projects and organizations (Abdelhafidh et al., 2019; Rane et al., 2023; Vinueza-Martinez et al., 2024). Without clear industry-wide guidelines, companies risk investing in solutions that may become obsolete or incompatible with future technologies.

11.5 Societal and Environmental Barriers

Public perception, regulatory frameworks, and environmental considerations also influence adoption. UAV-based monitoring, for example, may face restrictions due to airspace regulations or privacy concerns (Wang, Li, Yang, & Peng, 2025; Telli et al., 2023). Similarly, IoT and sensor deployment must consider energy consumption and environmental impact to align with sustainability goals (Wu et al., 2025; Wojciechowska et al., 2024).

Social acceptance of AI and autonomous systems in construction remains variable. Concerns about job displacement, safety, and human oversight can hinder stakeholder buy-in, particularly in regions with limited exposure to digital construction technologies (Ryalat, Almtireen, Al-refai, Elmoaqet, & Rawashdeh, 2025; Ali et al., 2024).

11.6 Strategies to Overcome Barriers

Addressing these challenges requires a multi-pronged approach:

- **Capacity Building and Training:** Upskilling the workforce to operate and interpret AI-driven systems (Adebayo et al., 2025; Liu et al., 2022).
- **Standards and Interoperability:** Developing industry-wide protocols for sensor networks, BIM integration, and AI model deployment (Abdelhafidh et al., 2019; Mazzetto, 2024).
- **Security and Privacy by Design:** Embedding robust cybersecurity measures and privacy-preserving architectures from the outset (Shende, Tembhurne, & Ansari, 2024; Aouedi et al., 2024).
- **Cost-Effective Solutions:** Leveraging scalable, modular, or cloud-based systems to reduce upfront investments (Rane, 2023; Rane et al., 2023).
- **Regulatory Alignment and Ethical Oversight:** Ensuring compliance with legal frameworks and ethical guidelines for human-centered AI deployment (Zangana & Omar, 2026a; Verma et al., 2022).

11.7 Summary

In summary, while intelligent monitoring technologies promise transformative benefits for construction and infrastructure management, adoption is hindered by technical limitations, workforce preparedness, high costs, data security concerns, regulatory challenges, and social acceptance issues. Overcoming these barriers requires integrated strategies combining technological innovation, human-centered design, ethical practices, and cross-stakeholder collaboration to ensure sustainable and responsible deployment (Ali et al., 2024; Lu, 2019; Mondal & Chen, 2022; Xu et al., 2018).

12. FUTURE RESEARCH DIRECTIONS AND OPEN CHALLENGES

The evolution of intelligent systems, AI, IoT, UAVs, and smart monitoring technologies in construction, healthcare, and cyber-physical domains has shown promising advancements. However, despite significant progress, several open research challenges and directions remain to be addressed to ensure effective, sustainable, and secure adoption (Lu, 2019; Mondal & Chen, 2022; Xu, Xu, & Li, 2018). This section identifies the emerging research opportunities and unresolved issues in these domains.

12.1 Integration of AI, IoT, and Digital Twins

The integration of AI with IoT-enabled monitoring systems and digital twin platforms is a key area for future research. While AI models can provide predictive insights, IoT sensors supply real-time data, and digital twins allow simulation-based decision-making, combining these elements remains complex (Aouedi et al., 2024; Pan & Zhang, 2023; Mazzetto, 2024). Future research should focus on developing unified frameworks for interoperability, real-time data fusion, and adaptive model updating to enable predictive and prescriptive analytics for construction, healthcare, and smart city applications (Saleem, Crespi, Rehmani, & Copeland, 2019; Shafique, Khawaja, Sabir, Qazi, & Mustaqim, 2020).

12.2 Advanced Machine Learning and Deep Learning Techniques

There is a growing need to improve machine learning and deep learning algorithms for autonomous monitoring, anomaly detection, and decision support. Techniques that can handle heterogeneous and incomplete datasets, operate under uncertainty, and adapt to dynamic environments are essential (Sarker, 2021; Taye, 2023; Shende, Tembhurne, & Ansari, 2024). Research on federated learning, reinforcement learning, and hybrid architectures is critical for developing resilient and context-aware AI systems capable of working in complex construction and infrastructure scenarios (Mahmud et al., 2025; Ray, 2025a).

12.3 Security, Privacy, and Ethical Frameworks

As data-driven technologies proliferate, cybersecurity and privacy concerns are becoming increasingly prominent. Future research must focus on robust encryption, secure data sharing, and privacy-preserving machine learning techniques to safeguard sensitive operational and personal data (Aouedi et al., 2024; Zangana & Omar, 2026a; Verma et al., 2022). Ethical frameworks for AI deployment, including accountability, fairness, and transparency, are also critical, particularly for human-centered systems in healthcare, construction, and urban management (Zangana & Omar, 2026b; Kostopoulos et al., 2025).

12.4 Emerging Communication Technologies

6G networks and advanced wireless communication technologies are expected to transform real-time monitoring and control in smart infrastructure systems (Chowdhury, Shahjalal, Ahmed, & Jang, 2020). Future research should explore ultra-low latency, high-throughput, and reliable communication protocols that can support autonomous vehicles, UAV-ground coordination, and distributed sensing networks in both urban and rural contexts (Wang, Li, Yang, & Peng, 2025; Telli et al., 2023).

12.5 Human-Centered and Socio-Technical Considerations

Despite technological advances, human factors remain a central challenge. Research is needed to ensure that AI and automation systems are human-centric, supporting workers rather than replacing them, and enhancing decision-making through explainable AI and intuitive interfaces (Adebayo et al., 2025; Irani & Kamal, 2014). Human-in-the-loop approaches, trust-aware AI, and socio-technical integration strategies should be explored to facilitate adoption and improve user acceptance (Li et al., 2022; Ryalat et al., 2025).

12.6 Quantum Computing and High-Performance Systems

Quantum computing presents new opportunities for solving complex optimization problems in construction, healthcare, and logistics (Bukkarayasamudram et al., 2025). Research should focus on developing quantum algorithms for predictive maintenance, traffic and delivery optimization, and large-scale simulations. Integration of classical AI with quantum-enhanced models can address challenges in computational speed and scalability, enabling real-time decision-making in critical applications (Wu et al., 2025; Xiang et al., 2023).

12.7 Standardization, Interoperability, and Regulatory Alignment

Lack of standardized protocols for sensors, data formats, and AI deployment remains a significant barrier. Future research should emphasize creating industry-wide standards for interoperability, certification, and benchmarking of intelligent systems (Abdelhafidh et al., 2019; Rane, 2023; Vinueza-Martinez et al., 2024). Additionally, aligning technological development with regulatory and ethical requirements is crucial to ensure safety, legal compliance, and societal trust (Verma et al., 2022; Rane, Choudhary, & Rane, 2023).

12.8 Sustainability and Energy-Efficient Solutions

With growing concerns about environmental impact, future research should explore energy-efficient sensor networks, AI models, and smart infrastructure systems. Strategies such as edge computing, low-power IoT devices, and green data centers can reduce operational costs and carbon footprint while maintaining high performance (Shafique et al., 2020; Preethichandra et al., 2023; Wojciechowska et al., 2024).

12.9 Domain-Specific Applications and Multi-Disciplinary Integration

Research should focus on domain-specific applications, including structural health monitoring, healthcare delivery, autonomous construction, and agri-food engineering, while promoting interdisciplinary integration (Barbhuiya & Das, 2025; Deka et al., 2024; Wu et al., 2025). Hybrid frameworks combining AI, IoT, robotics, and blockchain can provide robust solutions for complex, multi-faceted challenges in the AEC and healthcare industries (Verma et al., 2022; Rane, 2023; Mondal & Chen, 2022).

12.10 Open Challenges and Future Directions

Despite progress, multiple open challenges remain:

1. Data heterogeneity and quality issues limiting AI model accuracy (Ali et al., 2024; Liu et al., 2022).

2. Integration complexities across heterogeneous systems such as BIM, IoT, and UAVs (Pan & Zhang, 2023; Mazzetto, 2024).
3. Cybersecurity and privacy vulnerabilities, especially for IoT-enabled infrastructures (Shende et al., 2024; Aouedi et al., 2024).
4. Human-AI collaboration and workforce training, ensuring systems are interpretable, safe, and trusted (Kostopoulos et al., 2025; Adebayo et al., 2025).
5. Scalability and standardization, including data protocols, regulatory frameworks, and ethical guidelines (Abdelhafidh et al., 2019; Rane et al., 2023).
6. Energy efficiency and sustainability challenges for large-scale deployments (Wojciechowska et al., 2024; Shafique et al., 2020).
7. Exploiting emerging technologies like quantum computing, 6G networks, and industry 5.0 enabling technologies for real-world adoption (Bukkarayasamudram et al., 2025; Xiang et al., 2023; Chowdhury et al., 2020).

Addressing these challenges will require concerted efforts from academia, industry, and policymakers. Future research should emphasize human-centered, secure, sustainable, and scalable AI-enabled systems capable of transforming construction, healthcare, smart city, and industrial applications while ensuring ethical, social, and environmental responsibility (Lu, 2019; Mahmud et al., 2025; Zangana & Omar, 2026a; Zangana & Omar, 2026b).

13. CONCLUSION

The rapid advancement of artificial intelligence, IoT, UAVs, digital twins, and smart monitoring technologies has ushered in a new era for construction, healthcare, and cyber-physical systems. These technologies offer unprecedented opportunities to improve efficiency, safety, and decision-making while enabling predictive and adaptive solutions across complex domains. However, their full potential can only be realized when the challenges of integration, interoperability, security, and human-centered design are carefully addressed.

The adoption of intelligent systems requires a balance between technological innovation and practical implementation. Ensuring that AI and automation tools are transparent, ethical, and reliable is essential for fostering trust and promoting acceptance among end-users. Moreover, sustainable and energy-efficient solutions must be prioritized to mitigate environmental impacts while maintaining high performance in large-scale deployments.

Looking forward, the convergence of emerging technologies such as quantum computing, advanced wireless communication, and industry 5.0 innovations holds great promise for solving complex, real-world problems. Yet, this progress must be accompanied by robust standards, regulatory alignment, and strategies for human-AI collaboration to ensure safe, scalable, and socially responsible applications.

In conclusion, the future of intelligent systems lies not only in technological breakthroughs but also in their thoughtful integration into societal and industrial contexts. By addressing current challenges, fostering interdisciplinary research, and focusing on sustainable, secure, and user-centered approaches, these systems have the potential to transform industries, enhance quality of life, and drive innovation across multiple sectors.

REFERENCES

Abdelhafidh, M., Fourati, M., Fourati, L. C., & Chouaya, A. (2019). Wireless sensor network monitoring system: Architecture, applications and future directions. *International Journal of Communication Networks and Distributed Systems, 23*(4), 413–451.

Adebayo, Y., Udoh, P., Kamudyariwa, X. B., & Osobajo, O. A. (2025). Artificial intelligence in construction project management: A structured literature review of its evolution in application and future trends. *Digital, 5*(3), 26.

Akinloye, B. (2025). DESIGN AND CONSTRUCTION OF A SINGLE-PHASE SYNCHRONOUS RELUCTANCE MOTOR DRIVE. *Vokasi Unesa Bulletin of Engineering. Technology and Applied Science, 3*(1), 12–24. DOI: 10.26740/vubeta.v3i1.42969

Ali, Z., Saad, S., Rasheed, K., & Ammad, S. (2024). AI future perspectives and trends in construction. *AI in Material Science*, 239-261.

Aouedi, O., Vu, T. H., Sacco, A., Nguyen, D. C., Piamrat, K., Marchetto, G., & Pham, Q. V. (2024). A survey on intelligent Internet of Things: Applications, security, privacy, and future directions. *IEEE Communications Surveys and Tutorials.*

Aribowo, W. (2025). The Conceptual Understanding of Metaheuristic Algorithms: A Brief Reviews. *Vokasi Unesa Bulletin of Engineering. Technology and Applied Science, 3*(1), 1–11. DOI: 10.26740/vubeta.v3i1.46163

Barbhuiya, S., & Das, B. B. (2025). Artificial Intelligence in Damage Detection of Concrete Structures: Techniques, Integration and Future Directions. In *Damage Detection and Structural Health Monitoring of Concrete and Masonry Structures: Novel Techniques and Applications* (pp. 31–92). Springer Nature Singapore.

Bukkarayasamudram, V. K., Reddy, P. C. S., Arun Kumar, K., Jagadish, R. M., Sharma, S., Prasad, M. L., & Thakur, G. K. (2025). Quantum computing revolution in healthcare: A systematic review of applications, issues and future directions. *Artificial Intelligence Review, 58*(12), 1–35.

Chowdhury, M. Z., Shahjalal, M., Ahmed, S., & Jang, Y. M. (2020). 6G wireless communication systems: Applications, requirements, technologies, challenges, and research directions. *IEEE Open Journal of the Communications Society, 1*, 957–975.

Deka, R. K., Ghosh, A., Nanda, S., Barik, R. K., & Saikia, M. J. (2024). Smart Healthcare System in Server-Less Environment: Concepts, Architecture, Challenges, Future Directions. *Computers, 13*(4), 105.

Irani, Z., & Kamal, M. M. (2014). Intelligent systems research in the construction industry. *Expert Systems with Applications*, *41*(4), 934–950.

Juma, M., & Shaalan, K. (2020). Cyberphysical systems in the smart city: Challenges and future trends for strategic research. In *Swarm intelligence for resource management in Internet of things* (pp. 65–85). Academic Press.

Kostopoulos, G., Gkamas, V., Rigou, M., & Kotsiantis, S. (2025). Agentic AI in Education: State of the Art and Future Directions. *IEEE Access : Practical Innovations, Open Solutions*.

Li, Z., Koban, K. C., Schenck, T. L., Giunta, R. E., Li, Q., & Sun, Y. (2022). Artificial intelligence in dermatology image analysis: Current developments and future trends. *Journal of Clinical Medicine*, *11*(22), 6826. PMID: 36431301

Liu, K., Meng, Q., Kong, Q., & Zhang, X. (2022). Review on the developments of structure, construction automation, and monitoring of intelligent construction. *Buildings (Basel, Switzerland)*, *12*(11), 1890.

Lu, Y. (2019). Artificial intelligence: a survey on evolution, models, applications and future trends. *Journal of management analytics*, *6*(1), 1-29.

Mahmud, D., Hajmohamed, H., Almentheri, S., Alqaydi, S., Aldhaheri, L., Khalil, R. A., & Saeed, N. (2025). Integrating llms with its: Recent advances, potentials, challenges, and future directions. *IEEE Transactions on Intelligent Transportation Systems*.

Marizal, M., & Jannah, Z. (2026). Modeling of Multiple Statistical Distributions for Extreme Rainfall Data Using Maximum Likelihood Estimation Methods and Bayesian Methods. *Vokasi Unesa Bulletin of Engineering. Technology and Applied Science*, *3*(1). Advance online publication. DOI: 10.26740/vubeta.v3i1.44270

Mazzetto, S. (2024). A review of urban digital twins integration, challenges, and future directions in smart city development. *Sustainability*, *16*(19), 8337.

Mondal, T. G., & Chen, G. (2022). Artificial intelligence in civil infrastructure health monitoring—Historical perspectives, current trends, and future visions. *Frontiers in Built Environment*, *8*, 1007886.

Pan, Y., & Zhang, L. (2023). Integrating BIM and AI for smart construction management: Current status and future directions. *Archives of Computational Methods in Engineering*, *30*(2), 1081–1110.

Preethichandra, D. M. G., Suntharavadivel, T. G., Kalutara, P., Piyathilaka, L., & Izhar, U. (2023). Influence of smart sensors on structural health monitoring systems and future asset management practices. *Sensors (Basel), 23*(19), 8279. PMID: 37837109

Rane, N. (2023). Integrating leading-edge artificial intelligence (AI), internet of things (IOT), and big data technologies for smart and sustainable architecture, engineering and construction (AEC) industry: Challenges and future directions. *Engineering and Construction (AEC) Industry: Challenges and Future Directions (September 24, 2023).*

Rane, N., Choudhary, S., & Rane, J. (2023). Artificial Intelligence (Ai) and Internet of Things (Iot)–based sensors for monitoring and controlling in architecture, engineering, and construction: Applications, challenges, and opportunities. *Engineering, and Construction: Applications, Challenges, and Opportunities (November 20, 2023).*

Ray, P. P. (2025). A Review of TRiSM Frameworks in Artificial Intelligence Systems: Fundamentals, Taxonomy, Use Cases, Key Challenges and Future Directions. *Authorea Preprints.*

Ray, P. P. (2025). A survey on model context protocol: Architecture, state-of-the-art, challenges and future directions. *Authorea Preprints.*

Ryalat, M., Almtireen, N., Al-refai, G., Elmoaqet, H., & Rawashdeh, N. (2025). Research and education in robotics: A comprehensive review, trends, challenges, and future directions. *Journal of Sensor and Actuator Networks, 14*(4), 76.

Saleem, Y., Crespi, N., Rehmani, M. H., & Copeland, R. (2019). Internet of things-aided smart grid: Technologies, architectures, applications, prototypes, and future research directions. *IEEE Access : Practical Innovations, Open Solutions, 7*, 62962–63003.

Sarker, I. H. (2021). Deep learning: A comprehensive overview on techniques, taxonomy, applications and research directions. *SN Computer Science, 2*(6), 1–20. PMID: 34426802

Shafique, K., Khawaja, B. A., Sabir, F., Qazi, S., & Mustaqim, M. (2020). Internet of things (IoT) for next-generation smart systems: A review of current challenges, future trends and prospects for emerging 5G-IoT scenarios. *IEEE Access : Practical Innovations, Open Solutions, 8*, 23022–23040.

Shende, S. W., Tembhurne, J. V., & Ansari, N. A. (2024). Deep learning based authentication schemes for smart devices in different modalities: Progress, challenges, performance, datasets and future directions. *Multimedia Tools and Applications, 83*(28), 71451–71493.

Sonko, S., Etukudoh, E. A., Ibekwe, K. I., Ilojianya, V. I., & Daudu, C. D. (2024). A comprehensive review of embedded systems in autonomous vehicles: Trends, challenges, and future directions. *World Journal of Advanced Research and Reviews*, *21*(1), 2009–2020.

Taye, M. M. (2023). Understanding of machine learning with deep learning: Architectures, workflow, applications and future directions. *Computers*, *12*(5), 91.

Telli, K., Kraa, O., Himeur, Y., Ouamane, A., Boumehraz, M., Atalla, S., & Mansoor, W. (2023). A comprehensive review of recent research trends on unmanned aerial vehicles (uavs). *Systems*, *11*(8), 400.

Verma, A., Bhattacharya, P., Madhani, N., Trivedi, C., Bhushan, B., Tanwar, S., & Sharma, R. (2022). Blockchain for industry 5.0: Vision, opportunities, key enablers, and future directions. *IEEE Access : Practical Innovations, Open Solutions*, *10*, 69160–69199.

Vinueza-Martinez, J., Correa-Peralta, M., Ramirez-Anormaliza, R., Franco Arias, O., & Vera Paredes, D. (2024). Geographic information systems (giss) based on webgis architecture: Bibliometric analysis of the current status and research trends. *Sustainability*, *16*(15), 6439.

Wang, Y., Li, J., Yang, X., & Peng, Q. (2025). UAV–Ground Vehicle Collaborative Delivery in Emergency Response: A Review of Key Technologies and Future Trends. *Applied Sciences (Basel, Switzerland)*, *15*(17), 9803.

Willie, M. (2026). Exploring Health Advice Adoption in Digital Spaces: A Dual-Theory Approach. *Vokasi Unesa Bulletin of Engineering. Technology and Applied Science*, *3*(1), 25–36. DOI: 10.26740/vubeta.v3i1.39662

Wojciechowska, G., Bednarz, Ł. J., Dolińska, N., Opałka, P., Krupa, M., & Imnadze, N. (2024). Intelligent Monitoring System for Integrated Management of Historical Buildings. *Buildings (2075-5309)*, *14*(7).

Wu, K., Ji, Z., Wang, H., Shao, X., Li, H., Zhang, W., & Bao, X. (2025). A comprehensive review of AI methods in agri-food engineering: Applications, challenges, and future directions. *Electronics (Basel)*, *14*(20), 3994.

Xiang, W., Yu, K., Han, F., Fang, L., He, D., & Han, Q. L. (2023). Advanced manufacturing in industry 5.0: A survey of key enabling technologies and future trends. *IEEE Transactions on Industrial Informatics*, *20*(2), 1055–1068.

Xu, L. D., Xu, E. L., & Li, L. (2018). Industry 4.0: State of the art and future trends. *International Journal of Production Research*, *56*(8), 2941–2962.

Zangana, H., & Omar, M. (Eds.). (2026). *Securing Large Language Models Against Emerging Threats*. IGI Global Scientific Publishing., DOI: 10.4018/979-8-3373-7133-7

Zangana, H., & Omar, M. (Eds.). (2026). *AI in Digital Forensics and Cybercrime Investigation: Methods, Ethics, and Emerging Technologies*. IGI Global Scientific Publishing., DOI: 10.4018/979-8-3373-6536-7

Compilation of References

Abdelhafidh, M., Fourati, M., Fourati, L. C., & Chouaya, A. (2019). Wireless sensor network monitoring system: Architecture, applications and future directions. *International Journal of Communication Networks and Distributed Systems*, *23*(4), 413–451.

Adebayo, Y., Udoh, P., Kamudyariwa, X. B., & Osobajo, O. A. (2025). Artificial intelligence in construction project management: A structured literature review of its evolution in application and future trends. *Digital*, *5*(3), 26.

Ahmad, H. M., & Rahimi, A. (2024). SH17: A dataset for human safety and personal protective equipment detection in manufacturing industry. *arXiv preprint arXiv:2407.04590*. https://arxiv.org/abs/2407.04590

Akinloye, B. (2025). DESIGN AND CONSTRUCTION OF A SINGLE-PHASE SYNCHRONOUS RELUCTANCE MOTOR DRIVE. *Vokasi Unesa Bulletin of Engineering. Technology and Applied Science*, *3*(1), 12–24. DOI: 10.26740/vubeta.v3i1.42969

Al-Bloush, S., & Al-Khasawneh, M. (2022). IoT-based energy management system using Python and MQTT protocol. In *2022 International Conference on Information Technology (ICIT)* (pp. 456–461). IEEE.

Alhaila, S., Almelhem, I., Alsnafi, A., Alameerah, M., Alrumaidhi, K., & Nadeem, M. (2024). YOLO-based helmet detection system for safety compliance in oil and gas industry. In *2024 5th International Conference on Industrial Engineering and Artificial Intelligence (IEAI)* (pp. 16-23). https://doi.org/DOI: 10. 1109/IEAI62569.2024.00012

Ali, Z., Saad, S., Rasheed, K., & Ammad, S. (2024). AI future perspectives and trends in construction. *AI in Material Science*, 239-261.

Aouedi, O., Vu, T. H., Sacco, A., Nguyen, D. C., Piamrat, K., Marchetto, G., & Pham, Q. V. (2024). A survey on intelligent Internet of Things: Applications, security, privacy, and future directions. *IEEE Communications Surveys and Tutorials.*

Aribowo, W. (2025). The Conceptual Understanding of Metaheuristic Algorithms: A Brief Reviews. *Vokasi Unesa Bulletin of Engineering. Technology and Applied Science, 3*(1), 1–11. DOI: 10.26740/vubeta.v3i1.46163

Atadoga, A., Sodiya, E. O., Umoga, U. J., & Amoo, O. O. (2024). A comprehensive review of machine learning's role in enhancing network security and threat detection. *World Journal of Advanced Research and Reviews, 21*(2), 877–886.

Baldwin, C. L., & Lewis, B. A. (2014). Perceived urgency mapping across modalities within a driving context. *Applied Ergonomics, 45*(5), 1270–1277. DOI: 10.1016/j.apergo.2013.05.002 PMID: 23910716

Barbhuiya, S., & Das, B. B. (2025). Artificial Intelligence in Damage Detection of Concrete Structures: Techniques, Integration and Future Directions. In *Damage Detection and Structural Health Monitoring of Concrete and Masonry Structures: Novel Techniques and Applications* (pp. 31–92). Springer Nature Singapore.

Barlybayev, A., Amangeldy, N., Kurmetbek, B., Krak, I., Razakhova, B., & Tursynova, N. (2024). Personal protective equipment detection using YOLOv8 architecture on object detection benchmark datasets: A comparative study. *Cogent Engineering, 11*(1), 2333209. https://www.tandfonline.com/doi/full/10.1080/23311916.2024.2333209

Baseri, Y., Chouhan, V., & Ghorbani, A. (2024). Cybersecurity in the quantum era: Assessing the impact of quantum computing on infrastructure. *arXiv preprint arXiv:2404.10659.*

Bas, S., Apaydin, N. M., Ilki, A., & Catbas, F. N. (2018). Structural health monitoring system of the long-span bridges in Turkey. *Structure and Infrastructure Engineering, 14*(4), 425–444.

BeanAir. (2024). Structural health monitoring - sensors. https://www.beanair.com/structural-health-monitoring.html

Bellini, P., Nesi, P., & Pantaleo, G. (2022). Open-source MQTT-based end-to-end IoT system for smart city scenarios. *IEEE Access : Practical Innovations, Open Solutions, 10*, 1215–1229. DOI: 10.1109/ACCESS.2022.3150258

Benaouda, D., Murtagh, F., Starck, J. L., & Renaud, O. (2006). Wavelet-based nonlinear multiscale decomposition model for electricity load forecasting. *Neurocomputing, 70*(1-3), 139–154.

Bishwas, A. K., & Sen, M. (2024). Strategic roadmap for quantum-resistant security: A framework for preparing industries for the quantum threat. *arXiv preprint arXiv:2411.09995*.

Bi, W., Tian, L., Li, C., & Zhang, S. (2023). Multi-hazard joint probability distribution model for wind speed, wind direction and rain intensity. *Science China. Technological Sciences*, *66*(2), 336–345.

Block, F. (2024). Social policy and accumulation: a critique of the new consensus. In *Stagnation and Renewal in Social Policy* (pp. 13–32). Routledge.

Buildings, M. D. P. I. (2024). Vision-based construction safety monitoring utilizing temporal analysis to reduce false alarms, 14(6), Article 1878. https://www.mdpi.com/2075-5309/14/6/1878

Bukkarayasamudram, V. K., Reddy, P. C. S., Arun Kumar, K., Jagadish, R. M., Sharma, S., Prasad, M. L., & Thakur, G. K. (2025). Quantum computing revolution in healthcare: A systematic review of applications, issues and future directions. *Artificial Intelligence Review*, *58*(12), 1–35.

Bush, S. F., Challener, W. A., & Mantelet, G. (2021). A perspective on industrial quantum networks. *AVS Quantum Science*, *3*(3).

Cai, H., & Mohammed, A. M. (2025). *Vision-based automated detection of surface damage in sports facilities and equipment*. IEEE 1st World Congress on Sports Science and Technology, Hong Kong.

Campero-Jurado, I., Márquez-Sánchez, S., Quintanar-Gómez, J., Rodríguez, S., & Corchado, J. (2020). Smart Helmet 5.0 for Industrial Internet of Things Using Artificial Intelligence. *Sensors (Basel)*, *20*(21), 6241. DOI: 10.3390/s20216241 PMID: 33139608

Campero-Jurado, I., Márquez-Sánchez, S., Quintanar-Gómez, J., Rodríguez, S., & Corchado, J. M. (2020). Smart helmet 5.0 for industrial internet of things using artificial intelligence. *Sensors (Basel)*, *20*(21), 6241. PMID: 33139608

CCOHS. (2024). *Health and safety programs - Leading and lagging indicators*. Canadian Centre for Occupational Health and Safety. https://www.ccohs.ca/

Chaabene, S., Bouaziz, B., Boudaya, A., Hökelmann, A., & Ammar, A. (2021). Convolutional neural network for drowsiness detection using EEG signals. *Sensors (Basel)*, *21*(5), 1734. PMID: 33802357

Chae, J., Hwang, S., & Kang, Y. (2024). Effectiveness of alarm sounds in preventing operator habituation to auditory warnings in construction equipment. *Automation in Construction, 168*, 105784. DOI: 10.1016/j.autcon.2024.105784

Chang, X., & Liu, X. M. (2018). Fault tree analysis of unreasonably wearing helmets for builders. *Journal of Jilin Jianzhu University, 35*(6), 67-71.

Chen, Y., Wan, Y., Yang, Y., Liao, Y., Jin, J., & Mohammed, A. M. (2025). *Deep learning-based fabric classification for sportswear using ensemble methods under small dataset constraints*. IEEE 1st World Congress on Sports Science and Technology, Hong Kong.

Chen, H., Li, Y., Wen, H., & Hu, X. (2023a). YOLOv5s-gnConv: Detecting personal protective equipment for workers at height. *Frontiers in Public Health, 11*, 1225478. Advance online publication. DOI: 10.3389/fpubh.2023.1225478 PMID: 37841722

Chiverton, J. (2012). Helmet presence classification with motorcycle detection and tracking. *IET Intelligent Transport Systems, 6*(3), 259–269. DOI: 10.1049/iet-its.2011.0138

Chowdhury, M. Z., Shahjalal, M., Ahmed, S., & Jang, Y. M. (2020). 6G wireless communication systems: Applications, requirements, technologies, challenges, and research directions. *IEEE Open Journal of the Communications Society, 1*, 957–975.

Construction Industry Council. (2024). [Leveraging AI to enhance performance and ensure safety through smart site systems]. https://www.cic.hk/chi/main/aboutcic/leadership/ChairmanBlog/blog-151.html

Deka, R. K., Ghosh, A., Nanda, S., Barik, R. K., & Saikia, M. J. (2024). Smart Healthcare System in Server-Less Environment: Concepts, Architecture, Challenges, Future Directions. *Computers, 13*(4), 105.

Development Bureau Hong Kong. (2024). Annual report on accident statistics and analyses for public works construction. https://www.devb.gov.hk/filemanager/en/content 32/2024%20Annual%20Report.pdf

Ding, Y., Liang, N. Y., Zhang, X. S., Wang, J., & Zeng, C. Q. (2025). Long short-term memory with wavelet decomposition for wind speed predicting based on SHM data. *Smart Structures and Systems, 35*(2), 65–75.

Ding, Y., Ye, X. W., & Guo, Y. (2023). Copula-based JPDF of wind speed, wind direction, wind angle, and temperature with SHM data. *Probabilistic Engineering Mechanics, 73*, 103483.

Dzikri, A., Hariadi, M., & Rachmadi, R. F. (2025). A vision-based framework for automated personal protective equipment detection and workplace safety enhancement. In *2025 International Seminar on Intelligent Technology and Its Applications (ISITIA)* (pp. 379-384). https://doi.org/. 2025.11137551DOI: 10.1109/ISITIA66279

Evalan. (2024). ARMOR - Heat stress monitoring wearable. https://evalan.com/products/armor/

Gallagher Bassett. (2024). The role of wearables and the IoT in enhancing construction site safety. https://www.gallagherbassett.com/news-and-insights/the-roleof-wearables-and-the-iot-in-enhancing-construction-site-safety/

Gao, Q.. (2024). Smart building management system based on digital twin: A case study on real-time environmental monitoring and thermal comfort prediction. *In 2024 IEEE International Symposium on Parallel and Distributed Processing with Applications (ISPA)* (pp. 2157-2163). https://doi.org/DOI: 10.1109/ISPA63168.2024.00294

Gómez, J., Casas, J. R., & Villalba, S. (2020). Structural Health Monitoring with Distributed Optical Fiber Sensors of tunnel lining affected by nearby construction activity. *Automation in Construction*, *117*, 103261.

Häikiö, J., Kallio, J., Mäkelä, S. M., & Keränen, J. (2020). IoT-based safety monitoring from the perspective of construction site workers. *International Journal of Occupational and Environmental Safety*, *4*(1), 1–14. DOI: 10.24840/2184-0954_004.001_

Hashemi, M., & Beheshti, S. (2009). Adaptive noise variance estimation in BayesShrink. *IEEE Signal Processing Letters*, *17*(1), 12–15.

He, H., Li, Y., Guan, Y., & Tan, J. (2020). A real-time driver fatigue detection method based on two-stage convolutional neural network. *IFAC-PapersOnLine*, *53*(2), 6645–6650.

Hochreiter, S., & Schmidhuber, J. (1997). Long short-term memory. *Neural Computation*, *9*(8), 1735–1780.

Hong Kong Centre for Health Protection. (2016). *Keep your ears safe*. Department of Health. https://www.chp.gov.hk/files/pdf/ncd_watch_feb2016.pdf

Hong Kong Centre for Health Protection. (2025). *Beware of heat stroke*. Department of Health. https://www.chp.gov.hk/en/static/90064.html

Hong Kong Labour Department. (2024). Occupational safety and health statistics 2024. https://www.labour.gov.hk/common/osh/pdf/OSH Statistics 2024 en.pdf

How, M. L., & Cheah, S. M. (2023). Business Renaissance: Opportunities and challenges at the dawn of the Quantum Computing Era. *Businesses*, *3*(4), 585–605.

IEEE Transactions on Plasma Science. (2019). Table of contents. *IEEE Transactions on Plasma Science, 47*(7), C1-C1. https://doi.org/DOI: 10.1109/TPS.2019.2922521

Intel. (2024). *Solution white paper: Intel® Xeon® scalable processor, Intel® Core™ processor, Intel® Video AI box, Intel® Distribution of OpenVINO™ Toolkit, Intel® Smart Edge Open Eazytec constructs an efficient smart building operations management platform to enable smart parks.* Intel Corporation.

International Labour Organization. (2023, December). *Nearly 3 million people die of work-related accidents and diseases each year.* https://www.ilo. org/resource/news/nearly-3-million-people-die-work-related-accidents-and-diseases

International Labour Organization. (2024). *Safety and health at work.* https://www .ilo.org/topics-and-sectors/safety-and-health-work

Irani, Z., & Kamal, M. M. (2014). Intelligent systems research in the construction industry. *Expert Systems with Applications, 41*(4), 934–950.

Jayasree, V., & Kumari, M. N. (2020, July). IOT based smart helmet for construction workers. In *2020 7th International Conference on Smart Structures and Systems (ICSSS)* (pp. 1-5). IEEE.

Jia, M., Tang, L., Chen, B.-C., Cardie, C., Belongie, S., Hariharan, B., & Lim, S.-N. (2023). Visual prompt tuning for open-world localization and detection. *arXiv.*

Jia, W., Xu, S., Liang, Z., Zhao, Y., Min, H., Li, S., & Yu, Y. (2021). Real-time automatic helmet detection of motorcyclists in urban traffic using improved YOLOv5 detector. *IET Image Processing, 15*, 3623–3637. DOI: 10.1049/ipr2.12295

Jirakitpuwapat, W.. (2024). Convergence property of Nesterov-accelerated adaptive moment estimation with safety helmet detection and classification in smart industry application. *Mathematical Methods in the Applied Sciences, 47*(16), 12650–12667. DOI: 10.1002/mma.10174

Jocher, G., Chaurasia, A., & Qiu, J. (2020). *Ultralytics YOLOv5* [Computer software]. GitHub. https://github.com/ultralytics/yolov5

Juma, M., & Shaalan, K. (2020). Cyberphysical systems in the smart city: Challenges and future trends for strategic research. In *Swarm intelligence for resource management in Internet of things* (pp. 65–85). Academic Press.

Kelm, A., Laußat, L., Meins-Becker, A., Platz, D., Khazaee, M. J., Costin, A. M., Helmus, M., & Teizer, J. (2013). Mobile passive radio frequency identification (RFID) portal for automated and rapid control of personal protective equipment (PPE) on construction sites. *Automation in Construction*, *36*, 38–52. DOI: 10.1016/j.autcon.2013.08.009

Khan, A. M., Alrasheed, K. A., Waqar, A., Almujibah, H., & Benjeddou, O. (2024). Internet of things (IoT) for safety and efficiency in construction building site operations. *Scientific Reports*, *14*, 28914. https://www.nature.com/articles/s41598-024-78931-0 PMID: 39572581

Khanam, R., & Hussain, M. (2024). YOLOv11: An overview of the key architectural enhancements. *arXiv preprint arXiv:2410.17725*. https://arxiv.org/ abs/2410.17725

Khan, M. A.. (2023). A Python-MQTT-based smart construction site monitoring framework. In *2023 IEEE International Conference on Artificial Intelligence and Internet of Things (AIIoT)* (pp. 112–118). IEEE.

Kim, H.-S., Seong, J., & Jung, H.-J. (2023). Real-time struck-by hazards detection system for small- and medium-sized construction sites based on computer vision using far-field surveillance videos. *Journal of Computing in Civil Engineering*, *37*(6). https://ascelibrary.org/doi/10.1061/JCCEE5.CPENG-5238

Kim, J. Y., Kang, S., Cho, J., Jeong, S., Kim, S., Sung, Y., Lee, B., & Kim, G. H. (2025). Implementation of integrated smart construction monitoring system based on point cloud data and IoT technique. *Sensors (Basel)*, *25*(13), 3997. DOI: 10.3390/s25133997 PMID: 40648254

King, F. O., Kong, S. H., Liu, C., & Mohammed, A. M. (2024). *Forecasting population trends and socioeconomic impacts in Asian tiger economies using stochastic and machine learning models. 22nd IEEE International Conference on Networking, Sensing and Control*, Finland.

Kopec, R. W., Kania, K., Sypniewski, J., & Strumiłło, P. (2013). The use of keystroke dynamics in the assessment of cognitive load. In Kurosu, M. (Ed.), *Human-Computer Interaction. Interaction Modalities and Techniques* (pp. 431–440). Springer.

Kostopoulos, G., Gkamas, V., Rigou, M., & Kotsiantis, S. (2025). Agentic AI in Education: State of the Art and Future Directions. *IEEE Access : Practical Innovations, Open Solutions.*

Kwon, D., Vannucci, M., Song, J. J., Jeong, J., & Pfeiffer, R. M. (2008). A novel wavelet-based thresholding method for the pre-processing of mass spectrometry data that accounts for heterogeneous noise. *Proteomics*, *8*(15), 3019–3029.

Li, C. Z., Zhao, Y., Xiao, B., Yu, B., Tam, V. W., Chen, Z., & Ya, Y. (2020). Research trend of the application of information technologies in construction and demolition waste management. *Journal of Cleaner Production, 263*, 121458.

Lingard, H., Hallowell, M., Salas, R., & Pirzadeh, P. (2017). Leading or lagging? Temporal analysis of safety indicators on a large infrastructure construction project. *Safety Science, 91*, 206–220. DOI: 10.1016/j.ssci.2016.08.020

Lin, T.-Y., Goyal, P., Girshick, R., He, K., & Dollar, P. (2017). Focal loss for dense object detection. In´ *Proceedings of the IEEE International Conference on Computer Vision (ICCV)* (pp. 2980-2988). https://arxiv.org/abs/1708.02002

Liu, C., Du, H., Ren, Y., & Mohammed, A. M. (2024). *Optimizing construction site resources under uncertainty through stochastic simulation and advanced scheduling techniques. 22nd IEEE International Conference on Networking, Sensing and Control*, Finland.

Liu, H., Tian, H. Q., & Li, Y. F. (2015). An EMD-recursive ARIMA method to predict wind speed for railway strong wind warning system. *Journal of Wind Engineering and Industrial Aerodynamics, 141*, 27–38.

Liu, K., Meng, Q., Kong, Q., & Zhang, X. (2022). Review on the developments of structure, construction automation, and monitoring of intelligent construction. *Buildings (Basel, Switzerland), 12*(11), 1890.

Liu, L., Guo, Z., Liu, Z., Zhang, Y., Cai, R., Hu, X., Yang, R., & Wang, G. (2024). Multi-task intelligent monitoring of construction safety based on computer vision. *Buildings (Basel, Switzerland), 14*(8), 2429. DOI: 10.3390/buildings14082429

Liu, W., Anguelov, D., Erhan, D., Szegedy, C., Reed, S., Fu, C.-Y., & Berg, A. C. (2016). SSD: Single shot multibox detector. In *Proceedings of the European Conference on Computer Vision (ECCV)* (pp. 21-37). Springer. https://arxiv.org/abs/1512.02325

Li, Y., Zhang, J., Hu, Y., Zhao, Y., & Cao, Y. (2022). Real-time Safety Helmet-wearing Detection Based on Improved YOLOv5. *Computer Systems Science and Engineering, 43*(3).

Li, Z., Koban, K. C., Schenck, T. L., Giunta, R. E., Li, Q., & Sun, Y. (2022). Artificial intelligence in dermatology image analysis: Current developments and future trends. *Journal of Clinical Medicine, 11*(22), 6826. PMID: 36431301

Loubet, G., Takacs, A., Gardner, E., & De Luca, A. (2019). Lo-RaWAN battery-free wireless sensors network designed for structural health monitoring in the construction domain. *Sensors (Basel), 19*(7), 1510. https://www.researchgate.net/publication/332076485 PMID: 30925754

Lougovski, P., Parekh, O., Broz, J., Byrd, M., Chapman, J. C., Chembo, Y., . . . Zeng, W. J. (2023). *Report for the ASCR Workshop on Basic Research Needs in Quantum Computing and Networking-2023*. USDOE Office of Science (SC)(United States).

Lu, Y. (2019). Artificial intelligence: a survey on evolution, models, applications and future trends. *Journal of management analytics, 6*(1), 1-29.

Lukman, N. (2021). LoRaWAN vs NB-IoT: A comparison. *NEXPCB*. https://www.nexpcb.com/blog/lorawan-vs-nb-iot-a-comparison

Luo, X., Wang, A., Zhang, X., Huang, K., Wang, S., Chen, L., & Cu, Y. (2024). Toward intelligent AIoT: A comprehensive survey on digital twin and multimodal generative AI integration. *Mathematics, 13*(21), 3382. https://www.mdpi.com/2227-7390/13/21/3382

Mahmud, D., Hajmohamed, H., Almentheri, S., Alqaydi, S., Aldhaheri, L., Khalil, R. A., & Saeed, N. (2025). Integrating llms with its: Recent advances, potentials, challenges, and future directions. *IEEE Transactions on Intelligent Transportation Systems*.

Manakitsa, N., Maraslidis, G. S., Moysis, L., & Fragulis, G. F. (2024). A review of machine learning and deep learning for object detection, semantic segmentation, and human action recognition in machine and robotic vision. *Technologies, 12*(2), 15. Advance online publication. DOI: 10.3390/technologies12020015

Marizal, M., & Jannah, Z. (2026). Modeling of Multiple Statistical Distributions for Extreme Rainfall Data Using Maximum Likelihood Estimation Methods and Bayesian Methods. *Vokasi Unesa Bulletin of Engineering. Technology and Applied Science, 3*(1). Advance online publication. DOI: 10.26740/vubeta.v3i1.44270

Masaba, M. G., Sserwadda, G., & Mwebaze, E. N. (2022). SHEL5K: An extended dataset and benchmarking for safety helmet detection. *Sensors (Basel), 22*(6), 2315. Advance online publication. DOI: 10.3390/s22062315 PMID: 35336491

Massachusetts Institute of Technology School of Engineering. (2011). *Why do we sweat more in high humidity?* https://engineering.mit.edu/engage/ask-an-engineer/why-do-we-sweat-more-in-high-humidity/

Max, C. K., Wang, Y., Pazo Recio, L., Cheuk Tung, G., & Mohammed, A. M. (2024). *Energy-efficient and high-accuracy AIoT visual sensing using K-vote, transfer-spiking net, and reconstructive classifier. 22nd IEEE International Conference on Networking, Sensing and Control*, Finland.

Mazzetto, S. (2024). A review of urban digital twins integration, challenges, and future directions in smart city development. *Sustainability, 16*(19), 8337.

Minderer, M., Gritsenko, A., Stone, A., Neumann, M., Weissenborn, D., Dosovitskiy, A., . . . Houlsby, N. (2022, October). Simple open-vocabulary object detection. In *European conference on computer vision* (pp. 728-755). Cham: Springer Nature Switzerland.

Mohammed, M. A., & Garimella, R. M. (2009). *Independent directional localization in mobile wireless sensor networks: A contemporary approach.* Fifth IEEE Consumer Communications & Networking Conference Demonstration Proposals (CCNC), Las Vegas, USA.

Mohammed, M. A., Shakir, M. Z., & Slim-Alouini, M. (2012). *An active 2-dimensional localization scheme for femtocell subscribers using E-UTRAN. IEEE International Conference on Computing, Networking and Communications (ICNC)*, Maui, Hawaii, USA.

Mohammed, M. A., & Slim-Alouini, M. (2011). *A localized cooperative wideband spectrum sensing using RF sensor networks. 7th IEEE International Wireless Communications and Mobile Computing Conference (IWCMC)*, Istanbul, Turkey.

MokoSmart. (2024). LoRaWAN vs NB-IoT: How do they compare and differ. https://www.mokosmart.com/lorawan-vs-nb-iot-how-do-they-compare-and-differ/

Mondal, T. G., & Chen, G. (2022). Artificial intelligence in civil infrastructure health monitoring—Historical perspectives, current trends, and future visions. *Frontiers in Built Environment, 8*, 1007886.

Nath, N. D., & Behzadan, A. H. (2020). Deep learning detection of personal protective equipment to maintain safety compliance on construction sites. In *Construction Research Congress 2020* (pp. 181-190). Reston, VA: American Society of Civil Engineers.

Nazib, A. A., & Moh, D. (2021). Security and privacy in industrial Internet of Things: A survey. *IEEE Access : Practical Innovations, Open Solutions, 9*, 110926–110954. DOI: 10.1109/ACCESS.2021.3101243

Occupational Safety and Health Branch of Hong Kong Labour Department. (2018). *Guidance notes on the selection, use and maintenance of safety helmets.* https://www.labour.gov.hk/eng/public/os/C/GN_on_helmet_eng.pdf

Occupational Safety and Health Council. (2022). ( ) [Analysis of a research report on the four major types of industrial accidents in Hong Kong's construction industry]. Green Cross. https://www.oshc.org.hk/oshc_data/files/greencross/2023/GX2023JUL-8.pdf

Pan, Y., & Zhang, L. (2023). AIoT-based building operation and maintenance supervision system. *Buildings (Basel, Switzerland)*, *13*(5), 1215. DOI: 10.3390/buildings13051215

Pan, Y., & Zhang, L. (2023). Integrating BIM and AI for smart construction management: Current status and future directions. *Archives of Computational Methods in Engineering*, *30*(2), 1081–1110.

Phan, A.-C., Trieu, T.-N., & Phan, T.-C. (2023). Driver drowsiness detection and smart alerting using deep learning and IoT. *Internet of Things : Engineering Cyber Physical Human Systems*, *22*, 100785.

Phase IV Engineering. (2024). The top 5 sensors for structural health monitoring (SHM). https://www.phaseivengr.com/blog/the-top-5-sensors-for-structural-healthmonitoring-shm/

Pierleoni, P., Conti, M., Belli, A., Palma, L., Incipini, L., Sabbatini, L., Valenti, S., Mercuri, M., & Concetti, R. (2019). IoT solution based on MQTT protocol for real-time building monitoring. *2019 IEEE 23rd International Symposium on Consumer Technologies (ISCT)*, 57-62. DOI: <ALIGNMENT.qj></ALIGNMENT>10.1109/ISCE.2019.8901024

Preethichandra, D. M. G., Suntharavadivel, T. G., Kalutara, P., Piyathilaka, L., & Izhar, U. (2023). Influence of smart sensors on structural health monitoring systems and future asset management practices. *Sensors (Basel)*, *23*(19), 8279. PMID: 37837109

Procore. (2024). *Leading and lagging indicators in construction: How to improve safety performance.* https://www.procore.com/

Rahman, A., Alatallah, F. A., Almubarak, A. J., Alkhazal, H. A., Alzayer, H. A., Shaaban, Y. Z., & Aloup, K. (2025). PPE-EYE: A deep learning approach to personal protective equipment compliance detection. *Computers*, *15*(1), 45. DOI: 10.3390/computers15010045

Rahul, R., Sindhu, P., Sundar, G. N., & Venkatesan, R. (2024). Fusing Deep Learning Techniques for Intrusion Detection in Smart Grids. *Fusion: Practice & Applications, 16*(1).

Ramkumar, G.. (2021). AIOT based real time environment monitoring and tracking system for soldiers using SN-MQTT protocol. *Journal of Physics: Conference Series, 1917*(1), 012005. DOI: 10.1088/1742-6596/1917/1/012005

Ramos, A. L. A., Sikansi, F., Pisa, I. T., de Faria, A. C., de Camargo, E. D. L. B., & Spadon, G. (2019). Driver drowsiness detection based on eye movement and yawning using facial landmark analysis. International Journal of Simulation - Systems. *Science & Technology, 20*(S2). Advance online publication. DOI: 10.5013/IJSSST.a.20.S2.18

Rane, N. (2023). Integrating leading-edge artificial intelligence (AI), internet of things (IOT), and big data technologies for smart and sustainable architecture, engineering and construction (AEC) industry: Challenges and future directions. *Engineering and Construction (AEC) Industry: Challenges and Future Directions (September 24, 2023).*

Rane, N., Choudhary, S., & Rane, J. (2023). Artificial Intelligence (Ai) and Internet of Things (Iot)–based sensors for monitoring and controlling in architecture, engineering, and construction: Applications, challenges, and opportunities. *Engineering, and Construction: Applications, Challenges, and Opportunities (November 20, 2023).*

Rao, N. S., Sri, K. H., Lakshmi, B. J. S., Sastry, R. R., Eswar, D., Sri, T. B. N., & Swamy, Y. T. V. (2024). Smart warehouse monitoring system using IoT. *IJARIIE, 10*(2).

Ray, P. P. (2018). A survey on Internet of Things architectures. *Journal of King Saud University. Computer and Information Sciences, 30*(3), 291–319. DOI: 10.1016/j.jksuci.2016.10.003

Ray, P. P. (2025). A Review of TRiSM Frameworks in Artificial Intelligence Systems: Fundamentals, Taxonomy, Use Cases, Key Challenges and Future Directions. *Authorea Preprints.*

Ray, P. P. (2025). A survey on model context protocol: Architecture, state-of-the-art, challenges and future directions. *Authorea Preprints.*

Redmon, J., & Farhadi, A. (2018). YOLOv3: An incremental improvement. *arXiv preprint arXiv:1804.02767.* https://arxiv.org/abs/1804.02767

Redmon, J., Divvala, S., Girshick, R., & Farhadi, A. (2016). You only look once: Unified, real-time object detection. In *Proceedings of the IEEE Conference on Computer Vision and Pattern Recognition (CVPR)* (pp. 779-788). https://doi.org/DOI: 10.1109/CVPR.2016.91

Ren, S., He, K., Girshick, R., & Sun, J. (2015). Faster R-CNN: Towards real-time object detection with region proposal networks. In *Advances in Neural Information Processing Systems* (Vol. 28, pp. 91-99). https://arxiv.org/abs/1506.01497

Resensys. (2024). Wireless structural health monitoring system. https://resensys.com/r20/sensor-applications/wireless-structural-healthmonitoring.html

Rubaiyat, A. H., Toma, T. T., Kalantari-Khandani, M., Rahman, S. A., Chen, L., Ye, Y., & Pan, C. S. (2016, October). Automatic detection of helmet uses for construction safety. In 2016 IEEE/WIC/ACM International Conference on Web Intelligence Workshops (WIW) (pp. 135-142). IEEE.

Ryalat, M., Almtireen, N., Al-refai, G., Elmoaqet, H., & Rawashdeh, N. (2025). Research and education in robotics: A comprehensive review, trends, challenges, and future directions. *Journal of Sensor and Actuator Networks*, *14*(4), 76.

Sadeghi, J., Ahmadi, A., & Phipps, R. (2025). Internet of things in construction: Trends and adoption insights from a scientometric perspective. *Engineering, Construction and Architectural Management*. https://www.tandfonline.com/doi/full/10.1080/15623599.2025.2508905

Saleem, Y., Crespi, N., Rehmani, M. H., & Copeland, R. (2019). Internet of things-aided smart grid: Technologies, architectures, applications, prototypes, and future research directions. *IEEE Access : Practical Innovations, Open Solutions*, *7*, 62962–63003.

Sarker, I. H. (2021). Deep learning: A comprehensive overview on techniques, taxonomy, applications and research directions. *SN Computer Science*, *2*(6), 1–20. PMID: 34426802

Schuldt, S. J., Nicholson, M. R., Adams, Y. A., & Delorit, J. D. (2021). Weather-related construction delays in a changing climate: A systematic state-of-the-art review. *Sustainability*, *13*(5), 2861.

Sensors, M. D. P. I. (2020). Development of an IoT-based construction worker physiological data monitoring platform at high temperatures, 20(19), Article 5682. https://www.mdpi.com/1424-8220/20/19/5682

Shafique, K., Khawaja, B. A., Sabir, F., Qazi, S., & Mustaqim, M. (2020). Internet of things (IoT) for next-generation smart systems: A review of current challenges, future trends and prospects for emerging 5G-IoT scenarios. *IEEE Access : Practical Innovations, Open Solutions*, *8*, 23022–23040.

Sharma, R. (2024). Enhancing industrial automation and safety through real-time monitoring and control systems. *International Journal on Smart & Sustainable Intelligent Computing*, *1*, 1–20. DOI: 10.63503/j.ijssic.2024.30

Shen, L., Cui, M., Huang, L., & Wang, J. (2025). *Optimizing sports facility management using OS-level process scheduling: A smart parking system for sports venues.* IEEE 1st World Congress on Sports Science and Technology, Hong Kong.

Shende, S. W., Tembhurne, J. V., & Ansari, N. A. (2024). Deep learning based authentication schemes for smart devices in different modalities: Progress, challenges, performance, datasets and future directions. *Multimedia Tools and Applications*, *83*(28), 71451–71493.

Shivani, S., Aditya, S., & Ramalingam, V. (2020). Driver drowsiness detection system using machine learning algorithms. [IJRTE]. *International Journal of Recent Technology and Engineering*, *8*(6), 3551–3555.

Silva, R. R. V. E., Aires, K. R. T., & Veras, R. D. M. S. (2017). Detection of helmets on motorcyclists. *Multimedia Tools and Applications*, *77*(5), 5659–5683. DOI: 10.1007/s11042-017-4482-7

Silva, R., Aires, K., Santos, T., Abdala, K., Veras, R., & Soares, A. (2013). Automatic detection of motorcyclists without helmet. *In 2013 XXXIX Latin American Computing Conference (CLEI)* (pp. 1-7). https://doi.org/DOI: 10.1109/CLEI.2013.6670613

Silva, R., Do Carmo, R., Silva, J. S., Boavida, F., & Schaeffer-Filho, A. (2021). A performance analysis of Internet of Things networking protocols: Evaluating MQTT, CoAP, OPC UA. *Applied Sciences (Basel, Switzerland)*, *11*(11), 4879. DOI: 10.3390/app11114879

Singh, V., Pencina, M., Einstein, A. J., Liang, J. X., Berman, D. S., & Slomka, P. (2021). Impact of train/test sample regimen on performance estimate stability of machine learning in cardiovascular imaging. *Scientific Reports*, *11*(1), 14376. Advance online publication. DOI: 10.1038/s41598-021-93651-5 PMID: 34262098

Sivakumar, S. K., Sivan, V., & Antony, G. C. (2024). SafeSense: Real-time safety alerts for construction workers. *Proceedings of the 41st International Symposium on Automation and Robotics in Construction (ISARC)*, 428-435. https://www.iaarc.org/publications/2024_proceedings_of_the_41st_isarc_lille_france/safesense_real-time_ safety_alerts_for_construction_workers.html

Song, C., & Li, Y. (2025). A YOLOv8 algorithm for safety helmet wearing detection in complex environment. *Scientific Reports*, *15*, 08828. https://www.nature.com/articles/s41598-025-08828-z PMID: 40624039

Sonko, S., Etukudoh, E. A., Ibekwe, K. I., Ilojianya, V. I., & Daudu, C. D. (2024). A comprehensive review of embedded systems in autonomous vehicles: Trends, challenges, and future directions. *World Journal of Advanced Research and Reviews*, *21*(1), 2009–2020.

Soukupová, T., & Cech, J. (2016). Real-time eye blink detection using facial landmarks. *21st Computer Vision Winter Workshop*, Rimske Toplice, Slovenia.

Spachos, P., & Plataniotis, K. (2018). Environmental monitoring for smart buildings: Communication networks and services. In *Smart buildings: Communication networks and services*. John Wiley & Sons, Inc., DOI: 10.1002/9781119360124.ch12

Taye, M. M. (2023). Understanding of machine learning with deep learning: Architectures, workflow, applications and future directions. *Computers*, *12*(5), 91.

Telli, K., Kraa, O., Himeur, Y., Ouamane, A., Boumehraz, M., Atalla, S., & Mansoor, W. (2023). A comprehensive review of recent research trends on unmanned aerial vehicles (uavs). *Systems*, *11*(8), 400.

Terven, J., & Cordova-Esparza, D. (2023). A comprehensive review of YOLO: From YOLOv1 to YOLOv8 and beyond. *arXiv preprint arXiv:2304.00501*. https://arxiv.org/abs/ 2304.00501

Tsao, Y.-F., Wang, Y., Hsu, H.-T., & Desai, A. (2022). Planar beam switched antenna module for 5G new-radio FR2 applications. *In 2022 Asia-Pacific Microwave Conference (APMC)* (pp. 396-398). https://doi.org/DOI: 10.23919/APMC55665.2022.9999999

Udurume, M., Hwang, T., Uddin, R., & Koo, I. (2025). Developing a fire monitoring system based on MQTT, ESP-NOW, and a REM in industrial environments. *Applied Sciences (Basel, Switzerland)*, *15*(2), 500. DOI: 10.3390/app15020500

Ultralytics. (2024). Construction-PPE dataset - Ultralytics YOLO docs. https://docs.ultralytics.com/datasets/detect/construction-ppe/

Ultralytics. (2024). Ultralytics YOLO11. https://docs.ultralytics.com/models/yolo11/

Ultralytics. (2024). *YOLOv8 documentation.* https://docs.ultralytics.com/

Venkata, P. B., & Suchismitha, C. (2021). Automatic classification methods for detecting drowsiness using wavelet packet transform extracted time-domain features from single-channel EEG signal. *Journal of Neuroscience Methods, 347,* 108968. PMID: 32941920

Verma, A., Bhattacharya, P., Madhani, N., Trivedi, C., Bhushan, B., Tanwar, S., & Sharma, R. (2022). Blockchain for industry 5.0: Vision, opportunities, key enablers, and future directions. *IEEE Access : Practical Innovations, Open Solutions, 10,* 69160–69199.

Vinueza-Martinez, J., Correa-Peralta, M., Ramirez-Anormaliza, R., Franco Arias, O., & Vera Paredes, D. (2024). Geographic information systems (giss) based on webgis architecture: Bibliometric analysis of the current status and research trends. *Sustainability, 16*(15), 6439.

Wang, C.-Y., Yeh, I.-H., & Liao, H.-Y. M. (2024a). YOLOv9: Learning what you want to learn using programmable gradient information. *arXiv preprint arXiv:2402.13616.* https://arxiv.org/abs/2402.13616

Wang, A., Chen, H., Liu, L., Chen, K., Lin, Z., Han, J., & Ding, G. (2024b). YOLOv10: Real-time end-to-end object detection. In *Proceedings of the 38th International Conference on Neural Information Processing Systems (NeurIPS).* https://arxiv.org/abs/2405.14458

Wang, H., Hu, Z., Guo, Y., Yang, Z., Zhou, F., & Xu, P. (2020). A real-time safety helmet wearing detection approach based on CSYOLOv3. *Applied Sciences (Basel, Switzerland), 10*(19), 6732. DOI: 10.3390/app10196732

Wang, Y., Li, J., Yang, X., & Peng, Q. (2025). UAV–Ground Vehicle Collaborative Delivery in Emergency Response: A Review of Key Technologies and Future Trends. *Applied Sciences (Basel, Switzerland), 15*(17), 9803.

Wang, Z., Xiao, H., & Chen, C. (2025). Research on the development of a building model management system integrating MQTT sensing. *Sensors (Basel), 25*(19), 6069. DOI: 10.3390/s25196069 PMID: 41094892

Wan, H. P., & Ni, Y. Q. (2018). Bayesian modeling approach for forecast of structural stress response using structural health monitoring data. *Journal of Structural Engineering, 144*(9), 04018130.

Wierwille, W. W., Wreggit, S. S., Kim, C. L., Ellsworth, L. A., & Fairbanks, R. J. (1994). *Research on vehicle-based driver status/performance monitoring: Development, validation, and refinement of algorithms for detection of driver drowsiness*. National Highway Traffic Safety Administration.

Williams, C. K., & Rasmussen, C. E. (2006). *No. 3* (Vol. 2). Gaussian processes for machine learning. MIT press.

Willie, M. (2026). Exploring Health Advice Adoption in Digital Spaces: A Dual-Theory Approach. *Vokasi Unesa Bulletin of Engineering. Technology and Applied Science*, *3*(1), 25–36. DOI: 10.26740/vubeta.v3i1.39662

Wojciechowska, G., Bednarz, Ł. J., Dolińska, N., Opałka, P., Krupa, M., & Imnadze, N. (2024). Intelligent Monitoring System for Integrated Management of Historical Buildings. *Buildings (2075-5309), 14*(7).

Wong, J. Y., & Lau, P. Y. (2019). Real-time driver alert system using raspberry pi. *ECTI Transactions on Electrical Engineering, Electronics, and Communications*, *17*(2), 112–119.

Woz´niak, Z., Trybuszewski, K., Nowobilski, T., Stolarz, M., & Šmalec, F. (2025). Integrated construction-site hazard detection system using AI algorithms in support of sustainable occupational safety management. *Sustainability*, *17*(23), 10584. DOI: 10.3390/ su172310584

Wu, Y., & Lei, X. (2025). Advancing construction safety: YOLOv8-CGS helmet detection model. *PLOS ONE, 20*(5), e0321713. https://doi.org/. 0321713DOI: 10.1371/journal.pone

Wu, K., Ji, Z., Wang, H., Shao, X., Li, H., Zhang, W., & Bao, X. (2025). A comprehensive review of AI methods in agri-food engineering: Applications, challenges, and future directions. *Electronics (Basel), 14*(20), 3994.

Xiang, W., Yu, K., Han, F., Fang, L., He, D., & Han, Q. L. (2023). Advanced manufacturing in industry 5.0: A survey of key enabling technologies and future trends. *IEEE Transactions on Industrial Informatics*, *20*(2), 1055–1068.

Xie, P., Zhang, Z.-S., & Zhang, J. (2018). Topology transformation based on time division: An improved inter-satellite link structure. *In 2018 Chinese Control And Decision Conference (CCDC)* (pp. 3761-3766). https://doi.org/DOI: 10.1109/ CCDC.2018.8407776

Xu, L. D., Xu, E. L., & Li, L. (2018). Industry 4.0: State of the art and future trends. *International Journal of Production Research*, *56*(8), 2941–2962.

Yi, T. H., Li, H. N., & Gu, M. (2013). Recent research and applications of GPS-based monitoring technology for high-rise structures. *Structural Control and Health Monitoring*, *20*(5), 649–670.

Zangana, H. M. (Ed.). (2025). *Defense in depth: Modern cybersecurity strategies and evolving threats*. Wiley., DOI: 10.1002/9781394340750

Zangana, H., & Omar, M. (Eds.). (2026). *AI in Digital Forensics and Cybercrime Investigation: Methods, Ethics, and Emerging Technologies*. IGI Global Scientific Publishing., DOI: 10.4018/979-8-3373-6536-7

Zangana, H., & Omar, M. (Eds.). (2026). *Securing Large Language Models Against Emerging Threats*. IGI Global Scientific Publishing., DOI: 10.4018/979-8-3373-7133-7

Zhang, Q., Xin, C., Shen, F., Gong, Y., Zi, Y., Guo, H., & Wang, Z. L. (2022). Human body IoT systems based on the triboelectrification effect: Energy harvesting, sensing, interfacing and communication. *Energy & Environmental Science*, *15*(9), 3688–3721.

Zhang, Y., Wang, J., Chen, X., & Liu, L. (2025). ESPCN-YOLO: A high-accuracy framework for personal protective equipment detection under low-light and small object conditions. *Buildings (Basel, Switzerland)*, *15*(10), 1609. Advance online publication. DOI: 10.3390/buildings15101609

Zhao, Z., Zhou, N., Zhang, L., Yan, H., Xu, Y., & Zhang, Z. (2020). Driver fatigue detection based on convolutional neural networks using em-cnn. *Computational Intelligence and Neuroscience*, *2020*, 7251280. PMID: 33293943

About the Contributors

Mohammed Aquil Mirza is a senior member of IEEE, a researcher and educator specializing in AI, AIoT, robotics, sensor networks, and medical imaging. Since 2025, he has been Dean and Professor at Liaoning Technical University, China. Previously, he was a Senior Lecturer at PolyU from 2023, teaching courses like Programming Fundamentals and AIoT, with high student feedback. Between 2021 and 2023, he led projects on wall-climbing robots, autonomous drones, and smart monitoring systems. He began his career with research roles at IIT Delhi and BITS Pilani. He holds a Ph.D. in Computing from PolyU, with master's degrees from KAUST and IIIT-Hyderabad. His research has been funded by major grants, and his work has earned awards such as the IEEE Best Student Paper (2025). With a strong academic background and innovative research, he continues to advance the fields of intelligent systems and robotics, inspiring future engineers

Hewa Majeed Zangana is an Assistant Professor at Duhok Polytechnic University (DPU), Iraq, and a current PhD candidate in Information Technology Management. He has over 15 years of academic and administrative experience, serving in key roles such as Acting Dean of the College of Computer and IT at Nawroz University, Head of the Computer Science Department, Director of the Curriculum Division at DPU, and Manager of the Information Unit at DPU's Research Center. His research expertise spans network systems, cybersecurity, computer vision, intelligent systems, and Information Technology, with a strong focus on integrating classical methods and advanced AI-driven approaches. Dr. Zangana has authored and co-authored numerous peer-reviewed journal articles, conference papers, and more than ten academic books with IGI Global, several indexed in Scopus and Web of Science (Clarivate). His works appear in respected outlets such as IEEE, EAI Endorsed Transactions on AI and Robotics, Sistemasi, BAMJ, and others. In addition to his prolific research output, he actively contributes to the academic community as a reviewer, editorial board member, and scientific committee member. He currently serves on the Editorial Board of EAI Endorsed Transactions on AI and Robotics (Scopus Q2), alongside roles with other international journals and organizations. Dr. Zangana is also an active member of several professional associations, including the European Alliance for Innovation (EAI), International Association of Engineers (IAENG), and the Computer Science Teachers Association (CSTA). Fluent in Kurdish, Arabic, and English, Dr. Zangana continues to contribute to advancing knowledge in AI, cybersecurity, and IT management, while mentoring future generations of scholars and practitioners in Iraq and beyond.

***

Jamal N. Al-Karaki is currently a full professor at CIS, Zayed university. He has a prolonged and rich university career in education, service, and research. He holds a Ph.D. degree (with Distinction) in Computer Engineering from Iowa State University, USA, with the Research Excellence Award. He has a proven record of progressive responsibility, including leadership positions as a college Dean, Director of IT, Co-Founder, and Dept. Head at various institutes. He led the development of some national centres in cybersecurity as well as undergraduate and graduate computing programs related to cyber security and AI. As an active researcher, he develops plans to advance the research agenda at CIS. He has Published 100+ refereed technical articles in scholarly international journals and conference proceedings. He is a senior member of IEEE and Tau Beta Pi. He is an affiliate member of the Mohammad Bin Rashid Academy for Scientists in UAE and constantly being listed among the top 2% highly cited researchers in his field worldwide. He holds a plethora of professional certification in the domain of cyber security as well as being certified accreditation reviewer for CAA-UAE, Quality Matters, and Pearson EDI verifier/assessor. Prof. Al-Karaki has participated in many panels to advance cyber security education and capacity building in UAE and beyond. Prof. Al-Karaki research interests covers the wide spectrum of computing and its applications in AI, human life, education, economy, healthcare, Society, Cyber security, among many others.

Widi Aribowo as an Assistant Professor in the Department of Electrical Engineering at the Faculty of Vocational Studies, Surabaya State University. Known as a leading researcher, he has made significant contributions in the fields of metaheuristics, optimization, artificial intelligence, power systems and Microgrid. His scientific impact is evident through various publications in leading journals. In addition, Widi Aribowo has demonstrated his expertise as a peer reviewer for several prestigious Q1 journals, demonstrating his commitment to advancing academic discourse in the fields of metaheuristics, optimization, artificial intelligence, power systems and Microgrid.

Xinwei Cao is presently a full professor with School of Business, Jiangnan University China. Her research includes fintech, operational research, AI and machine learning with applications to management and finance, etc.

Muhammad Irawan is a lecturer at the Faculty of Science and Technology, Universitas Islam Negeri Sumatera Utara (UINSU), Indonesia. He earned his bachelor's degree in Informatcs from Universitas Asahan and his master's degree in Informatics from Universitas Putra Indonesia YPTK Padang. His research focuses on algorithms, artificial intelligence, decision support systems, and broader areas in computer science. He has held editorial and reviewer roles in various national journals and serves as Journal Manager at UINSU. He is a member of professional organizations such as CASE, HEBII, APTIKOM, and IndoCEISS.

Aquil Mohammed is a highly accomplished researcher and educator, this and they specialize in cutting-edge fields such as AI, AIoT, Robotics, and Deep Machine Learning, with a keen focus on complex sensor networks, medical imaging, surgical robotics, and structural health monitoring. Currently serving as Dean and Professor at Liaoning Technical University's School of Low-Altitude Digital Intelligence since 2025, they lead innovative initiatives blending digital intelligence with practical applications. Prior to this role, they have contributed significantly to the academic landscape in Hong Kong, working as a Senior Lecturer at PolyU from 2023, where they received exceptional student feedback for courses like Programming Fundamentals, Operating Systems, and AIoT. From 2021 to 2023, they held the position of Scientific Officer at PolyU, spearheading projects involving robotic design, airborne drones, and network monitoring, showcasing their versatile expertise. Their career began in India, where they served as Assistant Professor at Bharat Institute of Engineering and Technology, teaching Data Analytics, Big Data, and Machine Learning. Earlier, they worked as a Lecturer at BITS Pilani and a Research Associate at IIT Delhi, gaining extensive research experience. Educationally, they hold a PhD in Computing from PolyU, with advanced degrees from KAUST and IIIT Hyderabad. Recognized with prestigious awards, including IEEE best paper honors, they remain at the forefront of technological innovation and academic excellence.

Firas Mahmood Mustafa holds a Ph.D. in Computer Engineering from Mosul University, Iraq. His academic journey began with a B.Sc. in Electrical Engineering (Electronics and Communication), graduating in the top quarter of his class. He earned an M.Sc. in Computer Engineering from Mosul University in 2000. Mustafa joined the Computer Science Department at AlHadba University in 2003 and completed his Ph.D. in 2007. From 2013 to 2017, he was with DPU University, and from 2017 to 2020, he chaired the CCE Department at Nawroz University. An active participant in Erasmus+ and IREX programs, he now teaches at DPU University, shaping future computer engineering professionals.

Marwan Omar serves as a full time faculty member of computer science and IT at Nawroz university, Duhok, Iraq. Omar is recognized for his information security expertise and knowledge and holds a security + certification from Comptia. Research interests are: cyber security, mobile security, open source software, and cloud computing.

Sharyar Wani is an Assistant Professor in the Department of Computer Science at the Kulliyyah of Information & Communication Technology (KICT), International Islamic University Malaysia (IIUM). He completed his Doctor of Philosophy (PhD) in Computer Science at IIUM, and subsequently pursued post-doctoral research at the Centre of Cyber Security, National Defence University of Malaysia, advancing his expertise in cybersecurity and intelligent systems. Following his postdoctoral appointment, he served as a Senior Lecturer at the National Defence University of Malaysia before joining IIUM, where he is an active member of the Computational Intelligence Research Group and contributes to both undergraduate and postgraduate education. His research spans artificial intelligence, information retrieval, data analytics, natural language processing, machine learning, and cybersecurity, with a substantial body of publications in reputable journals and conferences and significant citation impact in his field. At IIUM, Dr. Wani plays a central role in research supervision and scholarly mentorship, guiding students through thesis development and research dissemination while participating in academic service and collaborative research initiatives. His work is recognized for bridging theoretical foundations with practical applications, particularly in areas that address contemporary challenges in intelligent computing and secure information systems. Through his teaching, research leadership, and engagement with interdisciplinary projects, he contributes to the advancement of knowledge within both the national and international academic communities.

Index

H

I

L

M

O

P

R

S

W

Y

www.ingramcontent.com/pod-product-compliance
Lightning Source LLC
LaVergne TN
LVHW080901300926
849162LV00003B/504